KALMAN FILTERS - THEORY FOR ADVANCED APPLICATIONS

Edited by **Ginalber Luiz de Oliveira Serra**

Kalman Filters - Theory for Advanced Applications
http://dx.doi.org/10.5772/intechopen.68249
Edited by Ginalber Luiz de Oliveira Serra

Contributors

Kenshi Saho, Kok-Yong Seng, Ying Chen, Jia Guo, Weiping Priscilla Fan, Si Hui Maureen Lee, Junxian Ong, Poh Ling Tan, Yu Li Lydia Law, Kai Wei Jason Lee, Mark Wielitzka, Alexander Busch, Matthias Dagen, Tobias Ortmaier, Ahmad Shukri Abu Hasim, Syed Mohd Fairuz Syed Mohd Dardin, Zulkifilie Ibrahim, Felix Govaers, Sarita Nanda, Mudambi R Ananthasayanam, Qian Zhuang, Elias David Nino-Ruiz, Lalitha Pakala, Bernhard Schmauss, Angel Juan Sanchez Garcia, Homero Rios Figueroa, Antonio Marin Hernandez, Gustavo Quintana Carapia, Guocan Wu, Amir Khodabandeh, Peter Teunissen, Safoora Zaminpardaz, Mauro Hernán Riva

Published by InTech

Janeza Trdine 9, 51000 Rijeka, Croatia

Publishing Process Manager Iva Simcic
Technical Editor SPi Global
Cover InTech Design team

First published February, 2018
Printed in Croatia
Legal deposit, Croatia: National and University Library in Zagreb

Additional hard copies can be obtained from orders@intechopen.com

Kalman Filters - Theory for Advanced Applications, Edited by Ginalber Luiz de Oliveira Serra
p. cm.
Print ISBN 978-953-51-3827-3
Online ISBN 978-953-51-3828-0

For EU product safety concerns:
IN TECH d.o.o., Prolaz Marije Krucifikse Kozulić 3, 51000 Rijeka, Croatia,
info@intechopen.com or visit our website at intechopen.com.

Meet the editor

Dr. Ginalber Luiz de Oliveira Serra was born in São Luis, Brazil, in 1976. He received his BSc and MSc degrees in electrical engineering from Federal University of Maranhão, São Luis, MA, Brazil, in 1999 and 2001, respectively. Dr. Serra received his PhD degree in Electrical Engineering from the State University of Campinas, Campinas, SP, Brazil, in September 2005. He finished his postdoctoral research on neuro-fuzzy adaptive control, with the Department of Machinery, Components and Intelligent Systems, at the State University of Campinas, Campinas, SP, Brazil, in September 2006. Dr. Serra developed researches on control and intelligent systems, with the Department of Electrical Engineering, at the University of Santiago, Santiago, Chile, from 2006 to 2007. He is currently the professor and head of the research group on Computational Intelligence Applied to Technology, at Federal Institute of Education, Science and Technology, São Luis, MA, Brazil. Dr. Serra has contributed as the editor for the book *Frontiers in Advanced Control Systems* (InTech, 2012) and the author/reviewer of several papers for many prestigious international journals and conferences. His research interest includes topics on computational intelligence, signal processing, automation, and industrial control applications.

Contents

Preface

The content of this book brings together the research results on state-of-the-art Kalman filtering theory for advanced applications to real-world problems. In **Chapter 1**, the philosophy and the historical development of Kalman filter from ancient times to the present are followed by the connection between randomness, probability, statistics, random process, estimation theory, and the Kalman filter. A reference recursive recipe (RRR) methodology is proposed, and the efficacy is demonstrated by its application to a simulated spring, mass and damper system, and a real airplane flight data having a larger number of unknown parameters and statistics. In **Chapter 2**, a new structure of the forecast error covariance matrix is proposed to mitigate the problems with limited ensemble size and model error in an ensemble Kalman filter (EnKF). An adaptive procedure equipped with a second-order least squares method is applied to estimate the inflation factors of forecast and observational error covariance matrices. The proposed method is tested on the well-known atmosphere like Lorenz-96 model with spatially correlated observational systems. In **Chapter 3**, state and parameter estimation in vehicle dynamics utilizing the unscented Kalman filter is presented. The estimation runs in real time based on a detailed vehicle model and standard measurements taken within the car. The results are validated using a Volkswagen Golf GTE Plug-In Hybrid for various dynamic test maneuvers and a Genesys ADMA measurement unit for high precision measurements of the vehicle's states. In **Chapter 4**, a sensitivity-based adaptive square-root unscented Kalman filter (SRUKF) is presented. This algorithm combines an unscented Kalman filter (UKF) and the Recursive Prediction Error Method to estimate system states, parameters, and covariances online. In **Chapter 5**, an adaptive Taylor Kalman filter with PSO tuning for tracking nonstationary signal parameters in a noisy environment with primary focus on time-varying power signals is presented. The proposed PSO-tuned Taylor Kalman filter exhibits robust tracking capabilities even under changing signal dynamics, is immune to critical noise conditions and harmonic contaminations, and reveals excellent convergence properties. In **Chapter 6**, the estimation of heart strain from noninvasive measurements, heart rate (HR), and chest skin temperature (ST), obtained "online" via wearable body sensors via Kalman filter, is investigated. The experiments are performed using data from laboratory and outfield-based heart strain profiling studies in which subjects performed a high-intensity military foot march. In **Chapter 7**, a method of predicting financial distress based on Kalman filtering is improved dynamically. Based on the state-space method, two models that are used to describe the dynamic process and discriminant rules of financial distress are established, respectively: a process model and a discriminant model. An empirical study for China's manufacturing industry is also conducted. In **Chapter 8**, an application of the Kalman filter to the navigation of mobile robots, specifically the time-to-contact problem, is presented. A monocular vision-based approach to detect potential obstacles and to follow them over time through their apparent size change is used. The approach collects information about obstacle data and models the behavior while the robot is approaching the obstacle, in order to predict collisions. In **Chapter 9**, the efficient and practical matrix-free implementations of the ensemble Kalman filter (EnKF) in order to account for localization during the assimilation of observations are discussed. Experimental tests are performed making use of the Lorenz 96 model. In **Chapter 10**, the design and implementation of a three-phase shunt active power filter (APF) employing Kalman filter estimator are presented. Details on investigation between conventional and proposed methods under simulation based on MATLAB/SIMULINK platform and experiment are made for two types of load, namely, three-phase rectifier with RC-load and three-phase induction motor. In **Chapter 11**, various applications of Kalman filtering for coherent optical communication systems are reviewed. The numerical analysis concludes that the Kalman filter-based approaches outperform the conventional methods with better tracking capability and faster convergence besides offering more feasibility for real-time implementations. In **Chapter 12**, Kalman filters for tracking moving objects and their efficient design strategy based on steady-state performance analysis are presented. Numerical simulations show the validity of the theoretical analysis and effectiveness of the proposed strategy in realistic situations. In **Chapter 13**, the challenges in distributed tracking are explained. Possible solutions are derived, which include the distributed Kalman filter (DKF) and a more recent methodology based on "accumulated state densities" (ASD), which augment the states from multiple time instances to overcome spatial cross correlation ASD approach. In **Chapter 14**, Kalman filtering in its distributed information form is reviewed and applied to a network of receivers tracking Global Navigation Satellite Systems (GNSS). The consensus-based Kalman filter (CKF) of individual receivers is applied to deliver GNSS parameter solutions with comparable precision performance as their network-derived, fusion center-dependent counterparts. This is relevant as in the near future the proliferation of low-cost receivers will give rise to a significant increase in the number of GNSS users.

This book *Kalman Filters - Theory for Advanced Applications* presents the following aspects of interest: provide the state of the art on advanced theoretical and practical research and facilitate the proposal of new techniques and implementations and educational importance as handbook to help students and researchers on Kalman filtering technologies.

The editor would like to thank all authors who were valuable sources of information on the state-of-the-art Kalman filter technology. Thanks are still due to Ms. Iva Simcic, Publishing Process Manager (InTech), for her patience and accuracy in all steps in the project of this book. The editor is very grateful to the professors with the Department of Electro-Electronics (DEE/IFMA), for their motivations, their stimulations, and providing a pleasant environment of scientific study and research and MSc/PhD Electricity Engineering Program (PPGEE/UFMA) for its encouragement, during the project of this book. Finally, the editor would like to dedicate this book to his family (especially for his father Walber Serra, mother Raidalva Serra, and daughter Ester Luiza) for their valuable presence in his life.

Prof. Dr. Ginalber Luiz de Oliveira Serra
Federal Institute of Education, Sciences and Technology
Department of Electro-Electronics
São Luis, MA, Brazil

Chapter 1

A Reference Recursive Recipe for Tuning the Statistics of the Kalman Filter

Mudambi R Ananthasayanam

Additional information is available at the end of the chapter

http://dx.doi.org/10.5772/intechopen.71961

Abstract

The philosophy and the historical development of Kalman filter from ancient times to the present is followed by the connection between randomness, probability, statistics, random process, estimation theory, and the Kalman filter. A brief derivation of the filter is followed by its appreciation, aesthetics, beauty, truth, perspectives, competence, and variants. The menacing and notorious problem of specifying the filter initial state, measurement, and process noise covariances and the unknown parameters remains in the filter even after more than five decades of enormous applications in science and technology. Manual approaches are not general and the adaptive ones are difficult. The proposed reference recursive recipe (RRR) is simple and general. The initial state covariance is the probability matching prior between the Frequentist approach via optimization and the Bayesian filtering. The filter updates the above statistics after every pass through the data to reach statistical equilibrium within a few passes without any optimization. Further many proposed cost functions help to compare the present and earlier approaches. The efficacy of the present RRR is demonstrated by its application to a simulated spring, mass, and damper system and a real airplane flight data having a larger number of unknown parameters and statistics.

Keywords: adaptive EKF, expectation maximisation, maximum likelihood, Cramer Rao bound, probability matching prior

1. Brief introduction to the historical development of Kalman filter (KF)

1.1. Conceptual beginning of KF in ancient Indian astronomy

As is well known if there is one thing that does not change in nature it is the change. Such a change has to be captured by some means. In general, neither the change nor the capture is exact. Hence based on some suitable criteria a combination can be derived to correct. Such an update process has to be repeated at suitable intervals. The above philosophy of 'change, capture, and correct' is the one that is followed in the Kalman filter. The ancient Indian astronomers had understood the above philosophy.

The ancient Indian astronomers, at least since AD 500, used the above concept to update the parameters for predicting the position of celestial objects for timing their Vedic rituals based on measurements carried out at various time intervals which can be stated as

$$\begin{aligned}&\text{Updated parameter} = \text{Earlier parameter} + (\text{Some quantity})\\&\times(\text{Measured} - \text{Predicted})\ \text{Position of the celestial object}\end{aligned} \tag{1}$$

The 'some quantity' as we will see later on is the Kalman gain. Further, note the measured longitude of the celestial object is different from the state that is updated, which is the number of revolutions in a yuga just as state and measurements are in general different in many Kalman filter applications!

They needed to calculate the position of the celestial objects like Sun, Moon, and other planets to carry out the Vedic rituals. But their predicted positions changed over many centuries due to unmodeled or unmodelable causes. The French historian Billard [1] noted that measurements were carried out (in fact extending over many years or even decades!) at various times starting from around AD 500 by Aryabhata to AD 1600 and the parameters were corrected to make the predicted position of the objects consistent with new observations. **Table 1** shows such revisions over a period of time from Sarma [2].

During the above period, Nilakantha (around AD 1443) had stated that the eclipses cited in Siddhantas can be computed and the details verified. Similarly, other known eclipses, as well as those currently observable, are to be studied. In the light of such experience, future ones can be computed and predicted (**extrapolation!**). Or eclipses occurring at other places can be studied taking into account the latitude and longitude of the places and on this basis the method for the true Sun, Moon… can be perfected **(data fusion!).** Based on these, the past and future eclipses of one's own place can be studied and verified with appropriate refinement of the technique. This is just the idea of **'smoothing'!** Billard had a problem for later Indian

Source, Number of days per yuga +	Date AD	Moon 5,77,53 +	Moon's Apogee 4,88,000 +	Moon's Node 2,32,000 +	Mars 22,96,000 +	Mercury 1,79,37,000 +	Jupiter 3,64,000 +	Venus 70,22,000 +	Saturn 1,46,000 +
Aryabhatiyam, 500	499	336	219	226	824	020	224	388	564
Parahita,500	563	314.2	121.9	307.3	862.3	378	184.0	257.8	581.0
Drgganita,517.019	1431	321.009	123.229	297.832	862.959	072.112	172.296	270.775	626.695
Tantrasangraha 517.019	1443	320	122	300	864	058	180	268	612
Golasara,517.019	1486	320	120	300	864	040	180	270	600
Siddhantadarpana, 839.500	1500	332.321	123.318	226.745	862.137	120.175	160.611	270.552	571.016
Drkkarana,839.500	1607	320	122	298	863	100	166	272	612

Table 1. Corrections of planetary parameters in ancient India. From Sarma [2] (Sun 43,20,000; Number of days per yuga = 1,57,79,17,000+).

astronomers as 'If the elements of Aryabhata are now wrong, they must have been accurate when he was living. Thus, we ought to **establish the astronomical elements upon both on his own at his time and the new observations of the present time**'. He noted that based on the above reasoning, some Indian canons were evolved and one such canon around AD 898 shows a very high accuracy valid over a larger number of centuries [1].

1.2. Development of the concepts in KF during the medieval period

On January 1, 1801, Piazzi, while searching for a wrong entry in a star catalogue, discovered the largest and the first of the asteroids Ceres. He tracked its position for the next 41 days, before it was lost in space. He wrote to Bode who felt it to be the missing planet in his formula. Its orbital elements could not be determined by the then available methods. Newton had stated it as the most difficult nonlinear problem then in astronomy. Piazzi's discovery was published in 1801. Gauss tackled the problem and estimated its orbit and sent it to Piazzi who found it again on the last day of 1801! Gauss published his orbit determination methods only in 1809 [3] describing his 1795 method of least squares (LS) used in his estimates of the Ceres orbit. Gauss had an ideal situation with a good system model and only the measurement noise, and thus with his least squares approach he could get an estimate and a qualitative measure for the uncertainty. Independently, the method of LS had been discovered and published by Legendre of France and Robert Adrian of the United States. In fact, even before Gauss was born, the physicist Lambert had used it! Gauss has provided almost all the essentials of present day estimation theory. He postulated that a system model should be available, minimum number of measurements for observability, the redundant data helping to reduce the influence of measurement errors, a cost function based on the difference between the measurement and that predicted by the model should be minimised. There should be some a priori knowledge concerning the unknowns to be estimated. Further, since the errors could be unknown or unknowable, he had given hints about probabilistic approach, normal distribution, and method of maximum likelihood estimation, linearization, and the Gaussian elimination procedure. Gauss did not balance the governing differential equation, but tried to fit the measurement with the prediction. If he had tried the former, he would have been led to a biased solution! Fortune favours the brave! This is where a proper mathematical framework helps to understand if an algorithm converges to the correct value with more and more measurements. The post Gaussian contributions in ET consists of the method of moments, method of maximum likelihood estimation, the Kalman filter and its variants, frequency domain approach, and handling time varying dynamical state and parameters. Further, the use of matrix theory, sequential instead of batch processing, and real-time processing by computers exist. We are dealing with more difficult situations, but the conceptual framework to solve these problems had been fully laid out by Gauss!

1.3. Mature KF during the twentieth century and beyond

Almost every concept in present day science and technology seem to have its root in ancient times as mentioned earlier. It must be understood that the classic papers by Kalman [4] and Kalman Bucy [5] was conceptually preceded earlier by Thiele as mentioned by Lauritzen [6], Swerling [7], Stratonovich [8] and some other researchers. The development of the Kalman filter is one of the most interesting and useful innovations of the twentieth century, and it owes its origin to the least squares solution proposed by Gauss. For Gauss, the state equations were the exact Newton's laws of motion and only measurement noise existed. The sequential least squares were rediscovered by Plackett [9] and Kalman [4]. Thus, in fact, Sprott [10] has questioned if the Kalman Filter is really a significant contribution when Gauss was far ahead! The point is that the frequency domain approach of the Wiener filter [11] has been improved to the natural time domain approach. Further the shift from batch to the sequential approach is very convenient to handle continuous measurement data flow. It is to the credit of Kalman apart from unifying earlier results that he introduced the concept of controllability and observability, which means the system to be identified has to be properly excited and observed thus making the estimation problem systematic and consistent. The only slight difference, but very momentous between the Recursive Least Squares (RLS) and the Kalman filter is the time propagation of the state and covariance estimates between measurements (see for example [12]). Presently, the scale and magnitude of many difficult and interesting problems that estimation theory (ET) is handling could not have been comprehended by Gauss or Kalman. Examples are airplane flight test data analysis [13, 14], target tracking [15], evolution of the space debris scenario [16], fusion of GPS and INS data [17], study of the tectonic plate movements [18], high energy physics [19], agriculture, biology, and medicine [20], dendroclimatology [21], finance [22], source separation problem in telecommunications biomedicine, audio, speech and in particular astrophysics [23], and atmospheric data assimilation for weather prediction [24].

2. Randomness, probability, statistics, random process, estimation theory, and Kalman filter

The connection among probability, statistics, random process, and estimation theory by Ananthasayanam [12] as shown in **Figure 1**, are ubiquitous in science, technology, and life.

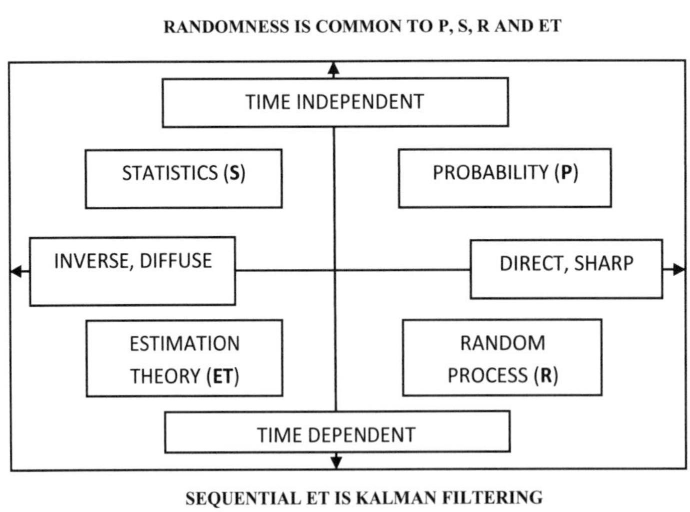

Figure 1. Relationship between probability, statistics, random process, and estimation theory.

The randomness is common to all and sequential time-dependent statistical analysis can be called as Kalman Filtering.

2.1. Randomness

We have moved unambiguously from determinism to randomness ever since ancient times to the present. We are compelled to understand, model, estimate and control nature which is random in a probabilistic way. Randomness occurs inevitably in all walks of life. The ontology (the true nature) is one thing and the epistemology (our understanding) is another thing. A computer generating a sequence of random numbers is deterministic ontology, but for the user who does not know how they are generated it is probabilistic epistemology. Randomness is patternless but not propertyless. Randomness could be our ignorance. Chance is no longer something to worry about or an expression of ignorance. On the contrary, it is the most logical way to present our knowledge. We are able to come to terms with uncertainty, to recognise its existence, to measure it and to show that advancement of knowledge and suitable action in the face of uncertainty are possible and rational. Random chance may be the antithesis of all law. We look for the alternatives and provide the probabilities of their happening as measures of their uncertainties. Knowing the consequences of each event and the probability of its happening, decision making under uncertainty can be reduced to an exercise in deductive logic. It is no longer a hit and miss affair. But, the way out is to discover the laws of chance and convert chance to choice in life. The aim of life is to make the earth a happy place to live.

Randomness could occur due to the uncertainty, variability, complexity, or enormity. A classic example is the deterministic coin tossing. There is enormity in dealing with a large number of air molecules, the complex interaction of air among themselves and with the coin, variability of the initial condition, and the uncertainty due to air currents. Of course, many deterministic mathematical problems could be handled using probabilistic approaches called Monte Carlo techniques. Quantum Mechanics seems to possess true randomness. One feels that randomness is a nuisance, and should be avoided. However we have to live with it and compulsively need it in many situations. In a multiple choice question paper, no examiner would dare to put all the correct answers at the same place! As another example the density, pressure, and temperature, or even many trace constituents in air can be measured with a confidence only due to the random mixing that invariably takes place over a suitable space and time scale. As we will see later, the introduction of random process noise into the kinematic or dynamical equations of motion of aircraft, missiles, launch vehicles, and satellite system helps to inhibit the onset of Kalman filter instability and thus track these vehicles. The well-known statistician Rao [25] states that statistics, as a method of learning from experience and decision making under uncertainty, must have been practiced from the beginning of mankind. The inductive reasoning in these processes has never been codified due to the uncertain nature of the conclusions drawn from any data. The break through occurred only at the beginning of the twentieth century with the realisation that inductive reasoning can be made precise by specifying additionally just the amount of uncertainty in the conclusions. This helped to work out an optimum course of action involving minimum risk in uncertain situations in the deductive process. This is what Rao [25] has stated as a new paradigm of thinking to handle uncertain situations as

$$\text{Uncertain knowledge} + \text{Knowledge of the amount of uncertainty in it} \\ = \text{Usable knowledge} \tag{2}$$

The Kalman filter dealing with only the estimates and their uncertainties thus has minimal additional information to handle.

2.2. General remarks on P, S, R, and ET

Though randomness exists in all the above, due to the undercurrent of some determinism wrapped by uncertainty the processes are not completely haphazard. The **P** and **S** are time independent, whereas **R** and **ET** are their time-dependent analogues. The **P** and **R** are direct using deductive logic; whereas **S** and **ET** are inverse problems using inductive logic, thus more difficult than the former. The **S** and **ET** deal with data that are insentient and do not speak. The analyst gives life to the data to find out the underlying model mechanism based on intuitive and subjective analysis to obtain results and conclusions which are meaningful and useful.

In an inverse problem with a limited sample size from the population with or without noise, the problem is considered, as well posed if (i) there exists a solution to the problem (existence), (ii) there is at least one solution to the problem (uniqueness), and (iii) the solution depends continuously on the data (stability). Generally, an acceptable and reasonable solution can be worked out by specifying some subjective criterion cast in terms of a quantitative cost function to match (in some reasonable way) the model and the measurements made on the system. Further, no matter whatever technique one adopts unless and until the model structure of the system is appropriate and the parameters in it are identified along with the noise sequence, one can never obtain the true value but only be around it all the time. When a system model characterises many effects, then the model structure should reflect each one appropriately and the parameters in them estimated for apportioning the different effects.

2.3. Probability (P)

The probabilistic approach subsumes in general the deterministic laws of physics (but one can include them also) or others postulated for any system. It utilises the axiomatic rules governing probability leading to outcomes regarding the behaviour of the ensemble. The probability itself can be broadly specified based on Classical, Bayesian, and the Frequentist approaches [26]. These are, respectively, based on the 'principle of indifference thus equiprobable', 'degree of belief', and 'limiting frequency' of the occurrence of the events. The law of large numbers and the central limit theorem under very general conditions (leading to a **Gaussian** distribution) assist in reaching practical conclusions.

2.4. Statistics (S)

In order to define statistics, it seems best to follow Feller [27] which in our language is to analyse the **measurement** data to develop a mathematical **model** with the intuitive **mind** providing the methodologies. The mind is nebulous, measurement is spotty, and model is smooth if one may say so. **Then, the simplest definition of statistics seems to be a good translation of measurements into a model by the mind.** Of course, the three are randomly used for progress in science, technology, and life. An intuitive subjective idea is cast in an objective form to help the analyst decide if the difference between measurement and model can be attributed to statistical fluctuation or deterministic change. Such a decision is always based on extra statistical (!) implications. All the concepts and problems of statistics go over to estimation theory, and in turn to Kalman filter which is to sequentially process the time-dependent data. Generally, in statistics, **linear** relationship among the variables is prolific since it is the simplest and anything else would have to be justified *a priori*.

The philosophical discussions about the Classical, Bayesian, and Frequentist approaches to probability and statistics can go on endlessly. The most prevalent view is Frequentists deal with data only, but Bayesians try to incorporate well established '*a priori*' information as well into the problem. But no matter whichever approach is followed, eventually one has to take a practical view and ensure the final results are credible and useful.

2.5. Random process (R)

Here, in **R**, the simplest characterisation of time-dependent noise is white in time following a Gaussian distribution, thus containing the least amount of information! However, due to the undercurrent of some amount of determinism wrapped by uncertainty, the processes are not completely haphazard. If the atmosphere and earth quakes behave like white noise then meteorologists, geologists would have no work! The underlying deterministic processes are wrapped in noise providing some underlying correlations. The white noise is the worst data that will fail any algorithm for prediction, but used most prolifically in **ET** since it is mathematically tractable!

2.6. Estimation theory (ET)

The basic framework of **ET** in analysing a given measurement data consists of qualitatively modelling the system, measurement and all the noise characteristics, a criterion to match or mix the model output with the measurements in some optimal sense, a numerical algorithm for the above task and consequently estimate the unknown parameters and the noise statistics together with their uncertainties and lastly an internal consistency check to ensure that the assumptions regarding model and measurement noise above are consistent and if not the above steps have to be modified and repeated.

There is a general feeling that ET solutions have to be objective with little scope for subjectivity. However, it is interesting to note that subjectivity cannot be avoided, but it is the one that helps all the way from the formulation of the problem to obtaining the final solutions. Deterministic approaches of the early period have given way to probabilistic approach that is neither a fashion nor the truth (!) The probabilistic rules aid in modelling the scenario and the statistical approach of analysing the data with all its subjectivity provides acceptable quantitative estimates together with their uncertainties.

3. Overview of Kalman filter

The simplest formulation of a Kalman filter is when the state and measurement equations are linear, a well-known fact as mentioned by Brown and Hwang [28]. For linear systems during evolution and update the form of the normal distribution is maintained, but with changed mean and covariance. However, the Kalman filter has found its greatest application for nonlinear systems. Particularly in many aerospace applications, the unknown parameters multiply the state variables to provide force or moments acting on the vehicles. If the unknown parameters are treated as additional states then the system of equations become nonlinear. Thus, the Extended Kalman filter (EKF) formalism can be used. The EKF formulation provides the simplest scenario to present the proposed Recurrence Recursive Recipe (**RRR**). Other filter formulations contain the effect of further approximations, discretizations and other features. A typical continuous state with discrete measurements in time forms a nonlinear filtering problem and can be written as

$$\mathbf{x}(k) = \mathbf{f}(\mathbf{x}(k-1), \boldsymbol{\Theta}, \mathbf{u}(k-1)) + \mathbf{w}(k) \tag{3}$$

$$\mathbf{Z}(k) = \mathbf{h}(\boldsymbol{x}(k), \boldsymbol{\Theta}) + \mathbf{v}(k), \ \ \mathrm{k} = 1, 2, 3, \ldots..\mathbf{N} \tag{4}$$

where '**x**' is the state vector of size (n × 1), '**u**' is the control input, **Θ** is the parameter vector of size (p × 1) and '**Z**' is the measurement vector of size (m × 1). The '**f**' and '**h**' are nonlinear functions of state and measurement equations, respectively. The process and measurement noise are assumed to be zero mean with covariance **Q** and **R**, respectively, and their sequences are uncorrelated with each other. The states are generally not directly observable, but the measurements are related to the states.

In EKF formulation, the parameter vector **Θ** is augmented as additional states. Thus,

$$\begin{bmatrix} x(k) \\ \boldsymbol{\Theta}(k) \end{bmatrix} = \begin{bmatrix} f(x(k-1), \ \boldsymbol{\Theta}(k-1), u(k-1)) \\ \boldsymbol{\Theta}(k-1) \end{bmatrix} + \begin{bmatrix} w(k) \\ 0 \end{bmatrix} \tag{5}$$

The nonlinear filtering problem is now redefined as

$$\mathbf{X}(k) = \mathbf{f}(\boldsymbol{X}(k-1)) + \mathbf{w}(k) \tag{6}$$

$$\mathbf{Z}(k) = \mathbf{h}(\boldsymbol{X}(k)) + \mathbf{v}(k), \mathrm{k} = 1, 2, \ldots..\mathbf{N} \tag{7}$$

where '**X**' and '**w**' are, respectively, the augmented state and process noise vector of size ((n + p) × 1). The control symbol '**u**' is not shown for brevity. The solution for the above filtering problem can be summarised as

$$\text{Initial State Estimate } \mathbf{X}(0|0) = \mathbf{X0} = E\,[\mathbf{X}(\mathrm{t}0)], \tag{8}$$

$$\text{Initial State Covariance Matrix } \mathbf{P}(0|0) = \mathbf{P0} = E\left[\,(\mathbf{X0}–\boldsymbol{X}(t0))(\mathbf{X0}–\boldsymbol{X}(t0))^{T}\right] \tag{9}$$

$$\text{Prediction Step}: \mathbf{X}(\mathrm{k}|\mathrm{k}-1) = \mathbf{f}(\mathbf{X}(\mathrm{k}-1|\mathrm{k}-1)), \tag{10}$$

$$\mathbf{P}(\mathrm{k}|\mathrm{k}-1) = \boldsymbol{F}(k-1)\boldsymbol{P}(k-1|k-1)\boldsymbol{F}(k-1)^{T} + \boldsymbol{Q}_{k} \tag{11}$$

We presume that $\mathbf{X}(\mathrm{k}|\mathrm{k}-1)$ and $\mathbf{P}(\mathrm{k}|\mathrm{k}-1)$ represent the estimates of the state and its covariance matrix at time index k, based on all information available up to and including time index k−1. Then, we update the state value from **X** (k|k − 1) to **X** (k|k) using the measurement **Z**(k) with uncertainty denoted by **R**(k) based on the value of **K**(k) called the Kalman gain such that the updated covariance **P**(k|k) has the individual terms along its major diagonal is a minimum leading to

$$\text{Update Step} : \mathbf{K}(\mathrm{k}) = \boldsymbol{P}(k|k-1)\,\boldsymbol{H}(k)^T\,[\boldsymbol{H}(k)\,\boldsymbol{P}(k|k-1)\,\boldsymbol{H}(k)T + \boldsymbol{R}(k)]^{-1} \tag{12}$$

$$\mathbf{X}\,(\mathrm{k}|\mathrm{k}) = \mathbf{X}\,(\mathrm{k}|\mathrm{k}-1) + \boldsymbol{K}(\mathrm{k})\,[\mathbf{Z}(\mathrm{k}) - \mathbf{h}(\mathbf{X}(\mathrm{k}|\mathrm{k}-1))] = \boldsymbol{K}(\boldsymbol{k})\mathbf{v}_{\mathbf{k}} \tag{13}$$

$$\mathbf{P}(\mathrm{k}|\mathrm{k}) = [\mathbf{I} - \mathbf{K}(\mathrm{k})\mathbf{H}(\mathrm{k})]\mathbf{P}(\mathrm{k}|\mathrm{k}-1) \tag{14}$$

where **P** denotes the uncertainty, **F**(k − 1) is the state Jacobian matrix (∂**f**/∂**X**) evaluated at **X** = **X**(k − 1|k − 1) and the measurement Jacobian **H**(k) evaluated at **X** = **X**(k). The **X**(k|k − 1) denotes the value at t(k) based on the process dynamics between t(k-1) and t(k), but before using the measurement information. The observation (measurement) of the process is at discrete time points in accordance with the local linearised relationship **H** = (∂**h**/∂**X**) evaluated at **X** = **X**(k) at the measurement time point. The quantity which is the difference between the actual measurement and the predicted model output

$$\mathbf{v}_{\mathbf{k}} = [\mathbf{Z}(\mathrm{k}) - \mathbf{h}(\mathbf{X}(\mathrm{k}|\mathrm{k}-1))] \tag{15}$$

is called the innovation. Further when the measurement is compared with the updated state **X**(k|k), then the quantity

$$\mathbf{v}_{\mathbf{f}} = [\mathbf{Z}(\mathrm{k}) - \mathbf{h}(\mathbf{X}(\mathrm{k}|\mathrm{k}))] \tag{16}$$

is called the filter residue. As the filter passes through the measurement data, the last measurement provides the best estimate using all the data points. In order to obtain similar estimates at all the intermediate time points using all the measurements, the filter can be operated backwards and with a proper blending provides the smoothed estimates such as by Rauch et al. [29]. The quantity

$$\mathbf{v}_{\mathbf{s}} = [\mathbf{Z}(\mathrm{k}) - \mathbf{h}(\mathbf{X}(\mathrm{k}|\mathbf{N}))] \tag{17}$$

is called as the smoothed residue where $\boldsymbol{X}(\mathrm{k}|\mathbf{N})$ is the smoothed state at time t(k) based on all the measurements **N**.

It may be noted that when the innovation is white it means all the information has been extracted from the data and no further information is left, with both the models and the algorithm have done their best job!

Note that we have combined the local state estimate and the measurement both at time t(k) to obtain an updated state. Further, the use of only the estimate and covariance all over the filter tacitly implies the state and measurement variables are all distributed or approximated as quasi Gaussian. Thus, if initially the state is assumed to be distributed normally with mean **X0** and covariance **P0** then the KF involves iterative operation of two steps: prediction and after an update (also called correction) step there is a subtle reset as a Gaussian. Thus, with all such subjective features, the final answer can only be an answer rather than a unique answer.

There are five steps in the Kalman filter, namely state and covariance propagation with time, Kalman gain calculation and the state and covariance updates by incorporating the measurement. The state propagation and update refer to the sample while the covariance propagation, update and the Kalman gain refer to the ensemble characteristics. It is possible that at times based on the measurements the updates may move locally away from the true values, but with increasing data it will move towards the true value. But this should be noted as the behaviour of the sample. What the filter gain denotes is that in an overall probabilistic **ensemble** sense the Kalman filter will outperform many other estimators. This is analogous in life to some righteous persons who appear to loose but that in the long run they will win! These steps that statistically combine two estimates at any given time point, one from state and the other from measurement equation, are formal if only their uncertainties denoted by their covariances are available. Thus, the states can be estimated given the initial **X0** and **P0** as well **Q** and **R** over time. Over a time span in order to match and minimise the difference called the innovation, in some best possible sense, a well-known criterion is the method of maximum likelihood estimation (MMLE). The innovation follows a white Gaussian distribution which is operationally equivalent to minimising the cost function.

$$\mathbf{J} = (\mathbf{1}/\boldsymbol{N})\Sigma\, \boldsymbol{\nu}_{\boldsymbol{k}}\, [\boldsymbol{H}(\boldsymbol{k})\boldsymbol{P}(\boldsymbol{k}|\boldsymbol{k}-\mathbf{1})\boldsymbol{H}(\boldsymbol{k})\boldsymbol{T} + \boldsymbol{R}(\boldsymbol{k})]^{-\mathbf{1}}\, \boldsymbol{\nu}_{\boldsymbol{k}}^{\boldsymbol{T}} = \mathbf{J}\,(\,\mathbf{X0}, \mathbf{P0}, \mathbf{Q}, \mathbf{R}, \boldsymbol{\Theta}\,) \tag{18}$$

based on summation over all the **N** measurements and has to be solved for **X0, P0, Q, R, Θ.** The importance of the innovation following white Gaussian for filter performance was brought out by Kailath [30]. Generally, mathematical treatment is terse (as the original paper of Kalman deriving the filter from the orthogonal projection principle) and refers to large data. But the sensitivity of the final results to the intermediate statistical quantities for filter consistency (at what confidence level?) is not apparent and sometimes lead to physically unacceptable results as noted in Shyam et al. [31]. Hence, for engineering applications, it is desirable to look for other statistics from the filter. When $\mathbf{Q} \equiv 0$, the MMLE is called as the output error method with the Kalman gain matrix being zero. When $\mathbf{Q} > 0$, the method is called as filter error method. In the usual Kalman filter implementation generally one does not solve for the statistics **P0, Q** and **R** but they are adjusted manually to obtain acceptable results. The numerical effort of minimising **J** has to appear in the estimation of the filter statistics. The Kalman filter is not a panacea to obtain better results when compared to simpler techniques of data analysis. The accuracy of the results using Kalman filter depends on its design based on the choice of **X0, P0, Θ, R** and **Q**. If the above values are not chosen properly then the filter results can be inferior to those from simpler techniques.

Other cost functions can and have been used in Kalman filter work such as the Integral of Time multiplied by Absolute Error with time as a scaling factor. This is meaningful since it is important to ensure a zero error after the filter has converged. This is given by

$$\mathrm{J_{ITAE}} = (\mathbf{1}/\mathbf{T}) \int (1/\mathbf{N}) \sum \mathbf{a_i v_i}\, \boldsymbol{dt} \tag{19}$$

where the $\mathbf{a_i}$ is suitable weight related to the innovation covariance. Another cost function useful to study the effects of inadequate modelling in state estimation problem that is very common in Kalman filter studies has been proposed and used in rendezvous and docking problem by Philip and Ananthasayanam [32] as

$$J_{MODEL} = (1/N)\sum \left(x(i) - x_{ref}(i)\right)\mathbf{P}^{-1}\left(x(i) - x_{ref}\right)^{T} \tag{20}$$

with the summation is over all the **N** time points and the suffix 'ref' refers to a desired reference trajectory to be followed and the argument in x (.) denotes the time step or point. The **P** is the covariance matrix obtained with nominal values for the unknown disturbances. If the variations or a deficiency in the modelling is beyond the statistical fluctuations as denoted by the covariance, then the above cost function changes substantially and indicates a degradation of the filter performance. Further cost functions are introduced here to obtain confidence in the results from the proposed **RRR** for tuning the filter statistics.

3.1. An appreciation of the Kalman filter

Science and technology has progressed through theories and experiments. It is only in Kalman filter that both theory and experiment are handled simultaneously almost all the time. Due to the seemingly unpretentious fact of splitting the state and measurement equations and switching between the state propagation and its update using the measurements, very interesting outcomes have been shown to be possible. This is similar to any amount of deep study and understanding of the state or the measurement equations (theory and experiment) separately may not be able to comprehend the exciting possibilities and abilities when both are combined together. This is similar to the components of a watch, or the cells in an organism leading, respectively, to the time keeping ability or life, which do not exist in the individual components. The GPS is another brilliant example of such a synergism. The competence of the Kalman filter is similar to the saying 'wholes are more than the sum of their parts' as stated by Minsky [33]. It is the above feature that can be called as synergistic, parallel, operator splitting that is the remarkable and profound aspect of the Kalman filter rather than describing it as a sequential least square estimator, or capable of handling time varying states and measurements.

3.2. Aesthetics, beauty, and truth of the Kalman filter

The aesthetics of the Kalman filter is to consider only the estimate and the covariance representing the uncertainty. Just only one additional quantity to move from a deterministic case to probabilistic scenario for describing the results is economical.

In Kalman Filter, at the initial time, the probability density function is assumed to be Gaussian. This need not be true if the state is not a measured quantity whence it could be true or an assumption is made. If the state equations are nonlinear then after the state propagation the probability density function becomes non-Gaussian. Next, in general, the combination of the propagated states and measurements (with any one being non-Gaussian) need not lead to an updated Gaussian. However, at a measurement update, only the estimate and the covariance of the density functions which could be non-Gaussian is used. Thus, even in nonlinear problems after an update when the distribution need not be Gaussian is subtly reset to be a Gaussian with the updated estimate and covariance.

These above are in some ways similar to many other problems in science and engineering wherein only the first and second derivatives or moments alone are considered. This is just the reason and fortuitously as well for using velocity and pressure or temperature in equilibrium thermodynamics, which depend on the first and second moments, respectively, of the distribution function governing the random velocity of the gas molecules. This leads to the consideration of fewer moments or states to describe the dynamics of the gas flow. What would happen if higher order moments had been relevant! Even in classical dynamics in the equations of motion, the linear and angular accelerations are only the second derivatives! As another example, take a rectangular distribution and with increase in sample size the lower order moments converge faster than the higher order moments. This is because away from the middle, the tail controls the higher order moments. For a very similar reason, the Boltzmann equation in kinetic theory deals only with single particle distribution as against multi particle distribution function.

The beauty in the Kalman filter is whether it is true or otherwise many random variables are assumed to follow a multivariate Gaussian distribution, and thus the derived joint and marginal density functions are all Gaussian. The Gaussian distribution provides an enormous amount of mathematical tractability exactly for linear systems and approximately for nonlinear systems.

The truth in the Kalman filter equations is that once it is derived in one way, it is possible to derive it in a variety of ways with slightly different assumptions, but mostly leading to similar set of basic equations as for linear problems. The author of each book has his own derivation! It is interesting to note that the simplest formulation of the Kalman filter is based on minimum amount of *a priori* knowledge or information in probability, statistics, and random process providing, respectively, the Gaussian distribution, linear relationship among the variables, and white noise [12]. Otherwise one has to justify them which may not be simple or easy. If necessary other suitable distributions, nonlinearity, and coloured noise can be introduced later into the filter framework.

3.3. Different perspectives and competence of Kalman filter

The Kalman filter can be viewed as an inverse problem, deterministic or probabilistic approach, reversing an 'irreversible' process. Also it can be considered as qualitative modelling and quantitative estimation, stochastic corrective process by Narasimha [34], data fusion and statistical estimation by probabilistic mixing, optimization in the time (or frequency) domain. The competence of the filter consists of estimating unknown or inaccurately known parameters (including deterministic errors) in the state and measurements, estimation of process and measurement noise, improved states and measurements by smoothed filter estimates, estimation of unmodelable inputs by modelling them in a probabilistic way, unobservables from observables, expansion of the scenario, handling computational errors by noise addition, consistency check of the whole process of modelling, convergence of the numerical algorithm, and the extraction of the complete information from the data by checking the correlation of the innovation sequence. All of the above are achieved based on the measurements and suitable modelling of the state and measurement equations. The above aspects are discussed in [12, 31].

3.4. Kalman filter and some of its variants

For nonlinear systems, even if the initial distribution is assumed normal, it gets distorted after propagation and so a suitable local approximation or quasi linearization has to be made. In the

http://dx.doi.org/10.5772/intechopen.71961

EKF the nonlinear systems and/or measurements equations are approximated by appropriate first order Taylor series expansion. The probability density is approximated by a Gaussian, which may distort the true structure and at times could lead to the divergence between the filter prediction and the measurements. In the Unscented Kalman filter (UKF) by Julier et al. [35] approach, instead of linearizing the functions, a set of chosen points are propagated through the nonlinear transformation. These points are so chosen such that the mean, covariance, and possibly also higher order moments match better with the propagated distribution. At an update, only the estimate and covariance matters.

The particle filtering by Gordon et al. [36] is a Monte Carlo technique for state estimation that can handle nonlinear models together with non-Gaussian noise. Here, the state probability density is approximated by using point particles having positive weights. Based on the initial distribution, the weights are chosen and then the particles are propagated following the system dynamics together with the state noise. Then using the measurement, their weights are adjusted and normalised among all the particles. The particles that can track the measurements gain weight and the ones far away lose their weights. However, after a while, all but one weight will become zero leading to degeneracy. A resampling scheme is introduced to solve the degeneracy problem that discard the particles with small weights and focus on the particles with more significant weights. Then, the procedure continues sequentially over the measurements.

For large size systems, such as those occurring in geophysical studies maintaining the covariance matrix computationally being difficult, in the ensemble KF (EnKF) for large problems Evensen [24], the estimate and the covariance matrix are replaced by the sample covariance from a large number of ensemble members similar to a particle in the particle filter. Each member of the ensemble is propagated including the process noise and later updated using a so-called virtual observation. Again, the procedure continues sequentially over the measurements.

One may note the evolution of the variants of the Kalman filter possesses some similarities as it progressed to handle simple, complex, to massive problems as in many other fields such as fluid dynamics or structural mechanics. In these cases, commencing from simple geometries one obtains closed form analytical solutions as in the linear KF, wherein the gains can be pre computed to process the data as and when they arrive. Then for involved nonlinear state and measurements one uses local linearization and numerical calculations as in EKF. When the geometry is complex and the boundary conditions are involved, it becomes necessary to discretise and form cells over appropriate space and time as in particle filtering to obtain the solution. Further when massively complex geometries and boundary conditions occur other innovative approaches like Ensemble Kalman Filter have been developed. An extensive bibliography of the nonlinear estimation is provided by Georgios [37] and an excellent review of nonlinear filters is given by Daum [38].

4. Tuning of the Kalman filter statistics

The solution for the linear filtering problem in discrete time was proposed in the famous 1960 paper by Kalman [4]. This was followed for continuous time in 1961 by Kalman and Bucy [5].

Not many know that the enthusiasm that followed soon after Kalman introduced his filter was damped, the solution was only formal and the statistics of the process **Q** and measurement noise **R** had to be specified to design and implement the filter. After the 1960 paper of Kalman by the time of Gelb's book [39] in 1974 most filter approaches, applications and numerical procedures except the tuning of the filter statistics were in place. Kalman when he proposed the filter dealt with only state estimation. In many present day applications, one does not even know the structure of the state and measurement equations as well as the parameters in them and the statistical characteristics of the state and measurement noise. One can also add the unknown initial state conditions **X0**. All the unknowns have to be estimated using the measurements only. The estimation of the system parameters **Θ**, **X0, P0**, together with **Q** and **R** is called filter design or filter tuning.

An interesting feature of KF is that one can use it to start with at least, without understanding the derivation can tune the filter by trial and error procedures for the statistics without carrying out an optimization as mentioned by Sorenson [40]. Even though the estimator performance may be satisfactory for some '*a priori*' reasonable choice of **P0**, **Q** and **R**, it could lead to unacceptable results in many cases. Rarely, sensitivity studies on **R** and **Q** like by Subbaraju et al. [41] while estimating the drag in the presence of thrust of Satellite Launch Vehicles are reported. Most reports and publications write out detailed filter equations but the tuning procedures are not spelt out. In fact, the ghost of filter tuning chases without exception every variant or formulation of the Kalman filter. If not tuned properly, it is difficult to infer if the performance of the filter is due to its formulation or filter tuning! It is surprising that most text books on Kalman filtering provide a scanty treatment of the problem of filter tuning that is at the heart of KF design.

One has to tune the statistics $\mathbf{P_0}$, **Q** and **R** for a satisfactory filter operation and even now this is generally done manually! Usually, the filter statistics are tuned off line using simulated data and subsequently used for on line and real time applications with some modifications. In spite of its immense applications for more than five decades in many problems of science and technology, the filter tuning has not matured to an easily implementable approach even to handle a constant signal with measurement noise! Generally the tuning is manual or with ad hoc quick fix solutions such as limiting **P** from going to zero, or adding **Q** to increase **P** before calculating the gain and multiplying **P** by a factor to limit **K**, all have obviously limitations in handling involved problems or scenarios. All the above introduce additional parameters to be adjusted that varies for every problem. The present work provides a generalised heuristic approach together with consistency check.

4.1. Qualitative features of P0, Q, and R the filter statistics

Should the $\mathbf{P0} \equiv \mathbf{Q} \equiv \mathbf{0}$ then the filter will not learn anything from the measurements at all which will be ignored. The **P0** is tricky and generally the off-diagonal elements are set to zero and the diagonal elements are set to large values. However their relative values are crucial for an optimum filter operation. The **R** can be determined from the measured data. In fact if one is satisfied with the measurement accuracy then no filter is needed. The main activity of the filter is to follow the measurements and further reduce or suppress the effect of noise. In spite of being labelled as 'notorious' it is only the **Q** that an analyst can estimate, account or offset for some deficiency, inaccuracy, or error in the following namely in the initial conditions, the unmodelled or unmodelable errors in the system and measurement equations, control or external input, measurement noise statistics, the errors in the numerical state and covariance propagation, a small increase in **Q** to offset modest errors in Taylor series approximations in EKF(!), or in the update operations of the Kalman filter. The **Q** helps to inject uncertainty into the state equations to assist the filter to learn from the measurements and also controls the steady state filter response. Too large a value of **Q** will lead to a short transient with large steady state uncertainty of the estimates and a small **Q** vice versa.

The **Q** is helpful to track systems whose dynamical equations are unknown. Some classic examples are the GPS receiver clocks, satellite, trajectory of aircraft, missiles, and re-entry objects. These are handled by using the kinematic relations as state equations among the position, velocity, acceleration and even jerk [42] driven by white Gaussian noise **Q** to enable the filter to track these systems. In general one can simulate any real world dynamical systems. Apart from simulating the dynamical system the process noise inhibits the onset of instability of the filter operation. Though **Q** is considered notorious it is the life line of the Kalman filter.

4.2. Choice of X0 and P0 for states and parameters

Since some of the states are generally measured either the first or the average of the first few measurements can be taken as the initial value **X0** for the state. The initial parameters values can be guessed if experimental or computational results are available. The **P0** is one of the important tuning parameters as stressed by very few like Maybeck [43], Candy [44], Gemson [45], Gemson and Ananthasayanam [46], and Sarkar et al. [47]. Generally a guess **P0** tends to become very low after some data points. In order to make the filter learn from the subsequent measurements an additional **Q** is introduced into the state equations even when there is no model uncertainty. In the present **RRR** a proper **P0** without any **Q** is shown to be possible for the above. The choice of **P0** can affect the final covariance from the filter operation, which can be crucial in certain state estimation problems such as impact point estimation and its uncertainty for target tracking. Even in parameter estimation problems the estimates and their uncertainties can be important in the design of control systems.

4.3. Tuning filter statistics with both R and Q

When the data contains the effect of both **R** and **Q** it becomes notorious for analysis. With no **R** and **Q** the system dynamics is exact. The process noise input at various times makes the system to wander randomly. When measurements are made on this wandering dynamical state it is blurred. The smoothing filter provides the best possible state estimate at all time points by suppressing the effect of measurement noise. Hence it is best to consider the smoothed state in order to estimate the process noise.

$$\left(\text{State dynamics with } \mathbf{R} \equiv \mathbf{Q} \equiv \mathbf{0}\right) + \text{Cumulative Effect of } \mathbf{Q} = \text{Smoothed State} \tag{21}$$

$$\text{Smoothed State} + \mathbf{R} = \text{Measured Signal or in other words} \tag{22}$$

$$\text{Measured Signal} + \text{Forward and Smoothing filter} = \text{Smoothed State with } \mathbf{R} \equiv \mathbf{0} \tag{23}$$

Shumway and Stoffer [48] did just this using the Estimation Maximisation (EM) approach. The book by McLachlan and Krishan [49] is well known on EM. Shyam et al. [31] introduced another approach to estimate **Q** based on the difference between the stochastic and dynamical trajectory called DSDT which provided statistically similar results to EM.

The interesting point is the filter by tracking the drifted dynamical behaviour with **Q**, it estimates the parameters controlling the original dynamics of the system without the effect of **R** and **Q**. Since **R** and **Q** occur, respectively, in the measurement and state equations their effects on the filter are negatively correlated as stated by Bohlin [50]. Thus during simultaneous recursive estimation if the statistics for estimating them are not properly chosen then **R** is overestimated and **Q** is underestimated and vice versa. This is just the reason in Gemson [45] and Gemson and Ananthasayanam [46] one has to update **R** and **Q** alternately. The filter operating on the data generates prior, posterior, and smoothed state estimates and their covariances thus helping to generate candidate 'statistic' to estimate **R** and **Q**. In EKF if the unknown noise covariances are incorrectly specified biased estimates can arise. Even when the **Θ** are known, with an inaccurate **R** and **Q** the filter may give poor estimates, or even diverge [51].

4.4. Adaptive Kalman filtering approaches

There are broadly four approaches for adaptive filtering namely Bayesian, Maximum Likelihood, Covariance Matching and Correlation Techniques (Mehra [52]) apart from other techniques. The present **RRR** falls in the category of covariance matching. Why there are so many formulations for solving an optimization problem? The reason is the unknowns do not occur in a simple way in the cost function, and there are many transformed variables with which one tries to solve for the basic unknowns, further the size and the required compatibility conditions among the transformed variables lead to the many difficulties not generally found in the classical optimization problems. Also many attempts have been made using probabilistic methods. However when the dimension, nonlinearity and the range of search space become large these could become computationally prohibitive and could lead to local minimum. One can summarise that either deterministic or probabilistic optimization approaches do not appear to be easy and general for solving the filter tuning problem. A simple recursive filtering approach was tried and fortuitously it did as will be shown subsequently.

Exact filtering solutions are very hard, approximate choice can lead to inappropriate results but heuristic approaches provide the middle path in designing the Kalman filter like the **RRR**. An adaptive heuristic approach in general updates the **X0, P0, Θ, R** and **Q** at a point, over a window, after a pass or after multiple passes by applying some corrections to them based on changes, iterations or sample statistics such that the numerical solution does not diverge but converges to the best possible estimates. Examples of heuristic approach is by Myers and Tapley (MT) in [53] for **R** and **Q** and in [45, 46] for **P0**, **Q**, and **R**.

5. Cost functions along with RRR for checking Kalman filter tuning

The present **RRR** contains no direct optimization of any cost function. We have purely iterated the filter on the measured data; but after each iteration, the unknowns **X0, P0**, **Q**, **R** and **Θ** have been updated. Such an iterative procedure leads to a converged solution for the estimate as well as the CRB that is very close to that obtained using any optimization method that minimises any suitable cost function. The estimates for the parameter, the covariance for uncertainty, and the noise statistics reach very near their final values in about five to ten iterations that can be extended for higher accuracy. Also, we introduce many cost functions as below to provide an indication about the filter performance and consistency.

Fundamentally, the estimation theory (ET) is an optimization problem. Hence a suitable cost function **J** is generally chosen. Essentially, there are two elements in ET: (i) defining a cost function and (ii) adopting a suitable algorithm to minimise the cost function. The general log likelihood cost **L** for normally distributed error 'e_k' summed over **N** time points is given by Gemson [45]

$$\mathbf{L}(\Theta) = (1/\mathbf{N}) \sum e_k \mathbf{A}_k{}^{-1} e_k{}^T + \log\ (\det(\mathbf{A}_k)) \tag{24}$$

where $\mathbf{A_k}$ is the error covariance matrix and det(**A**) represents determinant of matrix **A**. It may be noted the parameters **Θ** occur implicitly in the cost function **L**. From the $e_k$of the first term, one can see the principle of weighted least squares, and **L** for N-1 and N terms the sequential least squares with the constraint of the state equation as mentioned by Sorenson [40].

Further based on many available statistics from the filter operation it is possible to define many more cost functions (not written here for brevity but available in Shyam et al. [31]), but only describe them. The cost **J0** is based on squaring $[(\mathbf{X0}-\mathbf{X}(t0)]$ and scaled with respect to **P0** shows how well the initial conditions are balanced. Similarly the set of costs **(J1, J2, J3)** and **(J6, J7, J8)** are respectively derived from the sets $(\mathbf{v_k}, \mathbf{v_f}, \mathbf{v_s})$ and $(\mathbf{w}_1(k|\mathbf{N}), \mathbf{w}_2(k|\mathbf{N}), \mathbf{w}_3(k|\mathbf{N}))$ which are different estimates for local measurement and process noise samples. These indicate how well the measurement and state equations are balanced and should tend to the number of measurement and state equations for a good solution. When $\mathbf{Q} \equiv 0$, the cost **J4** is the difference between measurement and state dynamics based on the estimated parameter and is expected to tend towards the trace of **R**. The cost **J5** equals **J1** + log (**det**(cov($\mathbf{v_k}$)) is the negative of the log likelihood based on the innovation. All such costs indicate how well the state and measurement equations are balanced, and further the estimates and the covariances both given by the filter are consistent as well. One can formulate any number of cost functions to estimate the parameters and the filter statistics. But it is not possible to estimate the true value of the unknowns but be only around them due to statistical fluctuations percolating all over the unknowns.

5.1. Probability matching prior interpretation for P0

This interpretation was given earlier in the paper by Ananthasayanam et al. [54], and used by Shyam et al. [55]. As mentioned earlier the importance of **P0** has not been much appreciated in the literature on ET and more so in Kalman Filtering though statisticians have been discussing the philosophical and practical differences between Frequentist and Bayesian approach. The deterministic Newton-Raphson (NR) optimization of a cost function approach in Ananthasayanam et al. [56] provides Frequentist results and the Kalman filtering approach is the Bayesian route. Consider the simple case of a constant signal with measurement noise. In the Frequentist approach the calculation of the mean and standard deviation and the noise is simple. However in the Bayesian approach the above result is not reachable unless a proper **P0** is also chosen. The choice of appropriate **P0** is the probability matching prior (**PMP**) providing a bridge between the above approaches. With a large amount of data the differences in the results from the above approaches vanish. Since **PMP** is not unique its choice depends on the purpose. Presently **P0** is chosen to obtain proper estimates and CRBs for the unknown parameters as well as the noise statistics **R** and **Q**. The success of **RRR** has been due to the choice of **P0** by scaling and further trimming it. Further in addition the simultaneous choice of appropriate statistics for **R** and **Q** has been made using the many filter statistics available after every filter pass using the EM approach. When $\mathbf{Q} \equiv 0$ the choice for **R** is easy but when $\mathbf{Q} > 0$ since the Kalman filter is compulsory in both approaches we look for consistency based on simulated studies by comparing the statistical characteristics of the injected and estimated **R** and **Q** noise sample sequences. Further the various cost functions introduced earlier in **RRR** help to obtain confidence in the results and more so while analysing real flight test data. Since the present **RRR** is believed to provide near optimum but not an exact solution it is called as a 'reference' and not a 'standard'.

5.2. Choice of X0 and P0 in RRR

Commencing from an assumed reasonable initial choice for **X0**, **P0**, **Θ**, **R** and **Q** the first filter pass through the data is made. Then a backward smoothing is carried out using the Rauch et al. [29] smoother. The smoothing leads to the best possible state and parameter estimates and their covariances based on all the data. After smoothing the state estimates and their covariances change but not those of the parameters. If one uses the smoothed initial state covariance **P(0 | N)** and use it as the **P0** for the next pass then the final covariance will keep on decreasing with further filter passes and eventually tend towards zero. In order to overcome this, the final covariance at the end of the pass was scaled up by Shyam et al. [31] by **N** to provide **P0** at the beginning of the next pass:

$$\mathbf{P0} = \mathbf{N} \times \mathbf{P}(\mathbf{N}|\mathbf{N}) \tag{25}$$

A heuristic reasoning from statistics is that the mean from a sample has an uncertainty **P** that keeps decreasing with sample size as **P/N** where **P** is the population variance. Since, in the filter steps, the estimates and their update refer to the sample and the other covariance propagation and their update and the calculation of the Kalman gain refers to the ensemble characteristics before every filter pass, we carry out the above scale up to obtain the **P0** (after further trimming as well) for the next filter pass.

5.3. Estimation of R and Q using the EM/DSDT, MS and MT methods

We now provide some estimates for the measurement noise covariance **R.** Bavdekar et al. [57] use the extended EM given by:

$$\mathbf{R} = (1/\mathbf{N}) \sum_{1}^{N} \nu_k \, \nu_k^T + \mathbf{H}(k|\mathbf{N})\mathbf{P}(\mathrm{k}|\mathbf{N})\boldsymbol{H}(k|\boldsymbol{N})^T \tag{26}$$

The choice of Mohamed and Schwarz (MS) in [58] based on filter residue is

$$\mathbf{R} = (1/\mathbf{N}) \sum_{1}^{N} \boldsymbol{\nu}_f\, \boldsymbol{\nu}_f^T + \mathbf{H}(\mathrm{k}|\mathrm{k})\mathbf{P}(\mathrm{k}|\mathrm{k})\boldsymbol{H}(k|k)^T \tag{27}$$

The choice of Myers and Tapley (MT) in [50] based on innovation is

$$\mathbf{R} = (1/\mathbf{N}) \sum_{1}^{N} \boldsymbol{\nu}_s\, \boldsymbol{\nu}_s^T - \mathbf{H}(\mathrm{k}|\mathrm{k}-1)\mathbf{P}(\mathrm{k}|\mathrm{k}-1)\boldsymbol{H}(k|k-1)^T \tag{28}$$

Bavdekar et al. [57] use the smoothed statistic $w_1(\mathrm{k}|\mathbf{N}) = \mathbf{X}(\mathrm{k}|\mathbf{N})–\mathbf{f}(\mathbf{X}(\mathrm{k}-1|\mathbf{N}))$ in EM estimate as

$$\begin{aligned}\mathbf{Q} = (1/\mathbf{N}) \sum_{1}^{N} \Big\{ & w_1(k|\mathbf{N})\ w_1(k|\mathbf{N})^T + \mathbf{P}(\mathrm{k}|\mathbf{N}) + \mathbf{F}(\mathrm{k}-1|\mathbf{N})\mathbf{P}(k-1|N)\ F(k-1|\mathbf{N})^T \\ & -\mathbf{P}(k, k-1|N)F(k-1|\mathbf{N})^T - P(k, k-1|\mathbf{N})^T\mathbf{F}(\mathrm{k}-1|\mathbf{N}) \Big\}\end{aligned} \tag{29}$$

The DSDT statistic for $\mathbf{Q}$ is given in Shyam et al. [31]

$$\begin{aligned}\mathbf{Q} = \left(\frac{1}{N}\right) \sum_{1}^{N} \Big\{ & w_2(k|\mathbf{N})\, w_2(k|\mathbf{N})^T + \mathbf{P}(k|\mathbf{N}) + \mathbf{Fd}(k-1|\mathbf{N})\mathbf{P}(k-1|\mathbf{N})\boldsymbol{Fd}(k-1|\mathbf{N})^T \\ & -\mathbf{P}(k, k-1|N)\boldsymbol{Fd}(k-1|\mathbf{N})^T - P(k, k-1|\mathbf{N})^T\mathbf{Fd}(\mathrm{k}-1|\mathbf{N}) \Big\}\end{aligned} \tag{30}$$

where $\mathbf{w}_2(\mathrm{k}|\mathrm{N}) = \mathbf{X}(\mathrm{k}|\mathbf{N})–\mathbf{Xd}(\mathrm{k}|\mathbf{N})–\boldsymbol{Fd}(\mathrm{k}-1|\mathbf{N})(\mathbf{X}(\mathrm{k}-1|\mathbf{N})–\mathbf{Xd}(\mathrm{k}-1|\mathbf{N}))$ and with $\mathbf{Xd}(0|\mathbf{N}) = \mathbf{X}(0|\mathbf{N})$ is the predicted state trajectory without measurement and process noise using the estimated parameter $\boldsymbol{\Theta}(\mathbf{N}|\mathbf{N})$. The $\mathbf{P}(\mathrm{k}|\mathbf{N})$ is the smoothed covariance and $\mathbf{P}(\mathrm{k}, \mathrm{k}-1|\mathbf{N})$ is the lag-one covariance for k = **N**-1, **N**-2, …1.

The Mohamed and Schwarz [58] estimated $\mathbf{Q}$ in terms of the innovation and the smoothed gain $\mathbf{K}(\mathrm{k}|\mathbf{N})$ based on [29] by

$$\mathbf{Q} = \mathbf{K}(\mathbf{k}|\mathbf{N}) \left\{ \sum_{1}^{N} \boldsymbol{\nu}_s\ \boldsymbol{\nu}_s{}^T \right\} \boldsymbol{K}(k|N)^T \tag{31}$$

The choice of Myers and Tapley [53] for $\mathbf{Q}$ with $\mathbf{w}_3(\mathrm{k}|\mathbf{N}) = \mathbf{X}(\mathrm{k}|\mathrm{k}) - \mathbf{X}(\mathrm{k}|\mathrm{k}-1)$ is

$$\mathbf{Q} = \left(\frac{1}{N}\right) \sum_{1}^{N} \left\{ w_3(\mathrm{k}|\mathrm{k})\ w_3(k|k)^T - \left(\mathbf{F}(\mathrm{k}-1|\mathrm{k}-1)\mathbf{P}(\mathrm{k}|\mathrm{k}-1)\mathbf{F}(\mathrm{k}-1|\mathrm{k}-1)^{\mathrm{T}} - \mathbf{P}(\mathrm{k}|\mathrm{k}) \right\} \right. \tag{32}$$

All the process noise samples, $\mathbf{w}_1(\mathrm{k}|\mathrm{k})$, $\mathbf{w}_2(\mathrm{k}|\mathrm{k})$, and $\mathbf{w}_3(\mathrm{k}|\mathrm{k})$, are assumed to be of zero mean. It turns out that the smoothed statistics $\mathbf{w}_1(\mathrm{k}|\mathrm{k})$ and $\mathbf{w}_2(\mathrm{k}|\mathrm{k})$ based on EM and DSDT, respectively, are very close and either can be used for $\mathbf{Q}$ estimation.

5.4. The RRR method for the EKF

The following steps explain the recursive or iterative **RRR** algorithm for the EKF:

1. Given the system model and the measurements, the first filter pass through the data of EKF is carried out using guess values of **X0**, **P0**, **Θ**, **R** and **Q**.

2. The RTS smoother is used backwards to get smoothed state and covariance estimates.

3. If **X0** is unknown, then the smoothed state values can be used as the initial state values.

4. The estimated smoothed **P0** is scaled up by the number of time points **N** and further all elements except the diagonal terms corresponding to the parameters are set to zero. Due to the effect of statistical percolation effect, the estimated **R** and **Q** will in general be full. But, only the diagonal terms in **Q** need to be used in the basic state equations and not in the parameter states. Only the diagonal terms in **R** need to be used in the measurement equations. These are summarised as below. The quadrant on the upper left denotes the state, the bottom right the parameter states, and the others the cross terms. The (0) below shows the null and D the diagonal matrices. This is followed for all iterations.

$$\mathbf{P0} = \begin{bmatrix} 0 & 0 \\ 0 & D \end{bmatrix}; \ \mathbf{Q} = \begin{bmatrix} D & 0 \\ 0 & 0 \end{bmatrix}; \ \mathbf{R} = [D] \tag{33}$$

5. Then, the filter is run again using the above updates of **X0**, **P0**, **Θ**, **Q** and **R** till statistical equilibrium is reached.

6. Different cost functions (**J1–J8**) are checked for convergence.

The convergences of the following quantities after all the iterations are analysed:

1. The parameter estimates **Θ** and their covariances **P(Θ)**.

2. The sample noise sequences $\mathbf{v_k}$, $\mathbf{v_f}$, $\mathbf{v_s}$ and of $\mathbf{w}_1$, $\mathbf{w}_2$, and $\mathbf{w}_3$ with ± 1 sigma bounds, their autocorrelations and **Q** and **R**.

3. The state dynamics **Xd** (with $\mathbf{R} \equiv \mathbf{Q} \equiv 0$) based on the estimated **Θ**, smoothed **X**(k | **N**) and the measurement **Z**(k).

4. The various cost functions **J1–J8** after the final convergence.

5.5. Some remarks on running the RRR

If the value of **Q** for any state is known to be zero, then the value of **Q** is set at 10^{-10} or lower for all iterations to help the filter, which would otherwise generate a pseudo-**Q**, and then slowly grind it to zero in thousands of iterations. For $\mathbf{Q} \equiv 0$ case, one can estimate **R** even by ignoring the second order covariance terms. It is of interest to note that for $\mathbf{Q} > 0$ case unless the second order covariance terms are also included in the estimate for both **R** and **Q** the **RRR** does not converge.

http://dx.doi.org/10.5772/intechopen.71961

6. Simulation study of a spring, mass, and damper (SMD) system

The **RRR** is first applied to a very simple spring, mass, and damper system with $\mathbf{R} \equiv \mathbf{Q} \equiv 0$. For such a situation, the Newton Raphson optimization of the innovation cost [56] served as a bench mark for tuning the filter statistics in **RRR** (with no cost optimization!) to get the closest possible estimates and the CRB. Later, when **Q** is included, we looked for the consistency between the injected **R** and **Q** noise sequences and their statistics. The SMD system with weak nonlinear spring constant in time (t) is given by

$$\dot{x}_1(t) = x_2(t);\ \dot{x}_2(t) = -\Theta_1 x_1(t) - \Theta_2 x_2(t) - \Theta_3 x_1^3(t) \tag{34}$$

where x_1 and x_2 are the displacement and velocity state with initial conditions 1 and 0, respectively. The 'dot' represents differentiation with respect to time (t). The unknown parameter vector is $\boldsymbol{\Theta} = [\Theta_1, \Theta_2, \Theta_3]^{\mathbf{T}}$ with true values $\boldsymbol{\theta}_{\mathbf{true}} = [4.0, 0.4, 0.6]^{\mathbf{T}}$. Θ_3 is a weak parameter and does not affect the system dynamics much. The complete state vector $\mathbf{X} = [x_1, x_2, \Theta_1, \Theta_2, \Theta_3]^T$ is of size (5 x 1). The measurement equation is given by

$$\mathbf{Z}(\mathrm{k}) = \mathbf{H}\,\mathbf{X}(\mathrm{k}) \tag{35}$$

where $\mathbf{H} = \begin{bmatrix} 1 & 0 & 0 & 0 & 0 \\ 0 & 1 & 0 & 0 & 0 \end{bmatrix}$ is the measurement matrix. The noise covariances are **R** = diag (0.001, 0.004) and **Q** = diag (0.001, 0.002), thus keeping the noise levels of same order to test the robustness of the **RRR**. The initial guess value of **P0**, **Q**, were chosen, respectively, as diagonal with all values as 10^{-1}, and **R** = 2^{-1} for all measurement channels. The initial **Θ** was chosen within ±20% error of the true values. A total of **N** = 100 measurement data are simulated with the time varying from 0 to 10 s in very small steps of dt = 0.1 s. For zero process noise case, the maximum number of iterations is set to 20 over 50 simulations, and for nonzero process noise case, it is set to 100 over 50 simulations for obtaining generally four digits accuracy (though not necessary). The brief results as presented in **Figures 2** and **3**. In the present **RRR**, it was noticed that generally even if the initial **P0**, **Q**, and **R** were varied over a wide range of powers from −3 to +3 (or even more) together with all the initial **Θ** parameter set to zero leads to the same estimation results for a given data showing its robustness. Further, it may be noted that for all simulated and real data analysed and reported in [31] even if all the initial state and parameters were set to zero the solutions converged to the appropriate values!

6.1. Analysis of simulated SMD data

The **RRR**, when compared to [53, 57, 58] has hardly any instability when processing the simulated data, reaching statistical equilibrium in around 20 iterations. This can be seen from **Figure 2** for Θ**,** their CRBs (= **P0/N**), **R**, **Q**, **J1–J8.** The cost functions **J1–J8** provides confidence in the results and to compare **RRR** with other approaches. **Figure 3** shows the measurements wrap around the smoothed estimates. More details are available in Shyam et al. [31]. For further insight, the filter results with 500 iterations shown in **Figures 4** and **5** indicate three phases. Firstly, from a

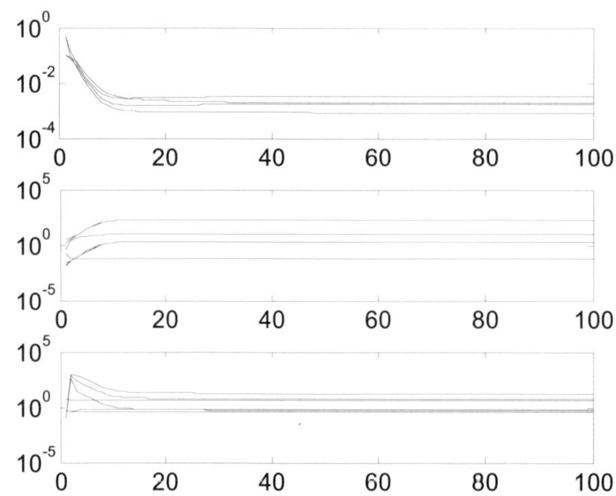

Figure 2. (i) Variation of **Q** and **R**, (ii) initial parameters **Θ**, and **P0**, and (iii) costs **J1–J8** with iterations.

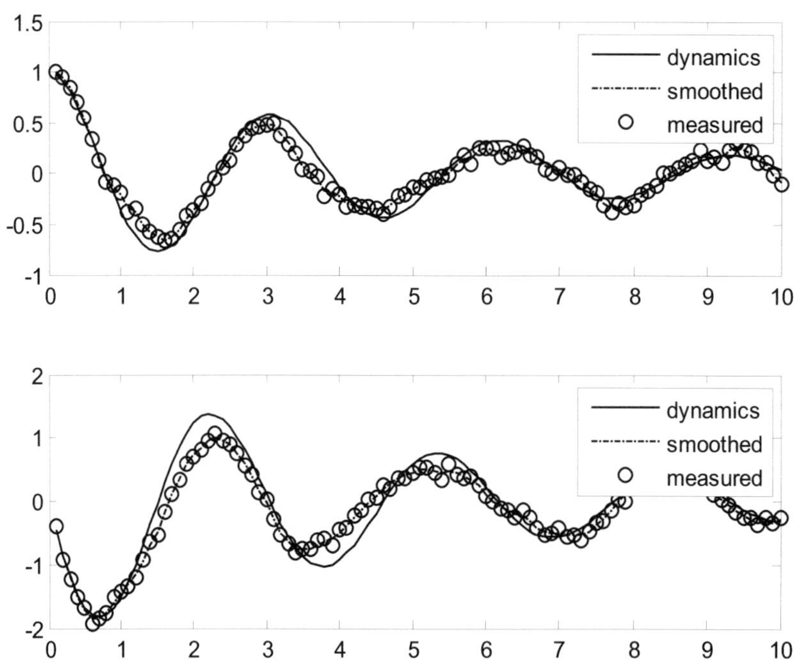

Figure 3. **Xd**, **X**(k|N), and **Z**(k) by (o) for displacement and velocity versus time (s).

nonequilibrium state the filter statistics reach statistical equilibrium in 10–20 iterations enough for use in practice. The next phase shows the second moments **R** and **Q** converge earliest followed by Θ and finally their CRBs. This is at variance with the fact mentioned in Section 3.2, namely lower order moments converge faster than higher order moments for samples from a distribution. But the statistics derived here from the Kalman filter are not simple. Finally, there is equilibrium with numerical fluctuations depending on computer accuracy.

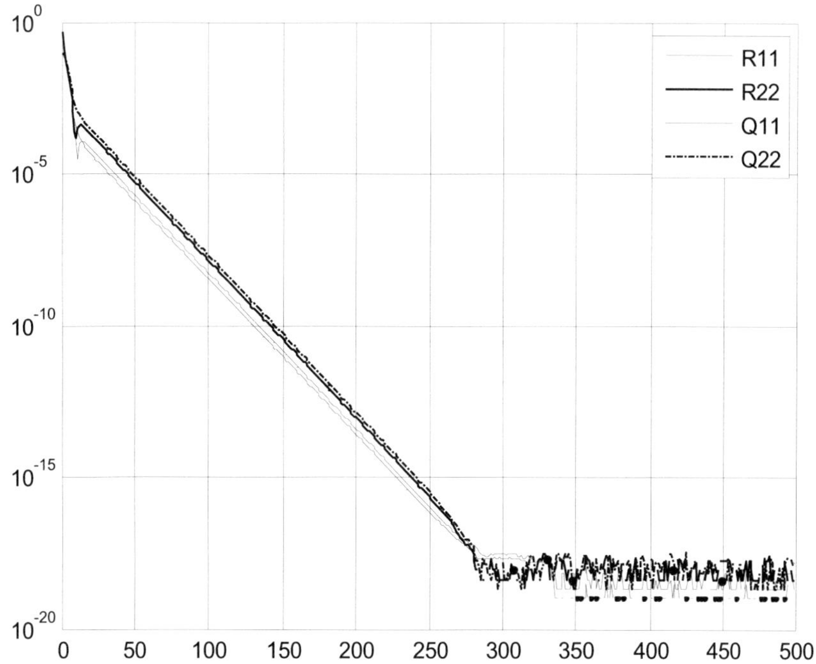

Figure 4. Difference of the diagonal elements of **R** and **Q** from their values after 500 iterations.

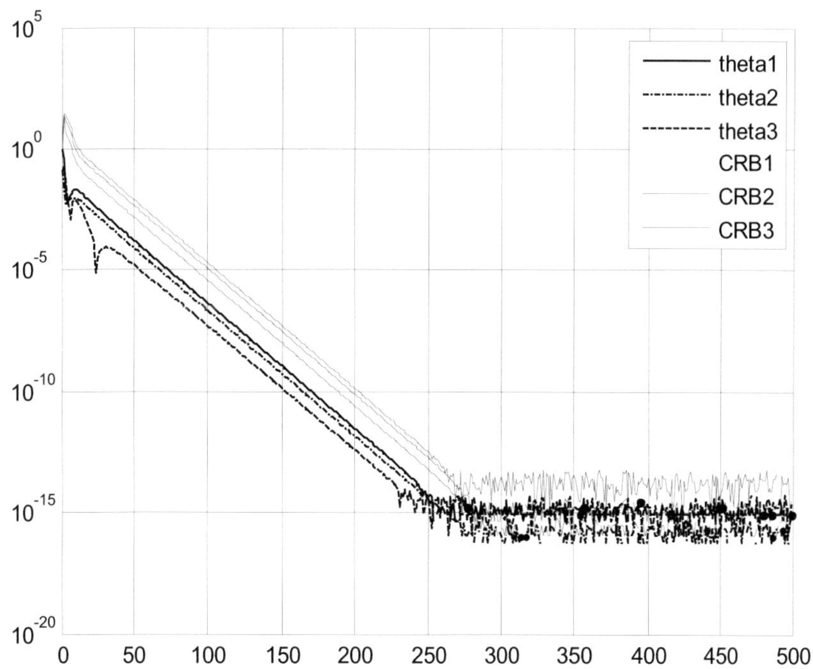

Figure 5. Difference of parameters **Θ** and their CRBs from their values after 500 iterations.

6.2. Analysis of real airplane flight test data

The real airplane flight tests cannot always be conducted in an ideal situation of **Q** and **R** being white Gaussian. The measurements may not be at the center of gravity, possess bias and scale factors, which have to be modelled and estimated. The coupling between the longitudinal and lateral motion brings in difficulty but makes the problem interesting. At times, the noisy measurements from the lateral motion are fed into the longitudinal states, and thus are input as process noise. This is an example of introducing subjectivity in estimation theory. The real data set is obtained along with notations from [59]. There is a peculiar manoeuvre, where the elevator angle (δ_e in deg) shown in **Figure 6** is imparted when the aircraft (T 37B) is rolling through a full rotation during aileron roll. The coupling between the longitudinal and lateral motion is replaced by their measured values, namely the roll angle (φ_m), sideslip (β_m), velocity (V_m), roll rate (p_m), yaw rate (r_m) and the angle of attack (α_m) as shown in **Figures 6–8**. The state equations (n = 3) for the angle of attack (α), pitch rate (q), and the pitch angle (0), respectively, are

$$\dot{\alpha} = -\frac{\bar{q}\,S}{mV_m\,Cos\left(\beta_m\right)}\left(C_{L_\alpha}\,\alpha + C_{L_{\delta_e}}\,\delta_e + C_{L_0}\right) + q + \frac{g}{V_m\,Cos\left(\beta_m\right)}\left(\cos\left(\phi_m\right)\left(\cos\left(\alpha_m\right)\cos\left(\theta\right) + \sin\left(\alpha_m\right)\,\sin\left(\theta\right)\right) - \tan\left(\beta_m\right)\left(p_m\cos\left(\alpha_m\right) + r_m\ \sin\alpha_m\right)\right) \tag{36}$$

$$\dot{q} = \frac{\bar{q}S\bar{c}}{I_{yy}}\left(C_{m_\alpha}\alpha + C_{m_q}\frac{\bar{c}}{2V}q + \ C_{m\dot{\alpha}}\ \frac{\bar{c}}{2V}\ \dot{\alpha} + C_{m_{\delta_e}}\ \delta_e + C_{m0}\right) + \frac{I_{zz} - I_{xx}}{I_{yy}}\ r_m\,p_m \tag{37}$$

$$\dot{\theta} = q\cos\left(\phi_m\right) - r_m\sin\left(\phi_m\right) + \theta_0 \tag{38}$$

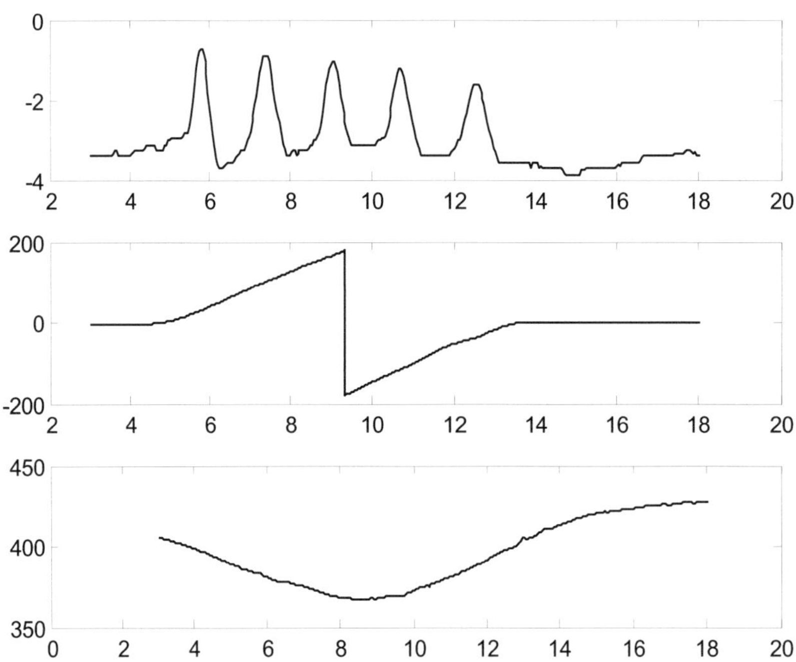

Figure 6. Control input δ_e in deg, roll angle ϕ_m in deg, velocity V_m in ft/sec versus time (s).

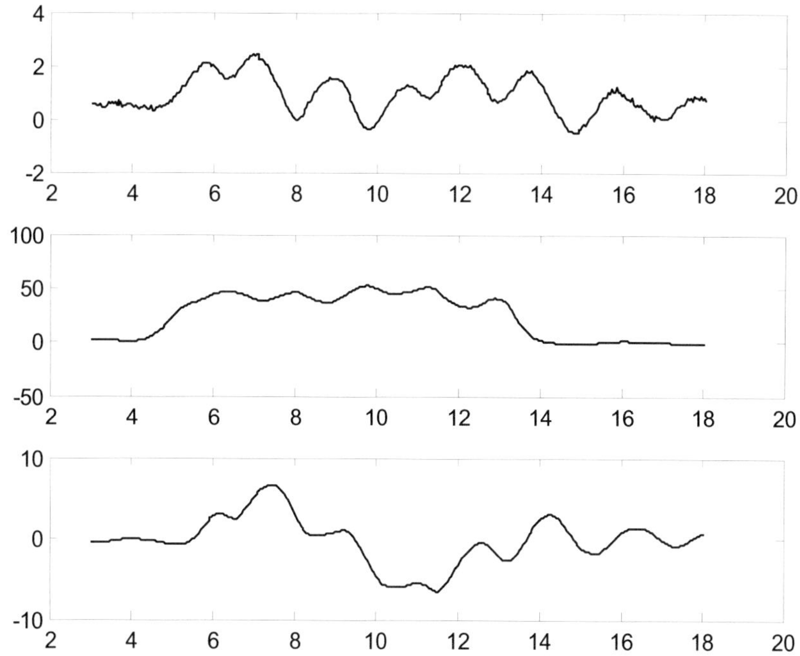

Figure 7. Sideslip angle β_m in deg, roll rate p_m in deg/sec, and yaw rate r_m in deg/sec, versus time (s).

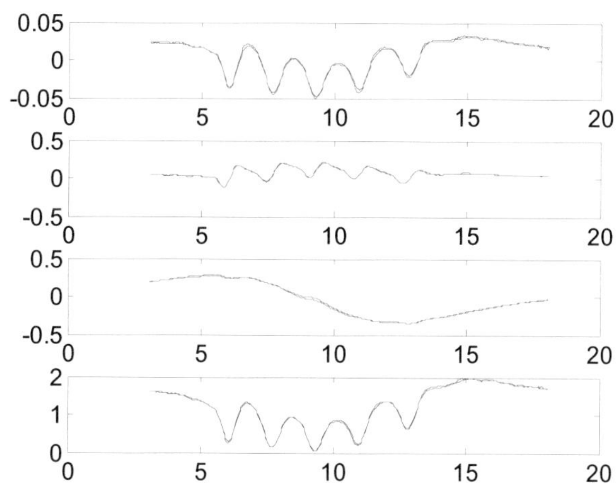

Figure 8. **Xd**(cont), and **Z**(dashdot) for α (deg), Q (deg/sec), θ (deg), and a_n with time (s).

The measurement equations (m = 4) for angle of attack, pitch rate, pitch, and normal acceleration are given by

$$\alpha_m = K_\alpha\, \alpha - K_\alpha x_\alpha \frac{q}{V};\ \ q_m = q\ ; \theta_m = \theta \tag{39}$$

$$a_{n_m} = \frac{\overline{q}S}{mg}\left(C_{N_\alpha}\alpha + C_{N_{\delta_e}}\delta_e + C_{N_0} \right) + \frac{x_{a_n}}{g}\ \dot{q} \tag{40}$$

The unknown parameters (p = 10) are $\left(C_{L_\alpha}, C_{L_{\delta_e}}, C_{L_0}, C_{m_\alpha}, C_{m_q}, C_{m\dot{\alpha}}, C_{m_{\delta_e}}, C_{m_0}, \theta_0, C_{N_0}\right)^T$ with the approximation $C_{N_\alpha} = C_{L_\alpha}$ *and* $C_{N_{\delta_e}} = C_{L_{\delta_e}}$. The suffix δ_e denotes control derivatives, and suffix zero refers to biases and all others are aerodynamic derivatives. The initial states are taken as the initial measurements and the initial parameter values are taken as (4, 0.15, 0.2, −0.5, −11.5, −5, −1.38, −0.06, −0.01, 0.2)^T. The other constant values used are S = 184, m = 196, I_{xx} = 6892.7, I_{yy} = 3952.3, I_{zz} = 10416.4, g = 32.2, $\bar{c}$ = 5.58, $K_\alpha x_\alpha$ = −0.0279, x_{a_n} = 0.101, and K_α = 1.

6.2.1. Remarks on the real data case results

All the real data studies were run for 100 iterations using the **RRR** with **Q** > 0, since the off-diagonal elements of the correlation coefficient matrix C for parameter estimates reduced substantially than for **Q** ≡ 0. **Figures 6–9** show the various input and output quantities from the **RRR**, and **Table 2** provides a comparison of the parameter estimates along with their CRBs

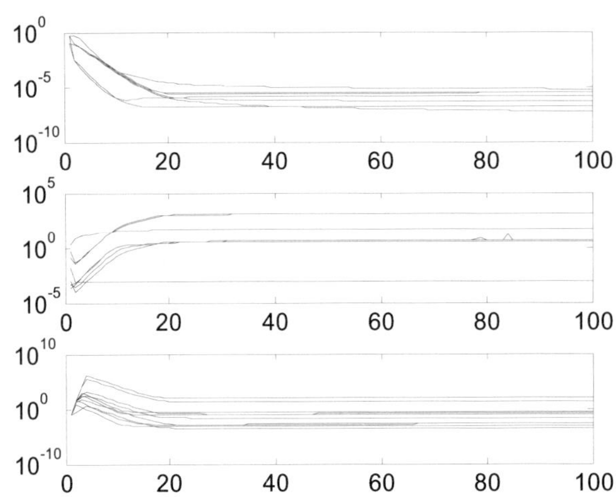

Figure 9. (i) **Q** and **R**, (ii) initial **Θ**, and **P0**, and (iii) the costs **J1–J8** with iterations.

Θ	RRR	NASA(Q≡0)	Gemson	MT	MS
C_{N_0}	0.2538 (0.0014)	0.2541 (0.008935)	0.2503 (0.0014)	0.2540 (0.0016)	0.2635 (0.0026)
C_{L_0}	0.2409 (0.0021)	0.2448 (0.009215)	0.2529 (0.0018)	0.2408 (0.0023)	0.2517 (0.0027)
C_{L_α}	4.9235 (0.0164)	5.1068 (0.1322)	4.9028 (0.0168)	4.9260 (0.0184)	5.0620 (0.0323)
$C_{L_{\delta_e}}$	0.1554 (0.0271)	0.1909 (0.1602)	0.0879 (0.0267)	0.1587 (0.0302)	0.3594 (0.0508)
C_{m_0}	−0.0425 (0.0009)	−.0505 (0.002655)	−0.0507 (0.0024)	−0.0424 (0.0009)	−0.0447 (0.0006)
C_{m_α}	−0.5293 (0.0079)	−0.6474 (0.02339)	−0.6174 (0.0211)	−0.5285 (0.0082)	−0.5590 (0.0055)
C_{m_q}	−11.8596 (0.2402)	−14.2600 (0.6528)	−18.8339 (0.8379)	−11.8255 (.2483)	−12.5965 (0.1400)
$C_{m\dot{\alpha}}$	−6.8959 (0.4891)	−8.2700 (1.296)	−7.1290 (1.544)	−6.8798 (0.5062)	−6.6713 (0.3021)
$C_{m_{\delta_e}}$	−0.9731 (0.0177)	−1.1614 (0.05371)	−1.1841 (0.471)	−0.9711 (0.0184)	−1.0247 (0.0129)
θ_0	0.0003 (0.0021)	−0.01177 (0.02528)	−0.0037 (0.001)	0.0002 (0.0011)	−0.0006 (0.0007)

Table 2. Real flight test data results $(\boldsymbol{\Theta}, \mathbf{CRB}(\boldsymbol{\Theta}))$.

Method		**Measurement noise covariance R × 10^{-6}**				**Process noise covariance Q × 10^{-6}**		
RRR		1.241	0.051	0.460	5.668	0.180	2.954	2.646
MT		1.614	0.240	2.316	2.929	0.203	3.153	0.667
MS		3.160	37.242	9.341	841.55	0.00005	0.0003	0.2386
Cost functions **J1–J8**								
RRR	3.934	4.223	3.616	0.0008	−44.1347	2.975	2.976	2.907
MT	3.766	4.519	3.838	0.0008	−43.7340	4.2266	4.228	2.949
MS	3.162	3.151	2.590	0.0007	−38.0517	8.477	8.466	3.022

Table 3. Real flight test data results **(R, Q, J)**.

(in parenthesis) for the real data from different approaches. Since MT and MS do not specify **P0**, the one in **RRR** is used. It turns out that even the parameters $C_{L_{\delta_e}}$ and C_{m_q} strongly affecting the airplane dynamics is estimated vary widely among the various approaches. Also, even if Θ are good the CRBs differ among the approaches such as for C_{N_0} and $C_{m\dot{\alpha}}$.

We noted earlier about the negatively correlated behaviour of **R** and **Q**. From **Table 3**, it may be noted that the **RRR** relative to other methods generally provides a larger **Q** and a smaller **R.** This implies that the larger Q assists in a better tracking of the wandering state. The smaller **R** tightly wraps around the smoothed state and thus providing generally smaller CRBs. **Figure 8** shows the close match of the measurements with the dynamics in this real case is due to the small **R** and **Q** than in the SMD case. The convergence remarks for Θ, their CRBs (= **P0/N**), **R**, **Q**, **J1–J8** as in SMD is valid here as well. The cost functions from **RRR** are generally closer to the number of states and measurements as expected.

The rounded 100(Correlation coefficient) matrix $\mathbf{C_{ij}} = \mathbf{P_{ij}}/sqrt(\mathbf{P_{ii}P_{jj}})$ with **P** denoting the covariance of parameter estimates is.

Θ	C_{N_0}	C_{L_0}	C_{L_α}	$C_{L_{\delta_e}}$	C_{m_0}	C_{m_α}	C_{m_q}	$C_{m\dot{\alpha}}$	$C_{m_{\delta_e}}$	θ_0
C_{N_0}	100	65	62	98	−13	−10	1	2	−12	0
C_{L_0}	65	100	41	64	−7	−4	−2	5	−5	0
C_{L_α}	62	41	100	67	−8	−19	1	1	−8	0
$C_{L_{\delta_e}}$	98	64	67	100	−13	−11	2	1	−12	0
C_{m_0}	−13	−7	−8	−13	100	88	9	84	99	0
C_{m_α}	−10	−4	−19	−11	88	100	25	70	91	0
C_{m_q}	1	−2	1	2	9	25	100	−27	21	1
$C_{m\dot{\alpha}}$	2	5	1	1	84	70	−27	100	80	−1
$C_{m_{\delta_e}}$	−12	−5	−8	−12	99	91	21	80	100	0
θ_0	0	0	0	0	0	0	1	−1	0	100

Based on the C from **RRR**, the weakest parameter can be inferred as θ_0, which is uncorrelated with all other parameters and its estimates and uncertainty can vary widely among the approaches. Next, it is possible to group the parameters as (C_{N_0}, C_{L_0}, C_{L_α}, $C_{L_{\delta_e}}$) and (C_{m_0}, C_{m_α}, C_{m_q}, $C_{m\dot{\alpha}}$, $C_{m_{\delta_e}}$). If a certain state is excited relatively more than others, then the estimated parameter that multiplies it will have lower correlation with other parameters in the set.

The downloadable MATLAB program available in Shyam et al. [31], which has many simulated and real flight test data analysis can be used all the way from teaching, learning to carry out research using Kalman filter.

7. Conclusions

A new approach called **RRR** has been proposed to handle the important problem of tuning the Kalman filter statistics. The importance of **P0** in EKF as the **PMP** is stressed. This along with suitably chosen **R** and **Q** after every filter pass through the data based on various filter statistics converges after few iterations without any optimization. Further, the many cost functions indicate the balance of the state and measurement equations and the consistency of the various filter statistics. These help the user to assess the results and compare **RRR** with other approaches. The efficacy of the **RRR** is demonstrated by application to a simple SMD system and a real airplane flight test data with a large number of unknown aerodynamic parameters.

Acknowledgements

Parts of this chapter are reproduced from the authors' previously published report [31]. There are three approaches to learn a subject namely based on intuitive concepts, practical calculations, and rigorous mathematics. This work has been written utilising the first two to help the readers and in particular learners so that they are not put off by the Kalman filter. The author sincerely thanks INTECH for providing such an opportunity. The present work is the outcome of the author's interaction with many colleagues and associates over many decades. It is difficult to list all their names but the ones in the references can be kindly taken as an acknowledgement. However, the author conveys his sincere apologies for any omissions relevant to the theme of the present work.

Author details

Mudambi R Ananthasayanam

Address all correspondence to: sayanam2005@yahoo.co.in

Department of Aerospace Engineering, Indian Institute of Science, Bangalore, India

References

[1] Billard R. Aryabhata and Indian astronomy. Indian Journal of History of Science. 1977;**12**(2): 207-224. Also R. Billard, "L'astronomie Indienne," Publications De Ecole Francaise D'-Extreme-Orient, Paris, 1971

[2] Sarma KV. Tradition of Aryabhatiya in Kerala: Revision of planetary parameters. Indian Journal of History of Science. 1977;**12**(2):194-199

[3] Gauss KF. Theoria Motus Corporum Coelestium in Sectionibus Conicis Solem Ambientium (Theory of the Motion of the Heavenly Bodies Moving about the Sun in Conic Sections [Translation]). New York: Dover; 1809. p. 1963

[4] Kalman RE. A new approach to linear filtering and prediction problems. Transactions of the ASME -Journal of Basic Engineering. 1960;**82**(Series D):35-45

[5] Kalman RE, Bucy RS. New results in linear filtering and prediction theory. Journal of Basic Engineering. 1961;**83**(1):95-108

[6] Lauritzen S. Thiele Pioneer in Statistics. Oxford University Press; 2002

[7] Swerling P. A Proposed Stagewise Differential Correction Procedure for Satellite Tracking and Prediction. Tech. Rep. P-1292. Rand Corporation; 1958

[8] Stratonovich RL. Application of the theory of Markov process in optimal signal discrimination. Radio Engineering and Electronic Physics. 1960;**1**:1-19

[9] Plackett RL. Some theorems in least squares. Biometrika. 1950;**37**:149157

[10] Sprott DA. Gauss's contributions to statistics. Historia Mathematica. 1978;**5**:183-203

[11] Wiener N. The Extrapolation, Interpolation and Smoothing of Stationary Time Series: With Engineering Applications. Cambridge, Massachusetts: MIT Press:; New York: Wiley and Sons; London: Chapman & Hall; 1949

[12] Ananthasayanam MR. A relook at the concepts and competence of the Kalman Filter. In: Proceedings of the 42nd AIAA Aerospace Sciences Meeting and Exhibit; 5–8 January 2004; Reno, Nevada. AIAA 2004-571

[13] Klein V, Morelli EA. Aircraft System Identification: Theory and Practice. AIAA Edn. Series; 2006

[14] Jategaonkar RV. Flight Vehicle System Identification: A Time Domain Methodology. Vol. 216. AIAA; 2006

[15] Bar-Shalom Y, Rong Li X, Kirubarajan T. Estimation with Applications to Tracking and Navigation, Theory, Algorithm and Software. John Wiley and Sons; 2000

[16] Ananthasayanam MR, Anilkumar AK, Subba Rao PV. New approach for the evolution and expansion of space debris scenario. Journal of Spacecraft and Rockets. 2006;**43**(6): 1271-1282

[17] Grewal MS, Lawrence RW, Andrews AP. Global Positioning Systems, Inertial Navigation, and Integration. 2nd ed. John Wiley & Sons Inc; 2007

[18] Kleusberg A, Teunissen PJG. GPS for Geodesy. 1st ed. Springer; 1996

[19] Fruhwirth R, Regier M, Bock RK, Grote H, Notz D. Data analysis techniques for high-energy physics. In: Cambridge Monographs on Particle Physics, Nuclear Physics and Cosmology; 2000

[20] Federer WT, Murthy BR. Kalman Filter Bibliography: Agriculture, Biology, and Medicine. Technical Report BU-1436-M. Department of Biometrics, Cornell University; 1998

[21] Visser H, Molenaar J. Kalman filter analysis in dendroclimatology. Biometrics. 1988;**44**: 929-940

[22] Wells C. The Kalman Filter in Finance. Springer-Science. Business Media, BV; 1996

[23] Costagli M, Kuruoglu EE. Image separation using particle filters. Digital Signal Processing. 2007;**17**:935-946

[24] Evensen G. Data Assimilation: The Ensemble Kalman Filter. Springer Verlag; 2009

[25] Rao CR. Statistics and Truth: Putting Chance to Work - Ramanujan Memorial Lectures. New Delhi, India: Council of Scientific and Industrial Research; 1987

[26] Samaniego FJ. A Comparison of the Bayesian and Frequentist Approaches to Estimation. Springer Science; Business Media, LLC; 2011

[27] Feller W. An Introduction to Probability Theory and its Applications. 3rd ed. Vol. 1. John Wiley and Sons; 1967

[28] Brown R, Hwang P. Introduction to Random Signals and Applied Kalman Filtering, with MATLAB Exercises. 4th ed. John Wiley and Sons; 2012

[29] Rauch HE, Tung F, Striebel CT. Maximum likelihood estimates of linear dynamic systems. AIAA Journal. 1965;**3**(8):1445-1450

[30] Kailath T. An innovations approach to least-squares estimation part I: Linear filtering in additive white noise. IEEE Transactions on Automatic Control. 2016;**AC-13**(6):646-655

[31] Shyam MM, Naik N, RMO G, Ananthasayanam MR. Introduction to the Kalman Filter and Tuning its Statistics for near Optimal Estimates and Cramer Rao Bound. TR/EE2015/ 401. Kanpur: IIT; 2015 http://arxiv.org/abs/1503.04313

[32] Philip NK, Ananthasayanam MR. Relative position and attitude estimation and control schemes for the final phase of an autonomous docking mission. Acta Astronautica. 2003; **52**(7):511-522

[33] Minsky M. The Society of the Mind. Picador Edition; 1988

[34] Narasimha R. Performance reliability of high maintenance systems. Journal of the Franklin Institute. 1975;**303**:15-29

[35] Julier SJ, Uhlmann JK, Durrant-Whyte HF. A new approach for filtering nonlinear systems. In: Proceedings of the 1995 American Control Conference. Vol. 1995. IEEE Press;

[36] Gordon NJ, Salmond DJ, Smith AFM. Novel approach to nonlinear/non-Gaussian Bayesian state estimation. IEE Proceedings. 1993;**140**(2):107-113

[37] Georgios BG, Erchin S. bibliography on nonlinear system identification. Signal Processing. 2001;**8**:533-580

[38] Daum F. Nonlinear filters: Beyond the Kalman filter. IEEE A&E Systems Magazine. 2005; **20**:8, Part 2: Tutorials-Daum

[39] Gelb A, editor. Applied Optimal Estimation. Cambridge: Massachusetts: MIT Press; 1974

[40] Sorenson HW. Least squares estimation from gauss to Kalman. IEEE Spectrum. 1970;**7**: 63-68

[41] Subbaraju PV, Ananthasayanam MR, Deshpande SM. Estimation of the Drag of a Satellite Launch Vehicle from Flight Data Using Extended Kalman Filter. Report 88 FM 3, Dept. of Aerospace Engineering. Indian Institute of Science, Bangalore; 1988. Accepted for presentation at the 8th IFAC/IFORS Symposium on Identification and System Parameter Estimation, August 27–31, 1988 at Beijing, China

[42] Mehrotra K, Mahapatra PR. A jerk model for tracking highly maneuvering targets. IEEE Transactions on Aerospace and Electronic Systems. 1997;**33**(4):1094-1105

[43] Maybeck PS. Stochastic Models, Estimation, and Control. Vol. 1. New York: Academic Press; 1979

[44] Candy JV. Signal Processing the Model Based Approach. McGraw Hill; 1986

[45] Gemson RMO. Estimation of aircraft aerodynamic derivatives accounting for measurement and process noise By EKF through adaptive filter tuning. [PhD thesis]. Department of Aerospace Engineering, Indian Institute of Science, Bangalore; 1991

[46] Gemson RMO, Ananthasayanam MR. Importance of initial state covariance matrix for the parameter estimation using adaptive extended Kalman filter. In: Proceedings of AIAA Atmospheric Flight Mechanics; 1998. AIAA-98-4153. pp. 94-104

[47] Sarkar AK, Ananthasayanam MR, Vathsal S. Sensitivity of initial state error covariance matrix in a practical adaptive EKF. In: Proceedings of AIAA Atmospheric Flight Mechanics Conference and Exhibit; 6–9 August 2001; Montreal, Canada. AIAA-2001-4202

[48] Shumway RH, Stoffer DS. Time Series Analysis and its Applications. New York: Springer Verlag; 2000

[49] McLachlan GJ, Krishan. The EM Algorithm and Extensions. 2nd ed. John Wiley and Sons; 2008

[50] Bohlin T. Four cases of identification of changing systems. In: System Identification: Advances and Case Studies. 1st ed. Academic Press; 1976

[51] Ljung L. Asymptotic behaviour of the EKF as a parameter estimator for linear systems. IEEE Transactions on Automatic Control. 1979;**24**:36-50

[52] Mehra R. Approaches to adaptive filtering. IEEE Transactions on Automatic Control. 1972;**17**:903-908

[53] Myers KA, Tapley BD. Adaptive sequential estimation with unknown noise statistics. IEEE Transactions on Automatic Control. 1976;**21**:520-525

[54] Ananthasayanam MR, Shyam MM, Naik N, Gemson RMO. A heuristic reference recursive recipe for adaptively tuning the Kalman filter statistics part-1: Formulation and simulation studies. Sadhana. 2016;**41**(12):1473-1490

[55] Shyam MM, Naik N, Gemson RMO, Ananthasayanam MR. A heuristic reference recursive recipe for adaptively tuning the Kalman filter statistics part-2: Real data studies. Sadhana. 2016;**41**(12):1491-1507

[56] Ananthasayanam MR, Suresh HS, Muralidharan MR. GUI Based Software for Teaching Parameter Estimation Technique Using MMLE. Report 2001 FM 1, Dept. of Aerospace Engineering. Bangalore: Indian Institute of Science; 2001

[57] Bavdekar VA, Deshpande AP, Patwardhan SC. Identification of process and measurement noise covariance for state and parameter estimation using extended Kalman filter. Journal of Process Control. 2011;**21**:585-601

[58] Mohamed AH, Schwarz KP. Adaptive Kalman filtering for INS/GPS. Journal of Geodesy. 1999;**73**(4):193-203

[59] Maine RE and Iliff KW. Programmer's Manual for MMLE3, A General Fortran Program for Maximum Likelihood Parameter Estimation. NASA TP-1690; 1981

Chapter 2

The Error Covariance Matrix Inflation in Ensemble Kalman Filter

Guocan Wu and Xiaogu Zheng

Additional information is available at the end of the chapter

http://dx.doi.org/10.5772/intechopen.71960

Abstract

The estimation accuracy of ensemble forecast errors is crucial to the assimilation results for all ensemble-based schemes. The ensemble Kalman filter (EnKF) is a widely used scheme in land surface data assimilation, without using the adjoint of a dynamical model. In EnKF, the forecast error covariance matrix is estimated as the sampling covariance matrix of the ensemble forecast states. However, past researches on EnKF have found that it can generally lead to an underestimate of the forecast error covariance matrix, due to the limited ensemble size, as well as the poor initial perturbations and model error. This can eventually result in filter divergence. Therefore, using inflation to further adjust the forecast error covariance matrix becomes increasingly important. In this chapter, a new structure of the forecast error covariance matrix is proposed to mitigate the problems with limited ensemble size and model error. An adaptive procedure equipped with a second-order least squares method is applied to estimate the inflation factors of forecast and observational error covariance matrices. The proposed method is tested on the well-known atmosphere-like Lorenz-96 model with spatially correlated observational systems. The experiment results show that the new structure of the forecast error covariance matrix and the adaptive estimation procedure lead to improvement of the analysis states.

Keywords: data assimilation, ensemble Kalman filter, error covariance inflation, observation-minus-forecast residual, least squares

1. Introduction

For state variables in geophysical research fields, a common assumption is that systems have "true" underlying states. Data assimilation is a powerful mechanism for estimating the true trajectory based on the effective combination of a dynamic forecast system (such as a numerical model) and observations [1]. It can produce an optimal combination of model outputs and observations [2]. The combined result is called analysis state, which should be closer to the true state than either the model forecast or the observation is. In fact, the analysis state can generally be treated as the weighted average of the model forecasts and observations, while the weights are approximately proportional to the inverse of the corresponding covariance matrices [3]. Therefore, the results of any data assimilation depend crucially on the estimation accuracy of the forecast and observational error covariance matrices [4]. If these matrices are estimated correctly, then the analysis states can be generated by minimizing an objective function which is technically straightforward and can be accomplished using existing engineering solutions [5], although finding the appropriate analysis state is still a quite difficult problem when the models are nonlinear [6, 7].

The ensemble Kalman filter (EnKF) is a widely used sequential data assimilation approach, which has been studied and applied since it is proposed by Evensen [8]. It is a practical ensemble-based assimilation scheme that estimates the forecast error covariance matrix using a Monte Carlo method with the short-term ensemble forecast states [9]. In EnKF, the forecast error covariance matrix is estimated as the sampling covariance matrix of the ensemble forecast states, which is usually underestimated due to the limited ensemble size and model error [10]. This finding indicates that the filter is over reliant on the model forecasts and excludes the observations. It may eventually lead to the divergence of the EnKF assimilation scheme [11, 12].

Therefore, the forecast error covariance inflation technique to address this problem becomes increasingly important. One of the error covariance matrix inflation techniques is additive inflation, in which a noise is added to the ensemble forecast states that sample the probability distribution of model error [13, 14]. Another widely used error covariance matrix inflation technique is multiplicative inflation, that is, to multiply the matrix by an appropriate factor. It can be used to mitigate filter divergence by inflating the empirical covariance and increasing the robustness of the filter [15].

In early studies of multiplicative inflation, researchers determine the inflation factor by repeated experimentation and choose a value according to their prior knowledge [11]. Hence, such experimental tuning is rather empirical and subjective. It is not appropriate to use the same inflation factor during all the assimilation procedure. Too small or too large an inflation factor will cause the analysis state to over rely on the model forecasts or observations and can seriously undermine the accuracy and stability of the filter. In later studies, the inflation factor is estimated online based on the observation-minus-forecast residual (innovation statistic) [16, 17] with different conditions.

Past work shows that moment estimation can facilitate the calculation by solving an equation of the observation-minus-forecast residual and its realization [18–20]. Maximum likelihood approach can obtain a better estimate of the inflation factor than moment approach, although it must calculate a high-dimensional matrix determinant [21–24]. Bayesian approach assumes a prior distribution for the inflation factor but is limited by spatially independent observational errors [25, 26]. Second-order least square estimation focus on minimizing the second-order least squares (SLS) [27] statistic of the squared observation-minus-forecast residual, which is not very expensive [28–30]. Generalized Cross Validation (GCV) [31, 32] can select a regularization parameter by minimizing the predictive data errors with rotation invariant in a least squares solution [33].

In practice, the observational error covariance matrix may also need to be adjusted, and an approach can be used to simultaneously optimize inflation factors of both forecast and observational error covariance matrices [21]. This approach is based on the optimization of the likelihood function of observation-minus-forecast residual. However, the likelihood function of observation-minus-forecast residual is nonlinear and involves the computationally expensive determinant and inverse of the residual covariance matrix. As compensation, the second-order least squares statistic of the squared observation-minus-forecast residual can be used as the cost function instead. The main advantage of the SLS cost function is that it is a quadratic function of the inflation factors, and therefore, the analytic forms of the estimators of the inflation factors can be easily obtained. Compared with the method based on maximum likelihood estimation method, the computational cost is significantly reduced.

Furthermore, unlike the sampling covariance matrix of the ensemble forecast states used in the conventional EnKF, a new structure of the forecast error covariance matrix is proposed in this chapter. In ideal situation, an ensemble forecast state is assumed as a random vector with the true state as its ensemble mean. Hence, it is should be defined that the ensemble forecast error is the ensemble forecast states minus true state rather than minus their ensemble mean [34]. This is because in a forecast model with large error and limited ensemble size, the ensemble mean of the forecast states can be very far from the true state. Therefore, the sampling covariance matrix of the ensemble forecast states can be very different from the true forecast error covariance matrix. As a result, the estimated analysis state can be substantially inaccurate. However, in reality, the true state is unknown, but the analysis state is a better estimate of the true state than the forecast state. Therefore, the information feedback from the analysis state can be used to revise the forecast error covariance matrix. In fact, the proposed forecast error covariance matrix is a combination of multiplicative and additive inflation. Bai and Li [14] also used the feedback from the analysis state to improve assimilation but in a different way.

This chapter consists of four sections. The EnKF scheme with a new structure of the forecast error covariance matrix and the adaptive estimation procedure is proposed in Section 2. The assimilation results on Lorenz model with a correlated observational system are presented in Section 3. Conclusions and discussion are given in Section 4.

2. Methodology

2.1. EnKF with SLS inflation scheme

Using the uniform notations for consistency, a nonlinear discrete-time forecast and linear observational system is written as [35]

$$\mathbf{x}_i^{\mathrm{t}} = M_{i-1}\left(\mathbf{x}_{i-1}^{\mathrm{a}}\right) + \boldsymbol{\eta}_i, \tag{1}$$

$$\mathbf{y}_i^{\mathrm{o}} = \mathbf{H}_i \mathbf{x}_i^{\mathrm{t}} + \boldsymbol{\varepsilon}_i, \tag{2}$$

where i is the time index; $\mathbf{x}_i^{\mathrm{t}} = \{\mathrm{x}_i^{\mathrm{t}}(1), \mathrm{x}_i^{\mathrm{t}}(2), \ldots, \mathrm{x}_i^{\mathrm{t}}(n)\}^{\mathrm{T}}$ is the n-dimensional true state vector at time step i; $\mathbf{x}_{i-1}^{\mathrm{a}} = \{\mathrm{x}_{i-1}^{\mathrm{a}}(1), \mathrm{x}_{i-1}^{\mathrm{a}}(2), \ldots, \mathrm{x}_{i-1}^{\mathrm{a}}(n)\}^{\mathrm{T}}$ is the n-dimensional analysis state vector which is an estimate of $\mathbf{x}_{i-1}^{\mathrm{t}}$, M_{i-1} is a nonlinear forecast operator such as a weather forecast model; $\mathbf{y}_i^{\mathrm{o}}$ is an observational vector with dimension p_i; $\mathbf{H}_i$ is an observational matrix of dimension $p_i \times n$ that maps model states to the observational space; $\boldsymbol{\eta}_i$ and $\boldsymbol{\varepsilon}_i$ are the forecast error vector and the observational error vector respectively, which are assumed to be statistically independent of each other, time-uncorrelated, and have mean zero and covariance matrices $\mathbf{P}_i$ and $\mathbf{R}_i$, respectively. The goal of the EnKF assimilation is to find a series of analysis states $\mathbf{x}_i^{\mathrm{a}}$ that are sufficiently close to the corresponding true states $\mathbf{x}_i^{\mathrm{t}}$, using the information provided by M_i and $\mathbf{y}_i^{\mathrm{o}}$.

It is well-known that any EnKF assimilation scheme should include a forecast error inflation scheme. Otherwise, the EnKF may diverge [11]. A procedure for estimating multiplicative inflation factor of $\mathbf{P}_i$ and adjustment factor of $\mathbf{R}_i$ can be carried out based on the SLS principle. The basic filter algorithm uses perturbed observations [9], but without localization [36]. The estimation steps of this algorithm equipped with SLS inflation are as follows.

Step 1. Calculate the perturbed forecast states

$$\mathbf{x}_{i,j}^{\mathrm{f}} = M_{i-1}\left(\mathbf{x}_{i-1,j}^{\mathrm{a}}\right), \tag{3}$$

where $\mathbf{x}_{i-1,j}^{\mathrm{a}}$ is the perturbed analysis states derived from the previous time step ($1 \leq j \leq m$ and m is the ensemble size).

Step 2. Estimate the improved forecast and observational error covariance matrices.

The forecast state $\mathbf{x}_i^{\mathrm{f}}$ is defined as the ensemble mean of $\mathbf{x}_{i,j}^{\mathrm{f}}$ and the initial forecast error covariance matrix is expressed as

$$\widehat{\mathbf{P}}_i = \frac{1}{m-1}\sum_{j=1}^{m}\left(\mathbf{x}_{i,j}^{\mathrm{f}} - \mathbf{x}_i^{\mathrm{f}}\right) \cdot \left(\mathbf{x}_{i,j}^{\mathrm{f}} - \mathbf{x}_i^{\mathrm{f}}\right)^{\mathrm{T}}, \tag{4}$$

and the initial observational error covariance matrix is $\mathbf{R}_i$. Then, the adjusted forms of forecast and observational error covariance matrices are $\lambda_i \widehat{\mathbf{P}}_i$ and $\mu_i \mathbf{R}_i$, respectively.

There are several approaches for estimating the inflation factors λ_i and μ_i. Wang and Bishop [19], Li et al. [18], and Miyoshi [20] use the first-order least square of the squared observation-minus-forecast residual $\mathbf{d}_i \equiv \mathbf{y}_i^{\mathrm{o}} - \mathbf{H}_i \mathbf{x}_i^{\mathrm{f}}$ to estimate λ_i; Liang et al. [21] maximizes the likelihood of $\mathbf{d}_i$ to estimate λ_i and μ_i. Here, the SLS approach is applied for estimating λ_i and μ_i. That is, λ_i and μ_i are estimated by minimizing the objective function

$$L_i(\lambda, \mu) = \mathrm{Tr}\left[\left(\mathbf{d}_i\mathbf{d}_i^{\mathrm{T}} - \lambda \mathbf{H}_i \widehat{\mathbf{P}}_i \mathbf{H}_i^{\mathrm{T}} - \mu \mathbf{R}_i\right)\left(\mathbf{d}_i\mathbf{d}_i^{\mathrm{T}} - \lambda \mathbf{H}_i \widehat{\mathbf{P}}_i \mathbf{H}_i^{\mathrm{T}} - \mu \mathbf{R}_i\right)^{\mathrm{T}}\right]. \tag{5}$$

This leads to

$$\widehat{\lambda}_i = \frac{\mathrm{Tr}\left(\mathbf{d}_i^{\mathrm{T}}\mathbf{H}_i\widehat{\mathbf{P}}_i\mathbf{H}_i^{\mathrm{T}}\mathbf{d}_i\right)\mathrm{Tr}\left(\mathbf{R}_i^2\right) - \mathrm{Tr}\left(\mathbf{d}_i^{\mathrm{T}}\mathbf{R}_i\mathbf{d}_i\right)\mathrm{Tr}\left(\mathbf{H}_i\widehat{\mathbf{P}}_i\mathbf{H}_i^{\mathrm{T}}\mathbf{R}_i\right)}{\mathrm{Tr}\left(\mathbf{H}_i\widehat{\mathbf{P}}_i\mathbf{H}_i^{\mathrm{T}}\mathbf{H}_i\widehat{\mathbf{P}}_i\mathbf{H}_i^{\mathrm{T}}\right)\mathrm{Tr}\left(\mathbf{R}_i^2\right) - \mathrm{Tr}\left(\mathbf{H}_i\widehat{\mathbf{P}}_i\mathbf{H}_i^{\mathrm{T}}\mathbf{R}_i\right)^2}, \tag{6}$$

$$\widehat{\mu}_i = \frac{\mathrm{Tr}\left(\mathbf{H}_i\widehat{\mathbf{P}}_i\mathbf{H}_i^{\mathrm{T}}\mathbf{H}_i\widehat{\mathbf{P}}_i\mathbf{H}_i^{\mathrm{T}}\right)\mathrm{Tr}\left(\mathbf{d}_i^{\mathrm{T}}\mathbf{R}_i\mathbf{d}_i\right) - \mathrm{Tr}\left(\mathbf{d}_i^{\mathrm{T}}\mathbf{H}_i\mathbf{P}_i\mathbf{H}_i^{\mathrm{T}}\mathbf{d}_i\right)\mathrm{Tr}\left(\mathbf{H}_i\widehat{\mathbf{P}}_i\mathbf{H}_i^{\mathrm{T}}\mathbf{R}_i\right)}{\mathrm{Tr}\left(\mathbf{H}_i\widehat{\mathbf{P}}_i\mathbf{H}_i^{\mathrm{T}}\mathbf{H}_i\widehat{\mathbf{P}}_i\mathbf{H}_i^{\mathrm{T}}\right)\mathrm{Tr}\left(\mathbf{R}_i^2\right) - \mathrm{Tr}\left(\mathbf{H}_i\widehat{\mathbf{P}}_i\mathbf{H}_i^{\mathrm{T}}\mathbf{R}_i\right)^2}. \tag{7}$$

(See Appendix A for detailed derivation). Similar to Wang and Bishop [19] and Li et al. [18], this procedure does not use Bayesian approach [20, 25, 26].

Step 3. Compute the perturbed analysis states.

$$\mathbf{x}_{i,j}^{\mathrm{a}} = \mathbf{x}_{i,j}^{\mathrm{f}} + \widehat{\lambda}_i\widehat{\mathbf{P}}_i\mathbf{H}_i^{\mathrm{T}}\left(\mathbf{H}_i\widehat{\lambda}_i\widehat{\mathbf{P}}_i\mathbf{H}_i^{\mathrm{T}} + \widehat{\mu}_i\mathbf{R}_i\right)^{-1}\left(\mathbf{y}_i^{\mathrm{o}} + \boldsymbol{\varepsilon}_{i,j}' - \mathbf{H}_i\mathbf{x}_i^{\mathrm{f}}\right), \tag{8}$$

where $\boldsymbol{\varepsilon}_{i,j}'$ is a normal random variable with mean zero and covariance matrix $\widehat{\mu}_i\mathbf{R}_i$ [9]. Here $\left(\mathbf{H}_i\widehat{\lambda}_i\widehat{\mathbf{P}}_i\mathbf{H}_i^{\mathrm{T}} + \widehat{\mu}_i\mathbf{R}_i\right)^{-1}$ can be effectively computed using the Sherman-Morrison-Woodbury formula [21, 37, 38]. Furthermore, the analysis state $\mathbf{x}_i^{\mathrm{a}}$ is estimated as the ensemble mean of $\mathbf{x}_{i,j}^{\mathrm{a}}$. Finally, set $i = i + 1$ and return to Step 1 for the assimilation at next time step.

2.2. EnKF with SLS inflation and new structure of forecast error covariance matrix

By Eqs. (1) and (3), the ensemble forecast error is defined as $\mathbf{x}_{i,j}^{\mathrm{f}} - \mathbf{x}_i^{\mathrm{t}}$. On the other hand, $\mathbf{x}_i^{\mathrm{f}}$ is an estimate of $\mathbf{x}_i^{\mathrm{t}}$ without knowing observations. The ensemble forecast error is initially estimated as $\mathbf{x}_{i,j}^{\mathrm{f}} - \mathbf{x}_i^{\mathrm{f}}$, which is used to construct the forecast error covariance matrix in Section 2.1. However, due to limited sample size and model error, $\mathbf{x}_i^{\mathrm{f}}$ can be far from $\mathbf{x}_i^{\mathrm{t}}$. Therefore, $\mathbf{x}_{i,j}^{\mathrm{f}} - \mathbf{x}_i^{\mathrm{f}}$ can be a biased estimate of $\mathbf{x}_{i,j}^{\mathrm{f}} - \mathbf{x}_i^{\mathrm{t}}$.

Here, the observations can be used for improving the estimation accuracy of ensemble forecast error. The basic sense is as follows: After the analysis state $\mathbf{x}_i^{\mathrm{a}}$ is derived, it should be a better estimate of $\mathbf{x}_i^{\mathrm{t}}$ than the forecast state $\mathbf{x}_i^{\mathrm{f}}$. Therefore, $\mathbf{x}_i^{\mathrm{f}}$ in Eq. (4) is substituted by $\mathbf{x}_i^{\mathrm{a}}$ for generating a revised forecast error covariance matrix. This procedure can be repeated iteratively until the corresponding objective function (Eq. (5)) converges. For the computational details, Step 2 in Section 2.1 is modified to the following adaptive procedure:

Step 2a. Use Step 2 in Section 2.1 to inflate the initial forecast error covariance matrix to ${}_0\widehat{\lambda}_i\,{}_0\widehat{\mathbf{P}}_i$ and adjust initial observational error covariance matrix to ${}_0\widehat{\mu}_i\mathbf{R}_i$. Then use Step 3 in Section 2.1 to estimate the initial analysis state ${}_0\mathbf{x}_i^{\mathrm{a}}$ and set $k = 1$.

Step 2b. Update the forecast error covariance matrix as

$$ {}_{k}\widehat{\mathbf{P}}_{i} = \frac{1}{m-1}\sum_{j=1}^{m}\left(\mathbf{x}_{i,j}^{\mathrm{f}} - {}_{k-1}\mathbf{x}_{i}^{\mathrm{a}}\right)\cdot\left(\mathbf{x}_{i,j}^{\mathrm{f}} - {}_{k-1}\mathbf{x}_{i}^{\mathrm{a}}\right)^{\mathrm{T}}. \tag{9} $$

Then, adjust the forecast and observational error covariance matrices to ${}_{k}\widehat{\lambda}_{ik}\widehat{\mathbf{P}}_{i}$ and ${}_{k}\widehat{\mu}_{i}\mathbf{R}_{i}$, where

$$ {}_{k}\widehat{\lambda}_{i} = \frac{\mathrm{Tr}\left(\mathbf{d}_{i}^{\mathrm{T}}\mathbf{H}_{ik}\widehat{\mathbf{P}}_{i}\mathbf{H}_{i}^{\mathrm{T}}\mathbf{d}_{i}\right)\mathrm{Tr}\left(\mathbf{R}_{i}^{2}\right) - \mathrm{Tr}\left(\mathbf{d}_{i}^{\mathrm{T}}\mathbf{R}_{i}\mathbf{d}_{i}\right)\mathrm{Tr}\left(\mathbf{H}_{ik}\widehat{\mathbf{P}}_{i}\mathbf{H}_{i}^{\mathrm{T}}\mathbf{R}_{i}\right)}{\mathrm{Tr}\left(\mathbf{H}_{ik}\widehat{\mathbf{P}}_{i}\mathbf{H}_{i}^{\mathrm{T}}\mathbf{H}_{ik}\widehat{\mathbf{P}}_{i}\mathbf{H}_{i}^{\mathrm{T}}\right)\mathrm{Tr}\left(\mathbf{R}_{i}^{2}\right) - \mathrm{Tr}\left(\mathbf{H}_{ik}\widehat{\mathbf{P}}_{i}\mathbf{H}_{i}^{\mathrm{T}}\mathbf{R}_{i}\right)^{2}}, \tag{10} $$

and

$$ {}_{k}\widehat{\mu}_{i} = \frac{\mathrm{Tr}\left(\mathbf{H}_{ik}\widehat{\mathbf{P}}_{i}\mathbf{H}_{i}^{\mathrm{T}}\mathbf{H}_{ik}\widehat{\mathbf{P}}_{i}\mathbf{H}_{i}^{\mathrm{T}}\right)\mathrm{Tr}\left(\mathbf{d}_{i}^{\mathrm{T}}\mathbf{R}_{i}\mathbf{d}_{i}\right) - \mathrm{Tr}\left(\mathbf{d}_{i}^{\mathrm{T}}\mathbf{H}_{ik}\widehat{\mathbf{P}}_{i}\mathbf{H}_{i}^{\mathrm{T}}\mathbf{d}_{i}\right)\mathrm{Tr}\left(\mathbf{H}_{ik}\widehat{\mathbf{P}}_{i}\mathbf{H}_{i}^{\mathrm{T}}\mathbf{R}_{i}\right)}{\mathrm{Tr}\left(\mathbf{H}_{ik}\widehat{\mathbf{P}}_{i}\mathbf{H}_{i}^{\mathrm{T}}\mathbf{H}_{ik}\widehat{\mathbf{P}}_{i}\mathbf{H}_{i}^{\mathrm{T}}\right)\mathrm{Tr}\left(\mathbf{R}_{i}^{2}\right) - \mathrm{Tr}\left(\mathbf{H}_{ik}\widehat{\mathbf{P}}_{i}\mathbf{H}_{i}^{\mathrm{T}}\mathbf{R}_{i}\right)^{2}}, \tag{11} $$

are estimated by minimizing the objective function.

$$ {}_{k}L_{i}(\lambda,\mu) = \mathrm{Tr}\left[\left(\mathbf{d}_{i}\mathbf{d}_{i}^{\mathrm{T}} - \lambda\mathbf{H}_{ik}\widehat{\mathbf{P}}_{i}\mathbf{H}_{i}^{\mathrm{T}} - \mu\mathbf{R}_{i}\right)\left(\mathbf{d}_{i}\mathbf{d}_{i}^{\mathrm{T}} - \lambda\mathbf{H}_{ik}\widehat{\mathbf{P}}_{i}\mathbf{H}_{i}^{\mathrm{T}} - \mu\mathbf{R}_{i}\right)^{\mathrm{T}}\right]. \tag{12} $$

If ${}_{k}\mathrm{L}_{i}\left({}_{k}\widehat{\lambda}_{i}, {}_{k}\widehat{\mu}_{i}\right) < {}_{k-1}\mathrm{L}_{i}\left({}_{k-1}\widehat{\lambda}_{i}, {}_{k-1}\widehat{\mu}_{i}\right) - \delta$, where δ is a pre-determined threshold to control the convergence of Eq. (12) and then estimate the k-th updated analysis state as

$$ {}_{k}\mathbf{x}_{i}^{\mathrm{a}} = \mathbf{x}_{i}^{\mathrm{f}} + {}_{k}\widehat{\lambda}_{ik}\widehat{\mathbf{P}}_{i}\mathbf{H}_{i}^{\mathrm{T}}\left(\mathbf{H}_{ik}\widehat{\lambda}_{ik}\widehat{\mathbf{P}}_{i}\mathbf{H}_{i}^{\mathrm{T}} + {}_{k}\widehat{\mu}_{ik}\mathbf{R}_{i}\right)^{-1}\left(\mathbf{y}_{i}^{\mathrm{o}} - \mathbf{H}_{i}\mathbf{x}_{i}^{\mathrm{f}}\right), \tag{13} $$

set $k = k + 1$ and return back to Eq. (9); otherwise, take ${}_{k-1}\widehat{\lambda}_{ik-1}\widehat{\mathbf{P}}_{i}$ and ${}_{k-1}\widehat{\mu}_{i}\mathbf{R}_{i}$ as the estimated forecast and observational error covariance matrices at i-th time step and go to Step 3 in Section 2.1.

A general flowchart of the proposed assimilation scheme is shown in **Figure 1**. Moreover, the proposed forecast error covariance matrix (Eq. (9)) can be expressed as.

$$ {}_{k}\lambda_{ik}\widehat{\mathbf{P}}_{i} = \frac{{}_{k}\lambda_{i}}{m-1}\sum_{j=1}^{m}\left(\mathbf{x}_{i,j}^{\mathrm{f}} - \mathbf{x}_{i}^{\mathrm{f}}\right)\left(\mathbf{x}_{i,j}^{\mathrm{f}} - \mathbf{x}_{i}^{\mathrm{f}}\right)^{\mathrm{T}} + \frac{{}_{k}\lambda_{i}m}{m-1}\left(\mathbf{x}_{i}^{\mathrm{f}} - {}_{k-1}\mathbf{x}_{i}^{\mathrm{a}}\right)\left(\mathbf{x}_{i}^{\mathrm{f}} - {}_{k-1}\mathbf{x}_{i}^{\mathrm{a}}\right)^{\mathrm{T}}, \tag{14} $$

which is a multiplicatively inflated sampling error covariance matrix plus an additive inflation matrix (see Appendix B for the proof).

2.3. Notes

2.3.1. Correctly specified observational error covariance matrix

If the observational error covariance matrix $\mathbf{R}_i$ is correctly known, then its adjustment is no longer required. In this case, the inflation factor ${}_{k}\widehat{\lambda}_{i}$ can be estimated by minimizing the following objective function

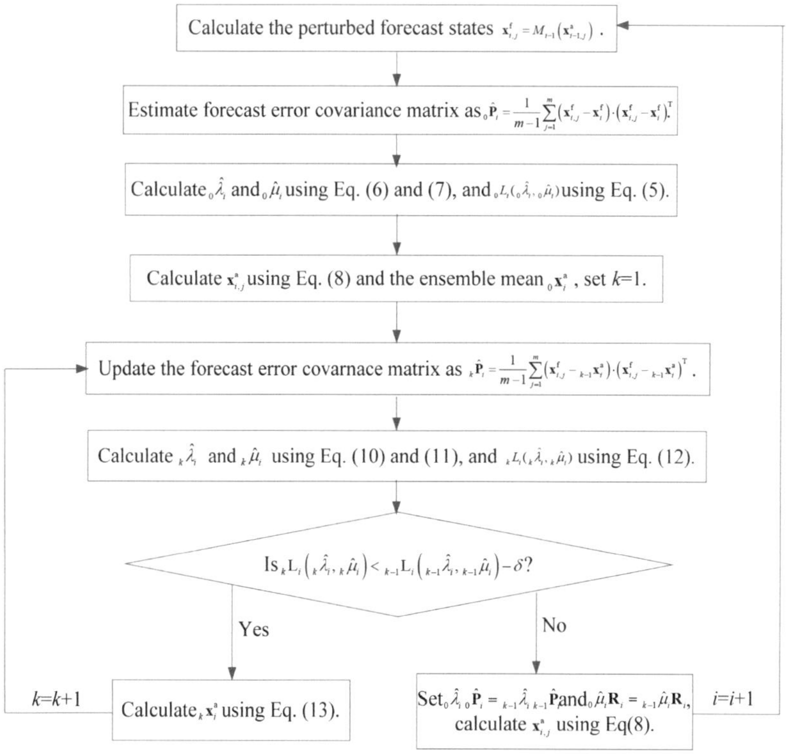

Figure 1. Flowchart of EnKF with SLS inflation scheme.

$$L_i(\lambda) = \mathrm{Tr}\left[\left(\mathbf{d}_i\mathbf{d}_i^T - \lambda\mathbf{H}_{ik}\widehat{\mathbf{P}}_i\mathbf{H}_i^T - \mathbf{R}_i\right)\left(\mathbf{d}_i\mathbf{d}_i^T - \lambda\mathbf{H}_{ik}\widehat{\mathbf{P}}_i\mathbf{H}_i^T - \mathbf{R}_i\right)^T\right]. \tag{15}$$

This leads to a simpler estimate

$${}_k\widehat{\lambda}_i = \frac{\mathrm{Tr}\left[\mathbf{H}_{ik}\widehat{\mathbf{P}}_i\mathbf{H}_i^T\left(\mathbf{d}_i\mathbf{d}_i^T - \mathbf{R}_i\right)\right]}{\mathrm{Tr}\left[\mathbf{H}_{ik}\widehat{\mathbf{P}}_i\mathbf{H}_i^T\mathbf{H}_{ik}\widehat{\mathbf{P}}_i\mathbf{H}_i^T\right]}. \tag{16}$$

2.3.2. *Validation statistics*

In any toy model, the "true" state $\mathbf{x}_i^t$ is known by experimental design. In this case, the root-mean-square error (RMSE) of the analysis state can be used to evaluate the accuracy of the assimilation results. The RMSE at *i*-th step is defined as

$$\mathrm{RMSE} = \sqrt{\frac{1}{n}\sum_{k=1}^{n}\left(\mathrm{x}_{ik}^a - \mathrm{x}_{ik}^t\right)^2}. \tag{17}$$

where x_{ik}^a and x_{ik}^t are the *k*-th components of the analysis state and true state at the *i*-th time step. In principle, a smaller RMSE indicates a better performance of the assimilation scheme.

3. Experiment on Lorenz 96 model

In this section, the EnKF with SLS inflation assimilation scheme is applied to a nonlinear dynamical system, which has properties relevant to realistic forecast problems: the Lorenz-96 model [39] with model error and a linear observational system. The performances of the assimilation schemes in Section 2 are evaluated through the following experiments.

3.1. Description of forecast and observational systems

The Lorenz-96 model [39] is a strongly nonlinear dynamical system with quadratic nonlinearity, which is governed by the equation.

$$\frac{d\mathbf{X}_k}{dt} = (\mathbf{X}_{k+1} - \mathbf{X}_{k-2})\mathbf{X}_{k-1} - \mathbf{X}_k + F \tag{18}$$

where $k = 1, 2, \cdots, K$ ($K = 40$; hence, there are 40 variables). For Eq. (18) to be well-defined for all values of k, it is defined that $\mathbf{X}_{-1} = \mathbf{X}_{K-1}, \mathbf{X}_0 = \mathbf{X}_K, \mathbf{X}_{K+1} = \mathbf{X}_1$. The dynamics of Eq. (18) are "atmosphere-like" in that the three terms on the right-hand side consist of a nonlinear advection-like term, a damping term, and an external forcing term respectively. These three terms can be thought of as some atmospheric quantity (e.g., zonal wind speed) distributed on a latitude circle. Therefore, the Lorenz-96 model has been widely used as a test bed to evaluate the performance of assimilation schemes in many studies [30].

The true state is derived by a fourth-order Runge–Kutta time integration scheme [40]. The time step for generating the numerical solution was set at 0.05 nondimensional units, which is roughly equivalent to 6 hours in real time, assuming that the characteristic time-scale of the dissipation in the atmosphere is 5 days [39]. The forcing term was set as $F = 8$, so that the leading Lyapunov exponent implies an error-doubling time of approximately 8 time steps, and the fractal dimension of the attractor was 27.1 [41]. The initial value was chosen to be $\mathbf{X}_k = F$ when $k \neq 20$ and $\mathbf{X}_{20} = 1.001F$.

In this study, the synthetic observations were assumed to be generated by adding random noises that were multivariate-normally distributed with mean zero and covariance matrix $\mathbf{R}_i$ to the true states. The frequency was four time steps, which can be used to mimic daily observations in practical problems, such as satellite data. The observation errors were assumed to be spatially correlated, which is common in applications involving remote sensing and radiance data. The variance of the observation at each grid point was set to $\sigma_o^2 = 1$, and the covariance of the observations between the j-th and k-th grid points was as follows:

$$\mathbf{R}_i(j,k) = \sigma_o^2 \times 0.5^{\min\{|j-k|, 40-|j-k|\}}. \tag{19}$$

Since it can deal with spatially correlated observational errors, the scheme may potentially be applied for assimilating remote sensing observations and radiances data.

The model errors by changing the forcing term are added in the forecast model because it is inevitable in real dynamic systems. Thus, different values of F are chose in the assimilation schemes while retaining $F = 8$ when generating the "true" state. The observations are simulated every four time steps analogizing 1 day in realistic problem for 2000 steps to ensure robust results. The ensemble size is used as 30. The pre-determined threshold δ to control the convergence of Eq. (12) is set to be 1, because the values of objective functions are in the order of 10^5. In most cases of the following experiment, the objective functions converge after 3–4 iterations, and the estimated analysis states also converge.

3.2. Comparison of assimilation schemes

In Section 2.1, the EnKF assimilation scheme with SLS error covariance matrix inflation is outlined. In Section 2.2, the improved EnKF assimilation scheme with the SLS error covariance matrix inflation and the new structure of the forecast error covariance matrix are summarized. In the following, the influences of these estimation methods on EnKF data assimilation schemes are assessed using Lorenz-96 model.

Lorenz-96 model is a forced dissipative model with a parameter F that controls the strength of the forcing (Eq. (18)). The model behaviors are quite different with different values of F, and chaotic systems are produced with integer values of F larger than 3. Therefore, several values of F are used to simulate a wide range of model errors. In all cases, the true states were generated by a model with $F = 8$. These observations were then assimilated into models with $F = 4, 5, \ldots, 12$.

3.2.1. *Correctly specified observational error covariance matrix*

Suppose the observational error covariance matrix $\mathbf{R}_i$ is correctly specified, the inflation adjustment on $\widehat{\mathbf{P}}_i$ is taken in each assimilation cycle and estimate the inflation factors λ_i by the methods described in Section 2.1. Then, the adaptive assimilation schemes with the new structure of the forecast error covariance matrix proposed in Section 2.2 are conducted.

Figure 2 shows the time-mean analysis RMSE of the two assimilation schemes averaged over 2000 time steps, as a function of F. Overall, the analysis RMSE of the two assimilation schemes gradually grows as increasing model error. When F is near the true value 8, the two assimilation schemes have almost indistinguishable values of the analysis RMSE. However, when F becomes increasingly distant from 8, the analysis RMSE of the assimilation scheme with the new structure of the forecast error covariance matrix becomes progressively smaller than that of the assimilation scheme with the forecast error covariance matrix inflation only.

For the Lorenz-96 model with large error (such as, the case with $F = 12$), the time-mean analysis RMSE of the two assimilation schemes is listed in **Table 1**, as well as the time-mean values of the objective functions. The conventional EnKF assimilation scheme is also included for comparison. These results show clearly that our two schemes have significantly smaller RMSE than

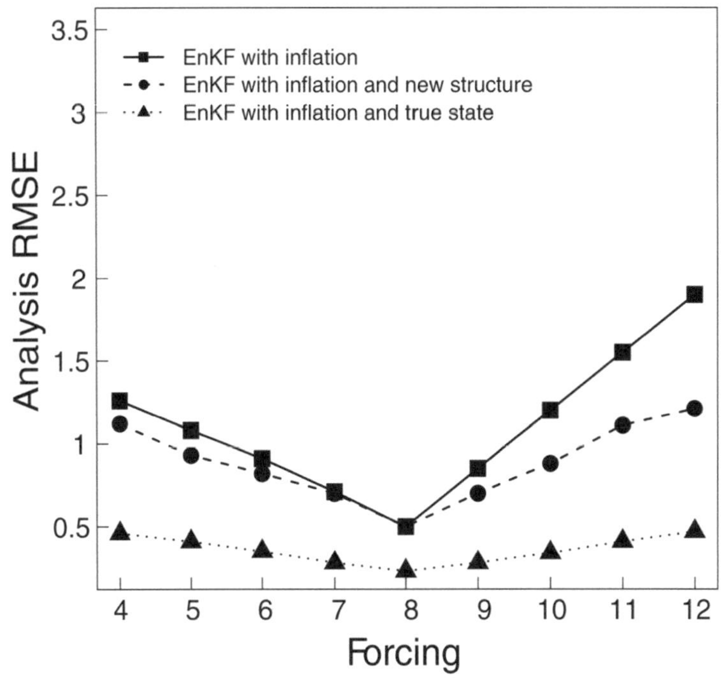

Figure 2. Time-mean values of the analysis RMSE as a function of forcing F when observational errors are spatially correlated and their covariance matrix is correctly specified, by using 3 EnKF schemes. 1) SLS only (solid line, described in Section 2.1); 2) SLS and new structure (dashed line, described in Section 2.2); and 3) SLS and true ensemble forecast error (dotted line, described in Section 5).

EnKF schemes	**Time-mean RMSE**	**Time-mean L**
Non-inflation	5.65	2,298,754
SLS	1.89	148,468
SLS and new structure	1.22	38,125
SLS and true ensemble forecast error	0.48	19,652

Table 1. The time-mean analysis RMSE and the time-mean objective function values in 4 EnKF schemes for Lorenz-96 model when observational errors are spatially correlated and their covariance matrix is correctly specified: (1) EnKF (non-inflation); (2) the SLS scheme in Section 2.1 (SLS); (3) the SLS scheme in Section 2.2 (SLS and new structure); (4) the SLS scheme in the discussion (SLS and true ensemble forecast error). The forcing term F = 12.

the EnKF assimilation scheme. Moreover, the assimilation scheme with the new structure of the forecast error covariance matrix performs much better than assimilation scheme with forecast error covariance matrix inflation only.

3.2.2. Incorrectly specified observational error covariance matrix

In this section, the observational error covariance matrix is supposed to be correct only up to a constant factor. The factor is estimated using different estimation methods, and the corresponding assimilation results are evaluated.

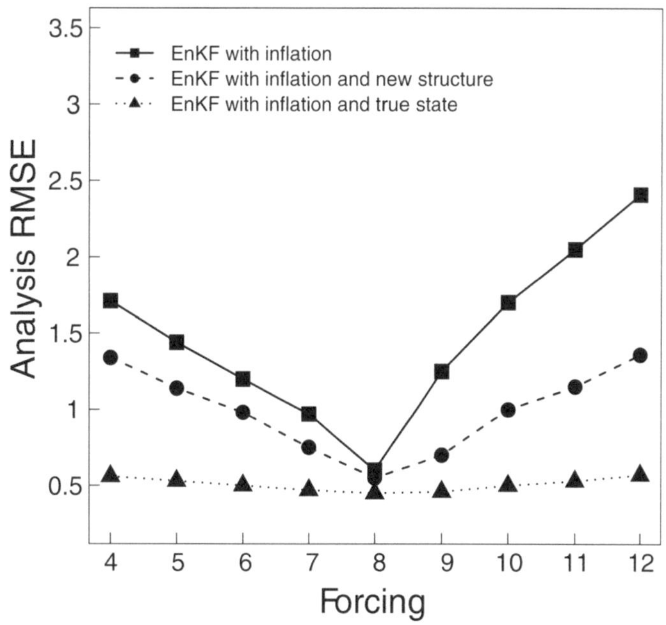

Figure 3. Time-mean values of the analysis RMSE as a function of forcing F when observational errors are spatially correlated and their covariance matrix is incorrectly specified, by using 3 EnKF schemes. 1) SLS only (solid line); 2) SLS and new structure (dashed line); and 3) SLS and true ensemble forecast error (dotted line).

The observational error covariance matrix $\mathbf{R}_i$ is set as four times of the true matrix and introduces another factor μ_i to adjust $\mathbf{R}_i$. The assimilation schemes are conducted in two cases: (1) inflate the forecast and observational error covariance matrices only (Section 2.1); (2) inflate the forecast and observational error covariance matrices and use the new structure of the forecast error covariance matrix (Section 2.2). Again, the forcing term F takes values 4, 5, …, 12 when assimilating observations, but $F = 8$ is used when generating the true states in all cases.

Figure 3 shows the time-mean analysis RMSE of the two cases averaged over 2000 time steps, as a function of forcing term. Generally speaking, the analysis RMSE of the two cases gradually

EnKF schemes	Ensemble size 30		Ensemble size 20	
	Time-mean RMSE	Time-mean L	Time-mean RMSE	Time-mean L
SLS	2.43	1,426,541	3.51	1,492,685
SLS and new structure	1.35	41,326	1.45	95,685
SLS and true ensemble forecast error	0.58	21,585	0.60	21,355

Table 2. The time-mean analysis RMSE and the time-mean objective function values in EnKF schemes for Lorenz-96 model when observational errors are spatially correlated and their covariance matrix is incorrectly specified: (1) SLS; (2) SLS and new structure; (4) SLS and true ensemble forecast error. The forcing term $F = 12$.

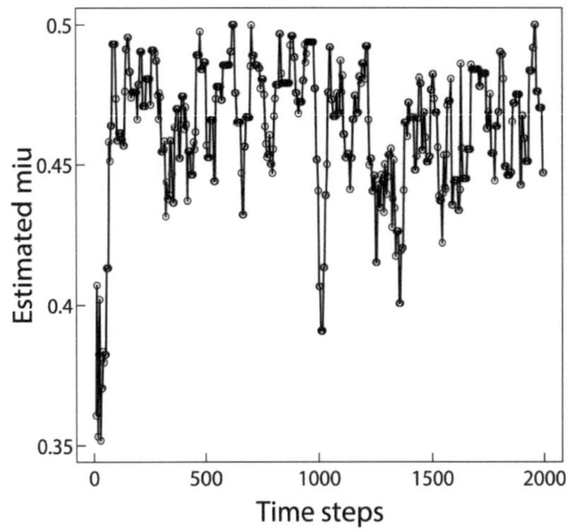

Figure 4. The times series of estimated $\widehat{\mu}_i$ when observational error covariance matrix is incorrectly specified.

grows as the increasing the model error. However, the analysis RMSE generated by using new structure of the forecast error covariance matrix (cases 2) is smaller than those by using the error covariance matrices inflation technique only (cases 1).

For the Lorenz-96 model with forcing term $F = 12$, the time-mean analysis RMSE of the two cases is listed in **Table 2**, along with the time-mean values of the objective functions. These

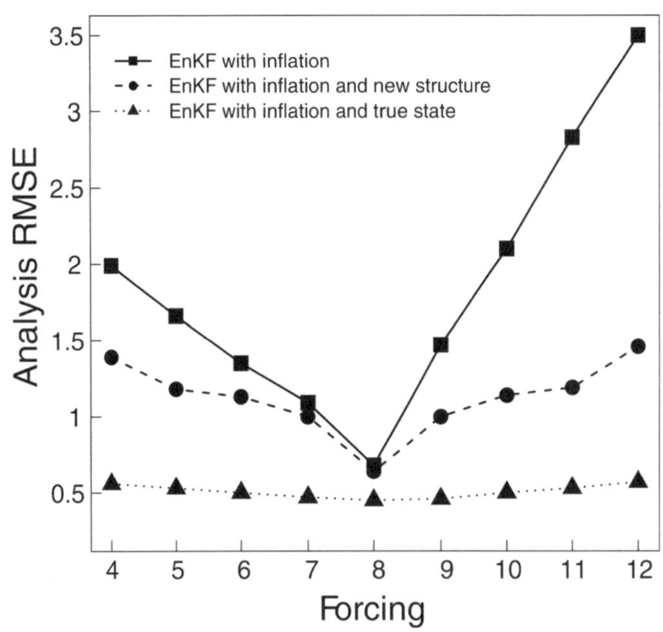

Figure 5. Similar to **Figure 3**, but ensemble size is 20.

results clearly show that when the observational error covariance matrix is incorrectly specified, the assimilation result is much better if the new structure of the forecast error covariance matrix is used (cases 2).

The estimated $\widehat{\mu}_i$ over 2000 time steps in the two cases of using the new structure of the forecast error covariance matrix (cases 2) are plotted in **Figure 4**. It can be seen that the time-mean value of estimated $\widehat{\mu}_i$ is 0.45, which is very close to the reciprocal of the constant that is multiplied to the observational error covariance matrix (0.25).

To further investigate the effect of ensemble size on the assimilation result, **Figure 3** is reproduced with the ensemble size 20. The results are shown in **Figure 5**, as well as in **Table 2**. Generally speaking, **Figures 5** is quite similar to **Figure 3** but with larger analysis error. This indicates that the smaller ensemble size can lead to the larger forecast error and analysis error. The analysis is also repeated with the ensemble size 10. However in this case, both the inflation and new structure are not effective. This could be due to that the ensemble size 10 is too small to generate robust covariance estimation.

4. Discussion and conclusions

It is well-known that accurately estimating the error covariance matrix is one of the most key steps in any ensemble-based data assimilation. In EnKF assimilation scheme, the forecast error covariance matrix is initially estimated as the sampling covariance matrix of the ensemble forecast states. But due to limited ensemble size and model error, the forecast error covariance matrix is usually an underestimation, which may lead to the divergence of the filter. Therefore, the initially estimated forecast error covariance matrix is multiplied by an inflation factor λ_i, and the SLS estimation is proposed to estimate this factor.

In fact, the true forecast error should be represented as the ensemble forecast states minus the true state. However, since in real problems, the true state is not available, the ensemble mean of the forecast states is used instead. Consequently, the forecast error covariance matrix is initially represented as the sampling covariance matrix of the ensemble forecast states. However, for the model with large error, the ensemble mean of the forecast states may be far from the true state. In this case, the estimated forecast error covariance matrix will also remain far from the truth, no matter which inflation technique is used.

To verify this point, a number of EnKF assimilation schemes with necessary error covariance matrix inflation are applied to the Lorenz-96 model but with the forecast state $\mathbf{x}_i^{\mathrm{f}}$ in the forecast error covariance matrix (Eq. (4)) substituted by the true state $\mathbf{x}_i^{\mathrm{t}}$. The corresponding RMSE are shown in **Figures 2–5** and **Tables 1** and **2**. All the figures and tables show that the analysis RMSE is significantly reduced.

However, since the true state $\mathbf{x}_i^{\mathrm{t}}$ is unknown, the analysis state $\mathbf{x}_i^{\mathrm{a}}$ is used to replace the forecast state $\mathbf{x}_i^{\mathrm{f}}$, because $\mathbf{x}_i^{\mathrm{a}}$ is closer to $\mathbf{x}_i^{\mathrm{t}}$ than $\mathbf{x}_i^{\mathrm{f}}$. To achieve this goal, a new structure of the forecast error covariance matrix and an adaptive procedure for estimating the new structure are proposed here to iteratively improve the estimation. As shown in this chapter, the RMSE of the corresponding analysis states are indeed smaller than those of the EnKF assimilation scheme with the error covariance matrix inflation only. For instance, in the experiment in Section 3.1, when the error covariance matrix inflation technique is applied, the RMSE is 1.89 which is much smaller than that for the original EnKF. When the new structure of the forecast error covariance matrix is used in addition to the inflation, the RMSE is reduced to 1.22 (see **Table 1**).

In the realistic problems, the observational error covariance matrix is not always correctly known, and hence it also needs to be adjusted too. Another factor μ_i is introduced to adjust the observational error covariance matrix in this chapter, which can be estimated simultaneously with λ_i by minimizing the second-order least squares function of the squared observation-minus-forecast residual.

The second-order least squares function of the squared observation-minus-forecast residual can be a good objective function to quantify the goodness of fit of the error covariance matrix. The SLS method proposed in this chapter can be used to estimate the factors for adjusting both the forecast and observational error covariance matrices, while the first order method can only estimate the inflation factor of the forecast error covariance matrix. The SLS can also provide a criterion for stopping the iteration in the adaptive estimation procedure when the new structure of the forecast error covariance matrix is used. This is important for preventing the proposed forecast error covariance matrix to depart from the truth in the iteration. In most cases in this study, the minimization algorithms converge after several iterations, and the objective function decreases sharply. On the other hand, the improved forecast error covariance matrix indeed leads to the improvement of analysis state. In fact, as shown in **Tables 1-2**, a small objective function value always corresponds to a small RMSE of the analysis state.

The difference of the EnKF assimilation scheme with SLS inflation is compared to that with maximum likelihood estimation (MLE) inflation [21]. Generally speaking, the RMSE of the analysis state derived using the MLE inflation scheme is a little smaller than that derived using the SLS inflation scheme only but is larger than that derived using the SLS inflation with the new structure of forecast error covariance matrix. For instance, for Lorenz-96 model with forcing term $F = 12$, the RMSE is 1.69 for MLE inflation, 1.89 for SLS inflation only, and 1.22 for SLS inflation and new structure (**Table 1**). Whether this is a general rule or not is still unclear and is subject to further investigation. However, in MLE inflation scheme, the objective function is nonlinear and especially involves the determinant of the observation-minus-forecast residual's covariance matrix, which is quite computationally expensive. The SLS objective function proposed in this chapter is quadratic, so its minimizer is analytic and can be easily calculated.

On the other hand, similar to other inflation schemes with single factor, this study also assumes the inflation factor to be constant in space. Apparently, this is not the case in many practical applications, especially for the cases that the observations are unevenly distributed. Persistently applying the same inflation values that are reasonably large to address problems in densely observed areas to all state variables can systematically overinflate the ensemble variances in sparsely observed areas [13, 26, 42]. Even when the adaptive procedure for estimating the error covariance matrix is applied, the problem may still exist in some extent. In the two case studies conducted here, the observational systems are relatively evenly distributed.

In the future study, we will investigate how to modify the adaptive procedure to suit the system with unevenly distributed observations. We also plan to apply our methodology to error covariance localization [43, 44] and to validate the proposed methodologies using more sophisticated dynamic and observational systems.

Acknowledgements

This work is supported by the National Natural Science Foundation of China (grant no. 91647202), the National Basic Research Program of China (grant no. 2015CB953703), the National Natural Science Foundation of China (grant no. 41405098) and the Fundamental Research Funds for the Central Universities. The authors gratefully acknowledge the anonymous reviewers for their constructive and relevant comments, which helped greatly in improving the quality of this manuscript. The authors are also grateful to the editors for their hard work and suggestions on this manuscript. Parts of this chapter are reproduced from the authors' previous publications [29, 30].

Appendix A

The forecast error covariance matrix $\widehat{\mathbf{P}}_i$ is inflated to $\lambda\widehat{\mathbf{P}}_i$. The estimation of the inflation factors λ is based on the observation-minus-forecast residual

$$\begin{aligned}\mathbf{d}_i &= \mathbf{y}_i^{\mathrm{o}} - \mathbf{H}_i\mathbf{x}_i^{\mathrm{f}} \\ &= \left(\mathbf{y}_i^{\mathrm{o}} - \mathbf{H}_i\mathbf{x}_i^{\mathrm{t}}\right) + \mathbf{H}_i\left(\mathbf{x}_i^{\mathrm{t}} - \mathbf{x}_i^{\mathrm{f}}\right)\end{aligned} \tag{A1}$$

The covariance matrix of the random vector $\mathbf{d}_i$ can be expressed as a second-order regression equation [27]:

$$E\left[\left(\left(\mathbf{y}_i^{\mathrm{o}} - \mathbf{H}_i\mathbf{x}_i^{\mathrm{t}}\right) + \mathbf{H}_i\left(\mathbf{x}_i^{\mathrm{t}} - \mathbf{x}_i^{\mathrm{f}}\right)\right)\left(\left(\mathbf{y}_i^{\mathrm{o}} - \mathbf{H}_i\mathbf{x}_i^{\mathrm{t}}\right) + \mathbf{H}_i\left(\mathbf{x}_i^{\mathrm{t}} - \mathbf{x}_i^{\mathrm{f}}\right)\right)^{\mathrm{T}}\right] = \mathbf{d}_i\mathbf{d}_i^{\mathrm{T}} + \Xi \tag{A2}$$

where E is the expectation operator and Ξ is the error matrix. The left-hand side of (A2) can be decomposed as

$$\begin{aligned}
&E\left[\left((\mathbf{y}_i^{\mathrm{o}} - \mathbf{H}_i\mathbf{x}_i^{\mathrm{t}}) + \mathbf{H}_i(\mathbf{x}_i^{\mathrm{t}} - \mathbf{x}_i^{\mathrm{f}})\right)\left((\mathbf{y}_i^{\mathrm{o}} - \mathbf{H}_i\mathbf{x}_i^{\mathrm{t}}) + \mathbf{H}_i(\mathbf{x}_i^{\mathrm{t}} - \mathbf{x}_i^{\mathrm{f}})\right)^{\mathrm{T}}\right] \\
&= E\left[(\mathbf{y}_i^{\mathrm{o}} - \mathbf{H}_i\mathbf{x}_i^{\mathrm{t}})(\mathbf{y}_i^{\mathrm{o}} - \mathbf{H}_i\mathbf{x}_i^{\mathrm{t}})^{\mathrm{T}}\right] + E\left[(\mathbf{H}_i(\mathbf{x}_i^{\mathrm{t}} - \mathbf{x}_i^{\mathrm{f}}))(\mathbf{H}_i(\mathbf{x}_i^{\mathrm{t}} - \mathbf{x}_i^{\mathrm{f}}))^{\mathrm{T}}\right] \\
&+ E\left[(\mathbf{y}_i^{\mathrm{o}} - \mathbf{H}_i\mathbf{x}_i^{\mathrm{t}})(\mathbf{H}_i(\mathbf{x}_i^{\mathrm{t}} - \mathbf{x}_i^{\mathrm{f}}))^{\mathrm{T}}\right] + E\left[(\mathbf{H}_i(\mathbf{x}_i^{\mathrm{t}} - \mathbf{x}_i^{\mathrm{f}}))(\mathbf{y}_i^{\mathrm{o}} - \mathbf{H}_i\mathbf{x}_i^{\mathrm{t}})^{\mathrm{T}}\right]
\end{aligned} \tag{A3}$$

Since the forecast and observational errors are statistically independent, we have

$$E\left[(\mathbf{H}_i(\mathbf{x}_i^{\mathrm{t}} - \mathbf{x}_i^{\mathrm{f}}))(\mathbf{y}_i^{\mathrm{o}} - \mathbf{H}_i\mathbf{x}_i^{\mathrm{t}})^{\mathrm{T}}\right] = \mathbf{H}_i E\left[(\mathbf{x}_i^{\mathrm{t}} - \mathbf{x}_i^{\mathrm{f}})(\mathbf{y}_i^{\mathrm{o}} - \mathbf{H}_i\mathbf{x}_i^{\mathrm{t}})^{\mathrm{T}}\right] = \mathbf{0}, \tag{A4}$$

$$E\left[(\mathbf{y}_i^{\mathrm{o}} - \mathbf{H}_i\mathbf{x}_i^{\mathrm{t}})(\mathbf{H}_i(\mathbf{x}_i^{\mathrm{t}} - \mathbf{x}_i^{\mathrm{f}}))^{\mathrm{T}}\right] = E\left[(\mathbf{y}_i^{\mathrm{o}} - \mathbf{H}_i\mathbf{x}_i^{\mathrm{t}})(\mathbf{x}_i^{\mathrm{t}} - \mathbf{x}_i^{\mathrm{f}})^{\mathrm{T}}\right]\mathbf{H}_i^{\mathrm{T}} = \mathbf{0}. \tag{A5}$$

From Eq. (2), $\mathbf{y}_i^{\mathrm{o}} - \mathbf{H}_i\mathbf{x}_i^{\mathrm{t}}$ is the observational error at i-th time step, and hence

$$E\left[(\mathbf{y}_i^{\mathrm{o}} - \mathbf{H}_i\mathbf{x}_i^{\mathrm{t}})(\mathbf{y}_i^{\mathrm{o}} - \mathbf{H}_i\mathbf{x}_i^{\mathrm{t}})^{\mathrm{T}}\right] = \mathbf{R}_i \tag{A6}$$

Further, since the forecast state $\mathbf{x}_{i,j}^{\mathrm{f}}$ is treated as a random vector with the true state $\mathbf{x}_i^{\mathrm{t}}$ as its population mean,

$$\begin{aligned}
&E\left[(\mathbf{H}_i(\mathbf{x}_i^{\mathrm{t}} - \mathbf{x}_i^{\mathrm{f}}))(\mathbf{H}_i(\mathbf{x}_i^{\mathrm{t}} - \mathbf{x}_i^{\mathrm{f}}))^{\mathrm{T}}\right] \\
&= \mathbf{H}_i E\left[(\mathbf{x}_i^{\mathrm{t}} - \mathbf{x}_i^{\mathrm{f}})(\mathbf{x}_i^{\mathrm{t}} - \mathbf{x}_i^{\mathrm{f}})^{\mathrm{T}}\right]\mathbf{H}_i^{\mathrm{T}} \\
&\approx \mathbf{H}_i \frac{\lambda}{m-1}\sum_{j=1}^{n}\left(\mathbf{x}_{i,j}^{\mathrm{f}} - \mathbf{x}_i^{\mathrm{f}}\right)\left(\mathbf{x}_{i,j}^{\mathrm{f}} - \mathbf{x}_i^{\mathrm{f}}\right)^{\mathrm{T}}\mathbf{H}_i^{\mathrm{T}} \\
&= \lambda \mathbf{H}_i\widehat{\mathbf{P}}_i\mathbf{H}_i^{\mathrm{T}}
\end{aligned} \tag{A7}$$

Substituting Eqs (A3)–(A7) into Eq. (A2), we have

$$\mathbf{R}_i + \lambda\mathbf{H}_i\widehat{\mathbf{P}}_i\mathbf{H}_i^{\mathrm{T}} \approx \mathbf{d}_i\mathbf{d}_i^{\mathrm{T}} + \Xi \tag{A8}$$

It follows that the second-order moment statistic of error Ξ can be expressed as

$$\begin{aligned}
&\mathrm{Tr}\left[\Xi\Xi^{\mathrm{T}}\right] \approx \mathrm{Tr}\left[\left(\mathbf{d}_i\mathbf{d}_i^{\mathrm{T}} - \mathbf{R}_i - \lambda\mathbf{H}_i\widehat{\mathbf{P}}_i\mathbf{H}_i^{\mathrm{T}}\right)\left(\mathbf{d}_i\mathbf{d}_i^{\mathrm{T}} - \mathbf{R}_i - \lambda\mathbf{H}_i\widehat{\mathbf{P}}_i\mathbf{H}_i^{\mathrm{T}}\right)^{\mathrm{T}}\right] \\
&\equiv L_i(\lambda)
\end{aligned} \tag{A9}$$

Therefore, λ can be estimated by minimizing objective function $L_i(\lambda)$. Since $L_i(\lambda)$ is a quadratic function of λ with positive quadratic coefficients, the inflation factor can be easily expressed as

$$\widehat{\lambda}_i = \frac{\mathrm{Tr}\left[\mathbf{H}_i\widehat{\mathbf{P}}_i\mathbf{H}_i^{\mathrm{T}}\left(\mathbf{d}_i\mathbf{d}_i^{\mathrm{T}} - \mathbf{R}_i\right)\right]}{\mathrm{Tr}\left[\mathbf{H}_i\widehat{\mathbf{P}}_i\mathbf{H}_i^{\mathrm{T}}\mathbf{H}_i\widehat{\mathbf{P}}_i\mathbf{H}_i^{\mathrm{T}}\right]} \tag{A10}$$

Similarly, if the amplitude of the observational error covariance matrix is not correct, we can adjust $\mathbf{R}_i$ to $\mu_i\mathbf{R}_i$ as well [21, 22]. Then, the objective function becomes

$$L_i(\lambda, \mu) = \mathrm{Tr}\left[\left(\mathbf{d}_i\mathbf{d}_i^{\mathrm{T}} - \mu\mathbf{R}_i - \lambda\mathbf{H}_i\widehat{\mathbf{P}}_i\mathbf{H}_i^{\mathrm{T}}\right)\left(\mathbf{d}_i\mathbf{d}_i^{\mathrm{T}} - \mu\mathbf{R}_i - \lambda\mathbf{H}_i\widehat{\mathbf{P}}_i\mathbf{H}_i^{\mathrm{T}}\right)^{\mathrm{T}}\right] \tag{A11}$$

As a bivariate function of λ and μ, the first partial derivative with respect to the two parameters respectively are

$$\lambda\mathrm{Tr}\left(\mathbf{H}_i\widehat{\mathbf{P}}_i\mathbf{H}_i^{\mathrm{T}}\mathbf{H}_i\widehat{\mathbf{P}}_i\mathbf{H}_i^{\mathrm{T}}\right) + \mu\mathrm{Tr}\left(\mathbf{H}_i\widehat{\mathbf{P}}_i\mathbf{H}_i^{\mathrm{T}}\mathbf{R}_i\right)\text{-}\mathrm{Tr}\left(\mathbf{d}_i\mathbf{d}_i^{\mathrm{T}}\mathbf{H}_i\widehat{\mathbf{P}}_i\mathbf{H}_i^{\mathrm{T}}\right) \tag{A12}$$

and

$$\lambda\mathrm{Tr}\left(\mathbf{H}_i\widehat{\mathbf{P}}_i\mathbf{H}_i^{\mathrm{T}}\mathbf{R}_i\right) + \mu\mathrm{Tr}\left(\mathbf{R}_i^{\mathrm{T}}\mathbf{R}_i\right)\text{-}\mathrm{Tr}\left(\mathbf{d}_i\mathbf{d}_i^{\mathrm{T}}\mathbf{R}_i\right) \tag{A13}$$

Setting Eqs (A12)–(A13) to zero and solving them lead to

$$\begin{aligned}\widehat{\lambda}_i &= \frac{\mathrm{Tr}\left(\mathbf{d}_i\mathbf{d}_i^{\mathrm{T}}\mathbf{H}_i\widehat{\mathbf{P}}_i\mathbf{H}_i^{\mathrm{T}}\right)\mathrm{Tr}\left(\mathbf{R}_i^2\right) - \mathrm{Tr}\left(\mathbf{d}_i\mathbf{d}_i^{\mathrm{T}}\mathbf{R}_i\right)\mathrm{Tr}\left(\mathbf{H}_i\widehat{\mathbf{P}}_i\mathbf{H}_i^{\mathrm{T}}\mathbf{R}_i\right)}{\mathrm{Tr}\left(\mathbf{H}_i\widehat{\mathbf{P}}_i\mathbf{H}_i^{\mathrm{T}}\mathbf{H}_i\widehat{\mathbf{P}}_i\mathbf{H}_i^{\mathrm{T}}\right)\mathrm{Tr}\left(\mathbf{R}_i^2\right) - \mathrm{Tr}\left(\mathbf{H}_i\widehat{\mathbf{P}}_i\mathbf{H}_i^{\mathrm{T}}\mathbf{R}_i\right)^2} \\ &= \frac{\mathrm{Tr}\left(\mathbf{d}_i^{\mathrm{T}}\mathbf{H}_i\widehat{\mathbf{P}}_i\mathbf{H}_i^{\mathrm{T}}\mathbf{d}_i\right)\mathrm{Tr}\left(\mathbf{R}_i^2\right) - \mathrm{Tr}\left(\mathbf{d}_i^{\mathrm{T}}\mathbf{R}_i\mathbf{d}_i\right)\mathrm{Tr}\left(\mathbf{H}_i\mathbf{P}_i\mathbf{H}_i^{\mathrm{T}}\mathbf{R}_i\right)}{\mathrm{Tr}\left(\mathbf{H}_i\widehat{\mathbf{P}}_i\mathbf{H}_i^{\mathrm{T}}\mathbf{H}_i\widehat{\mathbf{P}}_i\mathbf{H}_i^{\mathrm{T}}\right)\mathrm{Tr}\left(\mathbf{R}_i^2\right) - \mathrm{Tr}\left(\mathbf{H}_i\widehat{\mathbf{P}}_i\mathbf{H}_i^{\mathrm{T}}\mathbf{R}_i\right)^2}\end{aligned} \tag{A14}$$

$$\begin{aligned}\widehat{\mu}_i &= \frac{\mathrm{Tr}\left(\mathbf{H}_i\widehat{\mathbf{P}}_i\mathbf{H}_i^{\mathrm{T}}\mathbf{H}_i\widehat{\mathbf{P}}_i\mathbf{H}_i^{\mathrm{T}}\right)\mathrm{Tr}\left(\mathbf{d}_i\mathbf{d}_i^{\mathrm{T}}\mathbf{R}_i\right) - \mathrm{Tr}\left(\mathbf{d}_i\mathbf{d}_i^{\mathrm{T}}\mathbf{H}_i\widehat{\mathbf{P}}_i\mathbf{H}_i^{\mathrm{T}}\right)\mathrm{Tr}\left(\mathbf{H}_i\widehat{\mathbf{P}}_i\mathbf{H}_i^{\mathrm{T}}\mathbf{R}_i\right)}{\mathrm{Tr}\left(\mathbf{H}_i\widehat{\mathbf{P}}_i\mathbf{H}_i^{\mathrm{T}}\mathbf{H}_i\widehat{\mathbf{P}}_i\mathbf{H}_i^{\mathrm{T}}\right)\mathrm{Tr}\left(\mathbf{R}_i^2\right) - \mathrm{Tr}\left(\mathbf{H}_i\widehat{\mathbf{P}}_i\mathbf{H}_i^{\mathrm{T}}\mathbf{R}_i\right)^2} \\ &= \frac{\mathrm{Tr}\left(\mathbf{H}_i\widehat{\mathbf{P}}_i\mathbf{H}_i^{\mathrm{T}}\mathbf{H}_i\widehat{\mathbf{P}}_i\mathbf{H}_i^{\mathrm{T}}\right)\mathrm{Tr}\left(\mathbf{d}_i^{\mathrm{T}}\mathbf{R}_i\mathbf{d}_i\right) - \mathrm{Tr}\left(\mathbf{d}_i^{\mathrm{T}}\mathbf{H}_i\widehat{\mathbf{P}}_i\mathbf{H}_i^{\mathrm{T}}\mathbf{d}_i\right)\mathrm{Tr}\left(\mathbf{H}_i\widehat{\mathbf{P}}_i\mathbf{H}_i^{\mathrm{T}}\mathbf{R}_i\right)}{\mathrm{Tr}\left(\mathbf{H}_i\widehat{\mathbf{P}}_i\mathbf{H}_i^{\mathrm{T}}\mathbf{H}_i\widehat{\mathbf{P}}_i\mathbf{H}_i^{\mathrm{T}}\right)\mathrm{Tr}\left(\mathbf{R}_i^2\right) - \mathrm{Tr}\left(\mathbf{H}_i\widehat{\mathbf{P}}_i\mathbf{H}_i^{\mathrm{T}}\mathbf{R}_i\right)^2}\end{aligned} \tag{A15}$$

Appendix B

In fact,

$$
\begin{aligned}
&\frac{{}_k\lambda_i}{m-1}\sum_{j=1}^{m}\left(\mathbf{x}_{i,j}^{\mathrm{f}} - {}_{k-1}\mathbf{x}_i^{\mathrm{a}}\right)\left(\mathbf{x}_{i,j}^{\mathrm{f}} - {}_{k-1}\mathbf{x}_i^{\mathrm{a}}\right)^{\mathrm{T}} \\
&= \frac{{}_k\lambda_i}{m-1}\sum_{j=1}^{m}\left(\mathbf{x}_{i,j}^{\mathrm{f}} - \mathbf{x}_i^{\mathrm{f}} + \mathbf{x}_i^{\mathrm{f}} - {}_{k-1}\mathbf{x}_i^{\mathrm{a}}\right)\left(\mathbf{x}_{i,j}^{\mathrm{f}} - \mathbf{x}_i^{\mathrm{f}} + \mathbf{x}_i^{\mathrm{f}} - {}_{k-1}\mathbf{x}_i^{\mathrm{a}}\right)^{\mathrm{T}} \\
&= \frac{{}_k\lambda_i}{m-1}\left\{\sum_{j=1}^{m}\left(\mathbf{x}_{i,j}^{\mathrm{f}} - \mathbf{x}_i^{\mathrm{f}}\right)\left(\mathbf{x}_{i,j}^{\mathrm{f}} - \mathbf{x}_i^{\mathrm{f}}\right)^{\mathrm{T}} + \sum_{j=1}^{m}(\mathbf{x}_i^{\mathrm{f}} - {}_{k-1}\mathbf{x}_i^{\mathrm{a}})(\mathbf{x}_i^{\mathrm{f}} - {}_{k-1}\mathbf{x}_i^{\mathrm{a}})^{\mathrm{T}}\right. \\
&\left. + \sum_{j=1}^{m}\left(\mathbf{x}_{i,j}^{\mathrm{f}} - \mathbf{x}_i^{\mathrm{f}}\right)(\mathbf{x}_i^{\mathrm{f}} - {}_{k-1}\mathbf{x}_i^{k-1}\mathrm{a})^{\mathrm{T}} + \sum_{j=1}^{m}(\mathbf{x}_i^{\mathrm{f}} - {}_{k-1}\mathbf{x}_i^{k-1}\mathrm{a})\left(\mathbf{x}_{i,j}^{\mathrm{f}} - \mathbf{x}_i^{\mathrm{f}}\right)^{\mathrm{T}}\right\}
\end{aligned}
\tag{B1}
$$

Since $\mathbf{x}_i^{\mathrm{f}}$ is the ensemble mean forecast, we have

$$
\begin{aligned}
&\sum_{j=1}^{m}\left(\mathbf{x}_{i,j}^{\mathrm{f}} - \mathbf{x}_i^{\mathrm{f}}\right)(\mathbf{x}_i^{\mathrm{f}} - {}_{k-1}\mathbf{x}_i^{\mathrm{a}})^{\mathrm{T}} \\
&= \left[\sum_{j=1}^{m}\left(\mathbf{x}_{i,j}^{\mathrm{f}} - \mathbf{x}_i^{\mathrm{f}}\right)\right](\mathbf{x}_i^{\mathrm{f}} - {}_{k-1}\mathbf{x}_i^{\mathrm{a}})^{\mathrm{T}}c \\
&= \left[\sum_{j=1}^{m}\mathbf{x}_{i,j}^{\mathrm{f}} - m\frac{1}{m}\sum_{j=1}^{m}\mathbf{x}_{i,j}^{\mathrm{f}}\right](\mathbf{x}_i^{\mathrm{f}} - {}_{k-1}\mathbf{x}_i^{\mathrm{a}})^{\mathrm{T}} \\
&= \mathbf{0}
\end{aligned}
\tag{B2}
$$

and similarly.

$$
\sum_{j=1}^{m}(\mathbf{x}_i^{\mathrm{f}} - {}_{k-1}\mathbf{x}_i^{\mathrm{a}})\left(\mathbf{x}_{i,j}^{\mathrm{f}} - \mathbf{x}_i^{\mathrm{f}}\right)^{\mathrm{T}} = \mathbf{0}
\tag{B3}
$$

That is, the last two terms of Eq. (B1) vanish. Therefore, the proposed forecast error covariance matrix can be expressed as

$$
\begin{aligned}
&\frac{{}_k\lambda_i}{m-1}\sum_{j=1}^{m}\left(\mathbf{x}_{i,j}^{\mathrm{f}} - {}_{k-1}\mathbf{x}_i^{\mathrm{a}}\right)\left(\mathbf{x}_{i,j}^{\mathrm{f}} - {}_{k-1}\mathbf{x}_i^{\mathrm{a}}\right)^{\mathrm{T}} \\
&= \frac{{}_k\lambda_i}{m-1}\left\{\sum_{j=1}^{m}\left(\mathbf{x}_{i,j}^{\mathrm{f}} - \mathbf{x}_i^{\mathrm{f}}\right)\left(\mathbf{x}_{i,j}^{\mathrm{f}} - \mathbf{x}_i^{\mathrm{f}}\right)^{\mathrm{T}} + m(\mathbf{x}_i^{\mathrm{f}} - {}_{k-1}\mathbf{x}_i^{\mathrm{a}})(\mathbf{x}_i^{\mathrm{f}} - {}_{k-1}\mathbf{x}_i^{\mathrm{a}})^{\mathrm{T}}\right\} \\
&= \frac{{}_k\lambda_i}{m-1}\sum_{j=1}^{m}\left(\mathbf{x}_{i,j}^{\mathrm{f}} - \mathbf{x}_i^{\mathrm{f}}\right)\left(\mathbf{x}_{i,j}^{\mathrm{f}} - \mathbf{x}_i^{\mathrm{f}}\right)^{\mathrm{T}} + \frac{{}_k\lambda_i m}{m-1}(\mathbf{x}_i^{\mathrm{f}} - {}_{k-1}\mathbf{x}_i^{\mathrm{a}})(\mathbf{x}_i^{\mathrm{f}} - {}_{k-1}\mathbf{x}_i^{\mathrm{a}})^{\mathrm{T}}
\end{aligned}
\tag{B4}
$$

Author details

Guocan Wu[1,2]* and Xiaogu Zheng[3]

*Address all correspondence to: gcwu@bnu.edu.cn

1 College of Global Change and Earth System Science, Beijing Normal University, Beijing, China

2 Joint Center for Global Change Studies, Beijing, China

3 Key Laboratory of Regional Climate-Environment Research for East Asia, Institute of Atmospheric Physics, Chinese Academy of Sciences, Beijing, China

References

[1] Miller RN, Ghil M, Gauthiez F. Advanced data assimilation in strongly nonlinear dynamical systems. Journal of the Atmospheric Sciences. 1994;**51**:1037-1056

[2] Ravazzani G et al. Potentialities of ensemble strategies for flood forecasting over the Milano urban area. Journal of Hydrology. 2016;**539**:237-253

[3] Talagrand O. Assimilation of observations, an introduction. Journal of the Meteorological Society of Japan. 1997;**75**:191-209

[4] Wang Y et al. Improving precipitation forecast with hybrid 3DVar and time-lagged ensembles in a heavy rainfall event. Atmospheric Research. 2017;**183**:1-16

[5] Reichle RH. Data assimilation methods in the earth sciences. Advances in Water Resources. 2008;**31**:1411-1418

[6] Yang SC, Kalnay E, Hunt B. Handling nonlinearity in an ensemble Kalman filter experiments with the three-variable Lorenz model. Monthly Weather Review. 2012;**140**:2628-2645

[7] Sakov P, Oliver DS, Bertino L. An iterative EnKF for strongly nonlinear systems. Monthly Weather Review. 2012;**140**:1988-2004

[8] Evensen G. Sequential data assimilation with a nonlinear quasi-geostrophic model using Monte Carlo methods to forecast error statistics. Journal of Geophysical Research. 1994;**99**:10143-10162

[9] Burgers G, Leeuwen PJ, Evensen G. Analysis scheme in the ensemble kalman filter. Monthly Weather Review. 1998;**126**:1719-1724

[10] Luo X, Hoteit I. Ensemble Kalman filtering with residual nudging: An extension to state estimation problems with nonlinear observation operators. Monthly Weather Review. 2014;**142**:3696-3712

[11] Anderson JL, Anderson SL. A Monte Carlo implementation of the nonlinear fltering problem to produce ensemble assimilations and forecasts. Monthly Weather Review. 1999;**127**:2741-2758

[12] Constantinescu EM et al. Ensemble-based chemical data assimilation I: General approach. Quarterly Journal of the Royal Meteorological Society. 2007;**133**:1229-1243

[13] Hamill TM, Whitaker JS. Accounting for the error due to unresolved scales in ensemble data assimilation: A comparison of different approaches. Monthly Weather Review. 2005;**133**:3132-3147

[14] Bai Y, Li X. Evolutionary algorithm-based error parameterization methods for data assimilation. Monthly Weather Review. 2011;**139**:2668-2685

[15] Luo X, Hoteit I. Robust ensemble filtering and its relation to covariance inflation in the ensemble Kalman filter. Monthly Weather Review. 2011;**139**:3938-3953

[16] Dee DP. On-line estimation of error covariance parameters for atmospheric data assimilation. Monthly Weather Review. 1995;**123**:1128-1145

[17] Dee DP, Silva AM. Maximum-likelihood estimation of forecast and observation error covariance parameters part I: Methodology. Monthly Weather Review. 1999;**127**:1822-1834

[18] Li H, Kalnay E, Miyoshi T. Simultaneous estimation of covariance inflatioin and observation errors within an ensemble Kalman filter. Quarterly Journal of the Royal Meteorological Society. 2009;**135**:523-533

[19] Wang X, Bishop CH. A comparison of breeding and ensemble transform kalman filter ensemble forecast schemes. Journal of the Atmospheric Sciences. 2003;**60**:1140-1158

[20] Miyoshi T. The Gaussian approach to adaptive covariance inflation and its implementation with the local ensemble transform Kalman filter. Monthly Weather Review. 2011;**139**:1519-1534

[21] Liang X et al. Maximum likelihood estimation of inflation factors on error covariance matrices for ensemble Kalman filter assimilation. Quarterly Journal of the Royal Meteorological Society. 2012;**138**:263-273

[22] Zheng X. An adaptive estimation of forecast error statistic for Kalman filtering data assimilation. Advances in Atmospheric Sciences. 2009;**26**:154-160

[23] Zheng X et al. Using analysis state to construct forecast error covariance matrix in EnKF assimilation. Advances in Atmospheric Sciences. 2013;**30**:1303-1312

[24] Wu G, Dan B, Zheng X. Soil moisture assimilation using a modified ensemble transform Kalman filter based on station observations in the Hai River basin. Advances in Meteorology. 2016

[25] Anderson JL. An adaptive covariance inflation error correction algorithm for ensemble filters. Tellus. 2007;**59A**:210-224

[26] Anderson JL. Spatially and temporally varying adaptive covariance inflation for ensemble filters. Tellus. 2009;**61A**:72-83

[27] Wang L, Leblanc A. Second-order nonlinear least squares estimation. Annals of the Institute of Statistical Mathematics. 2008;**60**:883-900

[28] Huang C, Wu G, Zheng X. A new estimation method of ensemble forecast error in ETKF assimilation with nonlinear observation operator. SOLA. 2017;**13**:63-68

[29] Wu G et al. Improving the ensemble transform Kalman filter using a second-order Taylor approximation of the nonlinear observation operator. Nonlinear Processes in Geophysics. 2014;**21**:955-970

[30] Wu G et al. A new structure for error covariance matrices and their adaptive estimation in EnKF assimilation. Quarterly Journal of the Royal Meteorological Society. 2013;**139**:795-804

[31] Craven P, Wahba G. Smoothing noisy data with spline functions. Numerische Mathematik. 1979;**31**:377-403

[32] Wahba G et al. Adaptive tuning of numerical weather prediction models randomized GCV in three- and four-dimensional data assimilation. Monthly Weather Review. 1995;**123**:3358-3369

[33] Wu G, Zheng X. An estimate of the inflation factor and analysis sensitivity in the ensemble Kalman filter. Nonlinear Processes in Geophysics. 2017;**24**:329-341

[34] Evensen G. The ensemble Kalman filter theoretical formulation and practical implementation. Ocean Dynamics. 2003;**53**:343-367

[35] Ide K et al. Unified notation for data assimilation operational sequential and variational. Journal of the Meteorological Society of Japan. 1997;**75**:181-189

[36] Houtekamer PL, Mitchell HL. A sequential ensemble Kalman filter for atmospheric data assimilation. Monthly Weather Review. 2001;**129**:123-137

[37] Golub GH, Loan CFV. Matrix Computations. Baltimore: The Johns Hopkins University Press; 1996

[38] Tippett MK et al. Notes and correspondence ensemble square root filter. Monthly Weather Review. 2003;**131**:1485-1490

[39] Lorenz EN. Predictability a problem partly solved Paper presented at seminar on predictability. ECMWF: Reading UK; 1996

[40] Butcher JC. Numerical methods for ordinary differential equations. JohnWiley & Sons. 2003:425

[41] Lorenz EN, Emanuel KA. Optimal sites for supplementary weather observations simulation with a small model. Journal of the Atmospheric Sciences. 1998;**55**:399-414

[42] Yang S-C, Kalnay E, Enomoto T. Ensemble singular vectors and their use as additive inflation in EnKF. Tellus A. 2015;**67**

[43] Desroziers G, Arbogast E, Berre L. Improving spatial localization in 4DEnVar. Quarterly Journal of the Royal Meteorological Society. 2016;**142**:3171-3185

[44] Kirchgessner P, Berger L, Gerstner AB. On the choice of an optimal localization radius in ensemble Kalman filter methods. Monthly Weather Review. 2014;**142**:2165-2175

Unscented Kalman Filter for State and Parameter Estimation in Vehicle Dynamics

Mark Wielitzka, Alexander Busch,
Matthias Dagen and Tobias Ortmaier

Additional information is available at the end of the chapter

http://dx.doi.org/10.5772/intechopen.71900

Abstract

Automotive research and development passed through a vast evolution during past decades. Many passive and active driver assistance systems were developed, increasing the passengers' safety and comfort. This ongoing process is a main focus in current research and offers great potential for further systems, especially focusing on the task of autonomous and cooperative driving in the future. For that reason, information about the current stability in terms of dynamic behavior and vehicle environment are necessary for the systems to perform properly. Thus, model-based online state and parameter estimation have become important throughout the last years using a detailed vehicle model and standard sensors, gathering this information. In this chapter, state and parameter estimation in vehicle dynamics utilizing the unscented Kalman filter is presented. The estimation runs in real time based on a detailed vehicle model and standard measurements taken within the car. The results are validated using a Volkswagen Golf GTE Plug-In Hybrid for various dynamic test maneuvers and a Genesys Automotive Dynamic Motion Analyzer (ADMA) measurement unit for high-precision measurements of the vehicle's states. Online parameter estimation is shown for friction coefficient estimation performing maneuvers on different road surfaces.

Keywords: vehicle dynamics, state estimation, parameter estimation, unscented Kalman filter, dead-time compensation

1. Introduction

In the past decades, enormous developments in automotive research were achieved. Since the beginning of the twentieth century, a consistent search for solutions increasing vehicle's safety and comfort took place. Starting with passive safety systems in the early twentieth century, e.g., airbag, safety belt, and deformable zone, a vast improvement of the passenger's safety was accomplished. These systems reduce passenger injuries and or even death due to accidents.

Later, starting in the 1960th and 1970th active safety systems or advanced driver assistance systems (ADAS), preventing the vehicle from accidents by actively influencing the vehicle was developed. First implementations were traction control (TCS) and antilock braking systems (ABS), stabilizing the vehicle during longitudinal dynamic maneuvers. In 1995, the superordinate electronic stability control (ESC) was developed, combining stabilization during longitudinal and lateral dynamic maneuvers. Because of its success, ESC is mandatory in modern vehicles in Europe since the end of 2014. Other ADAS, e.g., adaptive cruise control (ACC), autonomous emergency braking (AEB), or lane detection system, have been developed in the meantime, further increasing safety and comfort. Future trends show an enormous potential for advanced systems, finally leading to the objective of autonomous and cooperative driving.

Many of these ADAS rely on parametric models, describing and predicting the vehicle's (future) behavior. Especially information about the vehicle's stability, characterized by dynamic states, is necessary. Furthermore, information about the vehicle's environment, e.g., by changing weather conditions and therefore changing friction conditions may influence the systems' performance drastically. Thus, online estimation of the vehicle's stability in terms of its dynamic behavior and online estimation of influential parameters, such as the friction coefficient, are challenging fields in modern automotive research.

Many methods for state estimation use simple models of vehicle dynamics to reduce the computational effort, e.g., considering a linear bicycle model [1–3] or a linear planar two-track model [4]. A further detailed description of the forces acting on the vehicle's tire is developed in Ref. [5] and utilized in Ref. [6]. A more detailed model, considering roll dynamics (cf. **Figure 1**), is utilized in [7, 8].

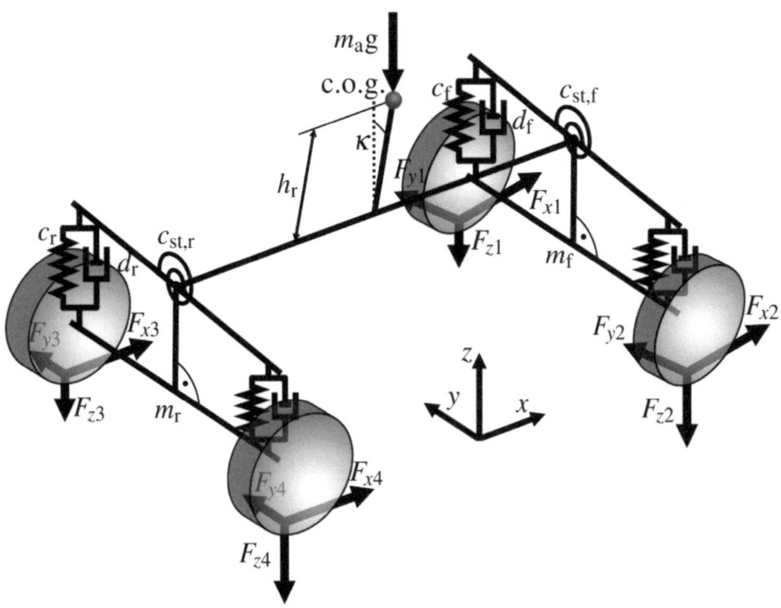

Figure 1. Two-track model including roll dynamics of a vehicle performing a left turn.

Similarly, various methods considering the estimation algorithm are utilized. For linear models, the Kalman filter serves as an optimal filter. Considering nonlinear models, Kalman filter derivatives, such as extended and unscented Kalman filters (EKF/UKF), are used. The EKF is used for tire road force and sideslip angle estimation in Ref. [9] and for sideslip angle estimation of low friction roads in Ref. [6]. The UKF for vehicle state estimation is presented in [7, 8].

Simultaneous state and parameter estimation with dual extended Kalman filter is presented in Ref. [10], estimating the vehicle mass and moment of inertia around the vertical axis. The same parameters are estimated in Ref. [11] using a joint UKF. Friction coefficient estimation using a joint UKF is realized in [12–14].

In this chapter, state estimation in vehicle dynamics utilizing an UKF is presented. The estimation is based on measurements taken with standard sensors, which are implemented in modern vehicles. Therefore, a nonlinear process and measurement model are introduced. Furthermore, dead times, due to CAN communication, are considered and compensated using model-based methods. Additionally, simultaneous state and parameter estimation considering the friction coefficient between tires and road is presented. All methods are validated online using a Volkswagen Golf GTE Plug-In Hybrid as the test vehicle, equipped with a Genesys ADMA inertial platform for precise reference measurements. The friction coefficient estimation is validated on a test track with two different surfaces.

The chapter is organized as follows: in Section 2, the nonlinear process and measurement models of the vehicle's longitudinal and lateral dynamics are introduced. Based on this, Section 3 addresses the state and parameter estimation utilizing the unscented Kalman filter. Furthermore, dead-time compensation and bounded parameter estimation are introduced. In Section 4, the estimation results, using this method, are presented. Thus, measurements taken on a test vehicle using a precise initial measurement unit are presented and discussed. The chapter is recapped with a Conclusion in Section 5.

2. Modeling

In this section, a detailed parametric model of the vehicle's dynamics is presented. Deriving this model for online application, a trade-off between accuracy and computational effort has to be faced. Starting from the contact patch of tires and road as a predominant source of forces acting on the vehicle, the full dynamics of the vehicle will be derived. Furthermore, a measurement model describing certain measurements, representing the vehicle's dynamics taken with standard sensors is presented. The resulting models form the basis for the later state and parameter estimation algorithm using the UKF.

2.1. Tire model

First, the vehicle's tires are considered, representing the contact patch between vehicle and road, consequently providing forces substantially influencing the vehicle's dynamics. These forces arise due to differences in relative motion between tire and road and therefore lead to a deformation of the tire due to friction, described by the friction coefficient μ. These differences in velocity can be expressed as tire slip

$$\lambda = \frac{\omega_t r_t - v_{t,x}}{\max(\omega_t r_t, v_{t,x})}, \tag{1}$$

for longitudinal and tire sideslip angle

$$\alpha = \delta_t - \arctan\left(\frac{v_{t,y}}{v_{t,x}}\right), \tag{2}$$

for lateral motion, respectively, where ω_t represents the rotational velocity of the tire, δ_t is the tire steering angle, $v_{t,x,y}$ is the components of the wheel's velocity v_t, and r_t is the effective tire radius, which is considered a constant. The tire steering angle results from the steering wheel angle δ as $\delta = i_{st}\delta_t$, with the assumption of constant steering transition i_{st}. Coming from the well-known Coulomb friction $F_{fric} = \mu F_z$, with normal force F_z, the associated stationary tire forces in longitudinal and lateral direction, F_x^S and F_y^S, are functions of the tire slip and sideslip angle, respectively, given as

$$F_x^S = f_x(\lambda, \mu_{max}, F_z)h(F_y), \tag{3}$$

$$F_y^S = f_y(\alpha, \mu_{max}, F_z)h(F_x), \tag{4}$$

assuming identical maximum friction coefficient for longitudinal and lateral forces. The nonlinear functions $f_{x,y}$ represent the (side)slip dependency of the tire forces by the magic formula tire model described by

$$f_{x,y}(\chi, \mu_{max}, F_z) = D \sin\left(C \arctan\left(B\frac{\chi}{\mu_{max}} - E\left(B\frac{\chi}{\mu_{max}} - \arctan\left(B\frac{\chi}{\mu_{max}}\right)\right)\right)\right), \tag{5}$$

with individual parameters for longitudinal and lateral dynamics $D = \mu_{max}F_z$, $B = C_F/CD$, and $C_F = c_1 \sin(2\arctan(F_z/c_2))$, while χ represents the slip λ or sideslip α [5]. Using this representation, the function maximum varies linearly over the (side)slip with changing maximum friction coefficient (cf. dashed black line in **Figure 2**). The function $f_y(\lambda, \mu_{max}, F_z)$ for one set of parameters C, E, c_1, and c_2, constant wheel load F_z and changing maximum friction coefficient μ_{max}, representing dry, wet, and icy conditions can be seen in **Figure 2**.

The function $h(F_{x,y}) = \cos(\arctan(B_{xy}F_{y,x}))$ accounts for the reduced forces in the presence of both lateral and longitudinal forces, with scaling factor B_{xy} [15]. Furthermore, the lateral forces F_y are modeled as PT_1-element as

$$F_y + \frac{l_t}{v_{t,x}}\dot{F}_y = f_y(\alpha, \mu_{max}, F_z)h(F_x), \tag{6}$$

with tire-delay constant l_t [16].

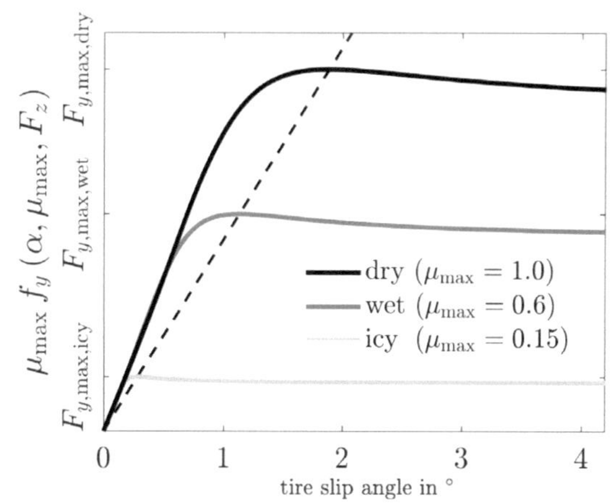

Figure 2. Schematic visualization of the used vehicle dynamics model.

The inputs to the system are the drive and break torques M_d and M_b, and the steering angle δ (cf. **Figure 3**). This leads to the equation of motion for the rotational velocity of one tire ω_t

$$J_\text{t}\dot{\omega}_\text{t} = M_\text{d} - M_\text{b} - M_\text{res} - F_x r_\text{t}, \tag{7}$$

with the tire's moment of inertia J_t, and a moment due to rolling resistance

$$M_\text{res} = F_z(c_{\text{res},1} + c_{\text{res},2}\omega_\text{t}), \tag{8}$$

with constant and velocity dependent part, represented by $c_{\text{res},1}$ and $c_{\text{res},2}$, respectively.

2.2. Vehicle body dynamics

Considering a two-track model with additional roll dynamics as displayed in **Figures 1** and **4**, the vehicle's dynamics under disregard of vertical dynamics can be described by the vehicle's yaw-rate $\dot{\psi}$, its sideslip angle β, the roll angle and rate κ and $\dot{\kappa}$, and its center of gravity (c.o.g.)

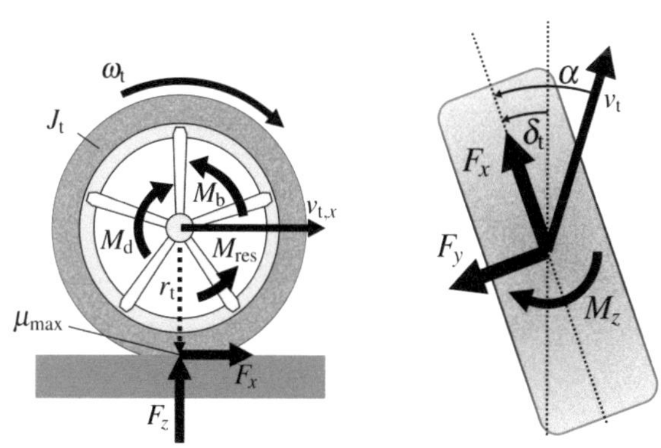

Figure 3. Torques and forces acting on the tire.

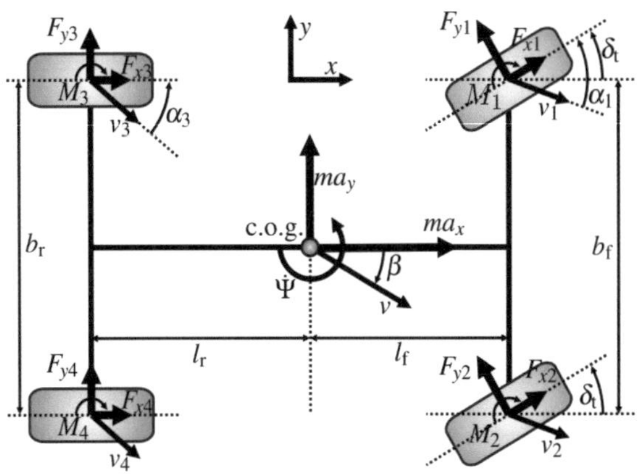

Figure 4. Top view of a two-track model including geometrical parameters.

velocity v. These quantities represent the angular velocity of rotation around the vertical axis, the angle between the vehicle's longitudinal axis and its velocity vector, the angle and rate between the vehicle's vertical axis and the stationary vertical axis, and the velocity of the center of gravity, respectively. This leads to the following equations of motion

$$J_z\ddot{\psi} = l_f(F_{y1} + F_{y2})\cos(\delta_t) - l_r(F_{y3} + F_{y4}) - \sum_{i=1}^{4} M_{z,i}, \tag{9}$$

$$mv(\dot{\beta} + \dot{\psi}) = (F_{y1} + F_{y2})\cos(\beta - \delta_t) + (F_{y3} + F_{y4})\cos(\beta), \tag{10}$$

$$m\dot{v} = (F_{x1} + F_{x2})\cos(\beta - \delta_t) + (F_{x3} + F_{x4})\cos(\beta) - F_{air}, \tag{11}$$

$$m_a h_r a_y = J_\kappa\ddot{\kappa} + d_\kappa(\dot{\kappa}) - c_\kappa(\kappa) - m_a g h_r \sin\kappa, \tag{12}$$

with l_f, l_r being the distance between front and rear axes to the c.o.g., respectively, mass m being moment of inertia with respect to the vertical axis J_z, velocity and acceleration of the c.o. g. v, $\dot{v}$, forces due to air resistance F_{air}, and self-aligning torques $M_{z,i}$.

Roll dynamics are represented in analogy to a spring-damper system with gravitational influence with chassis mass m_a, distance between roll axis and c.o.g. h_r, moment of inertia with respect to the roll axis J_κ, gravitational acceleration g, and nonlinear spring and damper coefficients $c_\kappa(\kappa)$ and $d\dot{\kappa}(\dot{\kappa})$, respectively. These characteristics are represented by

$$c_\kappa(\kappa) = c_{\kappa,1}\kappa + c_{\kappa,2}\kappa^3, \tag{13}$$

$$d_{\dot{\kappa}}(\dot{\kappa}) = d_{\dot{\kappa},1}\dot{\kappa} + d_{\dot{\kappa},2}\tanh(\dot{\kappa}). \tag{14}$$

These constants result from combinations of suspension and stabilization constants $c_{f,r}$, $d_{f,r}$, and $c_{st,f,r}$, respectively (cf. **Figure 1**). In conclusion, the resulting system state vector can be expressed as

$$\boldsymbol{x} = \left[\dot{\psi}, \beta, \kappa, \dot{\kappa}, F_{y,i}, \omega_i, v\right]^T. \tag{15}$$

2.3. Measurement model

Inside the vehicle, standard sensors are implemented to obtain information about the current driving state. These sensors measure the yaw-rate $\dot{\psi}$, longitudinal and lateral acceleration $a_{x,\mathrm{s}}$ and $a_{y,\mathrm{s}}$, and the four rotational velocities of the vehicle's tires ω_i. This leads to the measurement vector

$$y = \left[\dot{\psi}, a_{x,\mathrm{s}}, a_{y,\mathrm{s}}, \omega_{\mathrm{t},i}\right]^{\mathrm{T}}. \tag{16}$$

A measurement model representing these sensors is needed for the later implementation of KF derivatives. Since the measured yaw-rate and wheel velocities are states within the model, they are obtained directly and no further model is needed. The longitudinal and lateral accelerations in the sensor position $a_{x,\mathrm{s}}$ and $a_{y,\mathrm{s}}$, respectively, can be described by

$$a_{x,\mathrm{s}} = \dot{v}\cos\beta - v(\dot{\beta}+\dot{\psi})\sin\beta - l_y\ddot{\psi} + \dot{\psi}\left(l_z\dot{\kappa} - l_x\dot{\psi}\right), \tag{17}$$

$$a_{y,\mathrm{s}} = \dot{v}\sin\beta + v(\dot{\beta}+\dot{\psi})\cos\beta - l_z\ddot{\kappa} + l_x\ddot{\psi} - l_y\dot{\kappa}^2 - l_y\dot{\psi}^2, \tag{18}$$

with l_x, l_y, and l_z being the components of the distances from the c.o.g. to the sensor position.

Due to the sensors' sampling rate of 100 Hz, all later implementations on the control unit will be running at this frequency. Therefore, the continuous time differential Eqs. (1)–(18) are discretized using first-order Euler discretization.

Within the model, numerous parameters are utilized. These parameters can either be measured directly, e.g., geometrical parameters, or need to be identified using an offline identification algorithm. Since the model is strongly nonlinear, a particle swarm algorithm (PSO) is used. Therefore, measurements representing longitudinal and lateral dynamics, driven by a test vehicle, have to be performed. Hence, a sequential identification can be realized, first considering longitudinal excitation, neglecting lateral dynamics and subsequently lateral excitation. Further details can be found in Ref. [15].

In summary, the vehicle's dynamics can be expressed by the discrete time state space representation

$$\boldsymbol{x}_{k+1} = \boldsymbol{f}(\boldsymbol{x}_k, \boldsymbol{u}_k; \boldsymbol{p}), \tag{19}$$

$$\boldsymbol{y}_{k+1} = \boldsymbol{g}(\boldsymbol{x}_{k+1}, \boldsymbol{u}_{k+1}; \boldsymbol{p}), \tag{20}$$

at discrete time step $\boldsymbol{k}$ with all parameters included in $\boldsymbol{p}$. Thereby, $\boldsymbol{x}\in\mathbb{R}^{13\times 1}$ represents the system state and $\boldsymbol{y}\in\mathbb{R}^{7\times 1}$ represents the measurement vector.

3. State and parameter estimation

In this section, a brief overview over the used Kalman filter derivative will be given. At first, the algorithm for state estimation will be presented. Furthermore, a model-based dead-time compensation will be introduced. Secondly, the joint UKF for state and parameter estimation will be presented. Subsequently, an extension for the estimation of bounded parameters is introduced.

3.1. The unscented Kalman filter for state estimation

The process and measurement model presented in Section 2 are strongly nonlinear, especially considering the forces acting on the vehicle's tires (cf. Eq. (5)). Therefore, a Kalman filter derivative, capable of estimating nonlinear systems, the UKF is utilized. Since no information about the covariance is available, the additive form of the process and measurement equations

$$\boldsymbol{x}_{k+1} = \boldsymbol{f}(\boldsymbol{x}_k, \boldsymbol{u}_k; \boldsymbol{p}) + \boldsymbol{w}_k, \tag{21}$$

$$\boldsymbol{y}_{k+1} = \boldsymbol{g}(\boldsymbol{x}_{k+1}, \boldsymbol{u}_{k+1}; \boldsymbol{p}) + \boldsymbol{v}_{k+1}, \tag{22}$$

with $\boldsymbol{w}_k \propto \mathcal{N}(\boldsymbol{0}, \boldsymbol{Q})$, $\boldsymbol{Q} \in \mathbb{R}^{n_x \times n_x}$ and $\boldsymbol{v}_k \propto \mathcal{N}(\boldsymbol{0}, \boldsymbol{R})$, $\boldsymbol{R} \in \mathbb{R}^{n_y \times n_y}$ representing the process and measurement uncertainties by uncorrelated Gaussian random numbers, is assumed. The system state and measurement are described by the state and measurement vectors $\boldsymbol{x}_k \in \mathbb{R}^{n_x}$ and $\boldsymbol{y}_k \in \mathbb{R}^{n_y}$ with state dimension n_x and measurement dimension n_y. To initialize the filter, initial values for the state and covariance estimation, $\widehat{x}_0 \in \mathbb{R}^{n_x}$ and $\widehat{P}_0 \in \mathbb{R}^{n_x \times n_x}$, respectively, have to be set. Following this, the recursive estimation divided in two steps, i.e., the process and measurement update can be realized. Within the process update, an *a priori* state and covariance estimation utilizing the process model is executed. Using the unscented transformation [17], a carefully chosen set of $2n_x + 1$ sigma points for time step $k \in \{0, \ldots, \infty\}$

$$x_{k,0} = \widehat{x}_k, \tag{23}$$

$$x_{k,i} = \widehat{x}_k + \left(\sqrt{(n + \lambda_{\mathrm{ukf}})\widehat{P}_k}\right)_i \quad \text{for } i = 1, \ldots, n_x, \tag{24}$$

$$x_{k,i} = \widehat{x}_k - \left(\sqrt{(n + \lambda_{\mathrm{ukf}})\widehat{P}_k}\right)_{i-n_x} \quad \text{for } i = i + 1, \ldots, 2n_x, \tag{25}$$

with $\widehat{x}_k$ and $\widehat{P}_k$ representing the current state and covariance estimation, respectively, are calculated. Thereby, $\lambda_{\mathrm{ukf}} = \alpha_{\mathrm{ukf}} n_x + \kappa_{\mathrm{ukf}} - n_x$, with scaling parameters α_{ukf} and κ_{ukf}. Furthermore, $\left(\sqrt{(n + \lambda_{\mathrm{ukf}})\widehat{P}_k}\right)_i$ is the ith column of the matrix square root, e.g., using Cholesky decomposition. These sigma points characterize the current probability density function and undergo the real nonlinear transformation utilizing Eq. (21) to calculate the *a priori* estimation as

$$\boldsymbol{x}_{k+1,i}^- = \boldsymbol{f}(\boldsymbol{x}_{k,i}, \boldsymbol{u}_k; \boldsymbol{p}), \tag{26}$$

$$\boldsymbol{x}_{k+1}^- = \sum_{i=0}^{2n_x} W_i^{\mathrm{m}} \boldsymbol{x}_{k+1,i}^-, \tag{27}$$

$$\boldsymbol{P}_k^- - = \sum_{i=0}^{2n_x} W_i^{\mathrm{c}} \left(\boldsymbol{x}_{k+1,i}^- - \boldsymbol{x}_{k+1}^-\right)\left(\boldsymbol{x}_{k+1,i}^- - \boldsymbol{x}_{k+1}^-\right)^{\mathrm{T}}, \tag{28}$$

$$y^-_{k+1,i} = h(x_{k+1,i}, u_{k+1}; p), \tag{29}$$

$$y^-_{k+1} = \sum_{i=0}^{2n_x} W^{\mathrm{m}}_i y^-_{k+1,i}. \tag{30}$$

Following this, a measurement y_{k+1} is received and the measurement update

$$P_{xy,k+1} = \sum_{i=0}^{2n_x} W^{\mathrm{c}}_i \left(x^-_{k+1,i} - x^-_{k+1}\right)\left(y^-_{k+1,i} - y^-_{k+1}\right)^{\mathrm{T}}, \tag{31}$$

$$P_{yy,k+1} = \sum_{i=0}^{2n_x} W^{\mathrm{c}}_i \left(x^-_{k+1,i} - x^-_{k+1}\right)\left(y^-_{k+1,i} - y^-_{k+1}\right)^{\mathrm{T}}, \tag{32}$$

$$K_{k+1} = P_{xy,k+1} P^{-1}_{yy,k+1}, \tag{33}$$

$$\widehat{x}_{k+1} = x^-_{k+1} + K_{k+1}\left(y_{k+1} - y^-_{k+1}\right), \tag{34}$$

$$\widehat{P}_{k+1} = P^-_{k+1} - K_{k+1} P_{yy,k+1} K^{\mathrm{T}}_{k+1}, \tag{35}$$

with weighting factors $W^{\mathrm{m,c}}_i$ can be executed. This leads to the *a posteriori* estimations of the state and covariance, $\widehat{x}_{k+1}$ and $\widehat{P}_{k+1}$, respectively [18].

3.2. Dead-time compensation

When designing online methods for real-time applications, dead times are frequently to face. Especially, considering vehicular applications, the communication is realized via CAN-Bus, leading to dead times. In the following, a method to compensate for dead times within state estimation is presented.

Since the measurement update (Eqs. (31)–(35)) can only be processed, as soon as a measurement y_k is received, dead times t_{d} corrupt the UKF severely. Ignoring this dead time may lead to poor filter performance or even divergence. One solution is to accept the dead time and delay the estimation by exact this time. Alternatively, the system's state and covariance can be estimated by performing the process update (Eqs. (26)–(28)) during the dead time without doing the measurement update, based on the delayed filter estimation up to time step k, so that the state estimation at discrete time step $k + n_{t_{\mathrm{d}}}$, where $n_{t_{\mathrm{d}}}$ is the number of discrete time steps due to the dead time is

for j = 1 to $n_{t_{\mathrm{d}}}$ do

$$x_{k+j-1,0} = \widehat{x}_{k+j-1} \tag{36}$$

$$x_{k+j-1,i} = \widehat{x}_{k+j-1} + \left(\sqrt{(n + \lambda_{\mathrm{ukf}})\widehat{P}_{k+j-1}}\right)_i \text{ for } i = 1, \ldots, n_x \tag{37}$$

$$x_{k+j-1,i} = \widehat{x}_{k+j-1} - \left(\sqrt{(n + \lambda_{\mathrm{ukf}})\widehat{P}_{k+j-1}}\right)_{i-n_x} \text{ for } i = i+1, \ldots, 2n_x \tag{38}$$

$$x^-_{k+j,i} = f\left(x_{k+j-1}, u_{k-1}; p\right) \tag{39}$$

$$\widehat{x}_{k+j} = \sum_{i=0}^{2n_x} W_i^{\mathrm{m}} x_{k++j,i}^{-} \tag{40}$$

$$\widehat{P}_{k+j} = \sum_{i=0}^{2n_x} W_i^{\mathrm{c}} \left(x_{k+j,i}^{-} - \widehat{x}_{k+j}\right)\left(x_{k+j,i}^{-} - \widehat{x}_{k+j}\right)^{\mathrm{T}} \tag{41}$$

end

To reduce computational cost, the update steps can be reduced, so that only the mean is transformed and therefore no further sigma point needs to be calculated and transformed as

for j=1 to n_{t_d} do

$$\widehat{x}_{k+j} = f\left(\widehat{x}_{k+j-1}, u_{k-1}; p\right) \tag{42}$$

end

This may lead to reduced performance, depending on the length of the dead time, the process model complexity, and the uncertainties within the process.

3.3. Parameter estimation

Since parameters may vary within dynamic systems, simultaneous state and parameter estimation is considered. Various methods to solve this task have been developed in the past decades. In the following, the approach of joint state and parameter estimation is presented. Therefore, states and parameters are concentrated into one joint state vector as

$$x_k = \begin{bmatrix} x_k \\ p_{\mathrm{est},k} \end{bmatrix}, \tag{43}$$

with primary state $x_k \in \mathbb{R}^{n_x}$ and parameters to be estimated $p_{\mathrm{est},k} \in \mathbb{R}^{n_p}$. The model within the process update assumes constant parameters, i.e., $p_{\mathrm{est},k+1} = p_{\mathrm{est},k}$. The remaining UKF algorithm stays the same, with only the dimension of the estimated state $\widehat{\tilde{x}}_k \in \mathbb{R}^{n_x+n_p}$ changing.

If some parameters are bounded as $a_i \leq p_{\mathrm{est},i} \leq b_i$, these are not estimated directly, but using a substitute parameter $p_{\mathrm{sub},i}$ as

$$p_{\mathrm{est},i} = \frac{b_i - a_i}{2}\tanh\left(p_{\mathrm{sub},i}\right) + \frac{a_i + b_i}{2}. \tag{44}$$

Using this substitution, the estimated parameter $p_{\mathrm{sub},i}$ is not bounded and leads to the real parameters for $p_{\mathrm{est},i}$ in the intended range.

4. Estimation results

In this section, the results of the vehicle's states and parameter estimation using an unscented Kalman filter and the two-track model described Section 2 are presented. At

http://dx.doi.org/10.5772/intechopen.71900

first, the test vehicle and the measurement equipment that is necessary for the validation are introduced. Secondly, the results of the vehicle state estimation and the results by considering dead time within the estimation are shown. Furthermore, a simultaneous state and friction estimation are presented that exhibit improved estimation results for varying road conditions.

4.1. Measurement setup

The estimation results are verified by using a Volkswagen Golf GTE Plug-In Hybrid (**Figure 5(a)**) equipped with a Genesys ADMA-G-Eco + (**Figure 5(b)**). This system is developed especially for vehicle dynamics testing in the automotive sector. This inertial measurement unit (IMU) corrected by global positioning system (GPS) enables precise measurement of acceleration, speed, and position of the moving test vehicle in all three-dimensional axes. Furthermore, the pitch, roll and yaw angles, angular velocities as well as sideslip angle can be obtained. The GPS antenna is mounted on the roof of the test vehicle, whereas the IMU is placed in the footwell of the passenger seat. Ideally, the two sensors are placed in the center of gravity; unfortunately, in praxis, this requirement usually either cannot or can only be fulfilled with very high effort. Furthermore, it is hardly possible to exactly align the ADMA's measurement axes with the vehicle's axes. The errors caused by the distance between the installation position and the c.o.g as well as by misalignment angle can be mathematically compensated considering the lever arms and the angle offsets, respectively. The IMU's measurements are exclusively provided for the validation of the Kalman filter application. Additional measurements of the wheel speeds, accelerations and yaw rate as well as system inputs, i.e., steering angle, engine, and breaking torque, are taken from the vehicle's bus system. The onboard measurements are provided for the measurement update of the real-time Kalman filter application. Furthermore, an ES910 prototyping and interface module provides the connection to the vehicle bus and the computation of the filter application with system-level behavior.

4.2. Unscented Kalman filter setup

The UKF utilizes a two-track model including roll dynamics as described in Section 2. The estimator's state and measurement vector were therefore

(a) Volkswagen Golf GTE plug-in-Hybrid

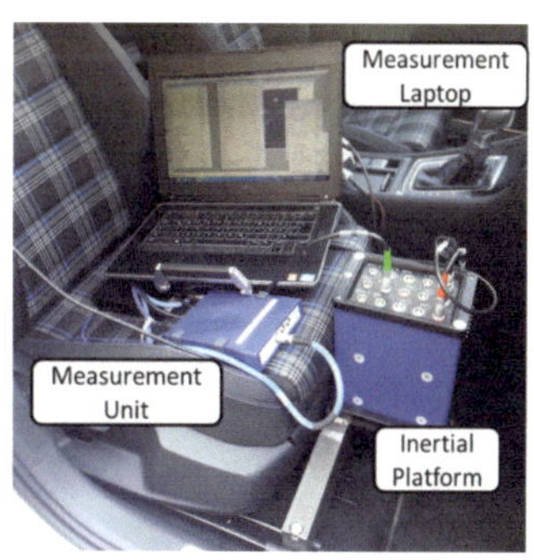

(b) Measurement setup

Figure 5. Test vehicle and measurement equipment.

$$x = \left(\dot{\psi}, \beta, \kappa, \dot{\kappa}, F_{y_i}, \omega_i, v_{\mathrm{cog}}\right)^{\mathrm{T}}, \tag{45}$$

$$y = \left(\dot{\psi}, a_{y,\mathrm{s}}, a_{x,\mathrm{s}}, \omega_i\right)^{\mathrm{T}}. \tag{46}$$

α_{UKF} determines the spread of the sigma points around the mean $\hat{x}$ and is set to 1, the scaling parameter κ_{UKF} is usually set to 0, and $\beta_{\mathrm{UKF}} = 2$ is an optimal value to approximate distribution of x as Gaussian distribution. The process and measurement covariances are empirically determined and set up to

$$Q = \mathrm{diag}\left(10^4, 1, 1, 1, 1, 1, 1, 1, 10^{10}, 10^{10}, 10^{10}, 10^{10}, 1\right)10^{-15} \tag{47}$$

$$R = \mathrm{diag}\left(10^{-6}, 10^{-3}, 10^{-2}, 10^{-3}, 10^{-3}, 10^{-3}, 10^{-3}\right). \tag{48}$$

The initial covariance matrix of the state distribution is initialized with $P_0 = Q$.

4.3. State estimation

In this section, the results of vehicle state estimation are shown. The UKF estimation results are displayed in light grey, while the reference measurements of the ADMA-G-Eco +are displayed in black and the vehicle`s onboard measurements are displayed in grey. Each displayed maneuver is subdivided into eight diagrams. The top line shows yaw-rate and sideslip angle representing lateral dynamics, and the second line shows roll angle and roll rate representing roll dynamics. Furthermore, the velocity in the c.o.g. and the wheel speeds representing longitudinal dynamics are displayed in the third line and at last longitudinal and lateral acceleration in the bottom line. The accelerations in the c.o.g are displayed with continuous lines and the accelerations measured with onboard acceleration sensor that is not mounted in the c.o.g. in the dashed lines. The distances of the sensor position from the c.o.g. are identified with $l_x = 1.07$ m, $l_y = -0.39$ m, and $l_z = 0.55$ m. In order to demonstrate the estimation quality, **Figure 6** shows a steering sweep maneuver with periodical steering angle input at nearly constant amplitude and increasing frequency and at a constant velocity of 37 km/h Despite the varying frequency, an accurate estimation of all dynamic states is evident, accompanied by improved estimation of the vehicles velocity compared to the onboard measurement of the vehicles' velocity. Only for the roll angle, a higher deviation can be recognized. This results from varying lateral inclination of the test road.

A steering sweep maneuver is optimal to validate the filter application and the integrated vehicle dynamic models, but it is not a practical example, whereas, for example, lane change maneuvers often occur. In addition, lane change maneuvers enable high lateral acceleration and high values of tire sideslip angles, which may lead, under certain conditions, to loss of stability due to nonlinear tire characteristics. Therefore, lane change maneuvers are suitable as practical driving situation.

http://dx.doi.org/10.5772/intechopen.71900

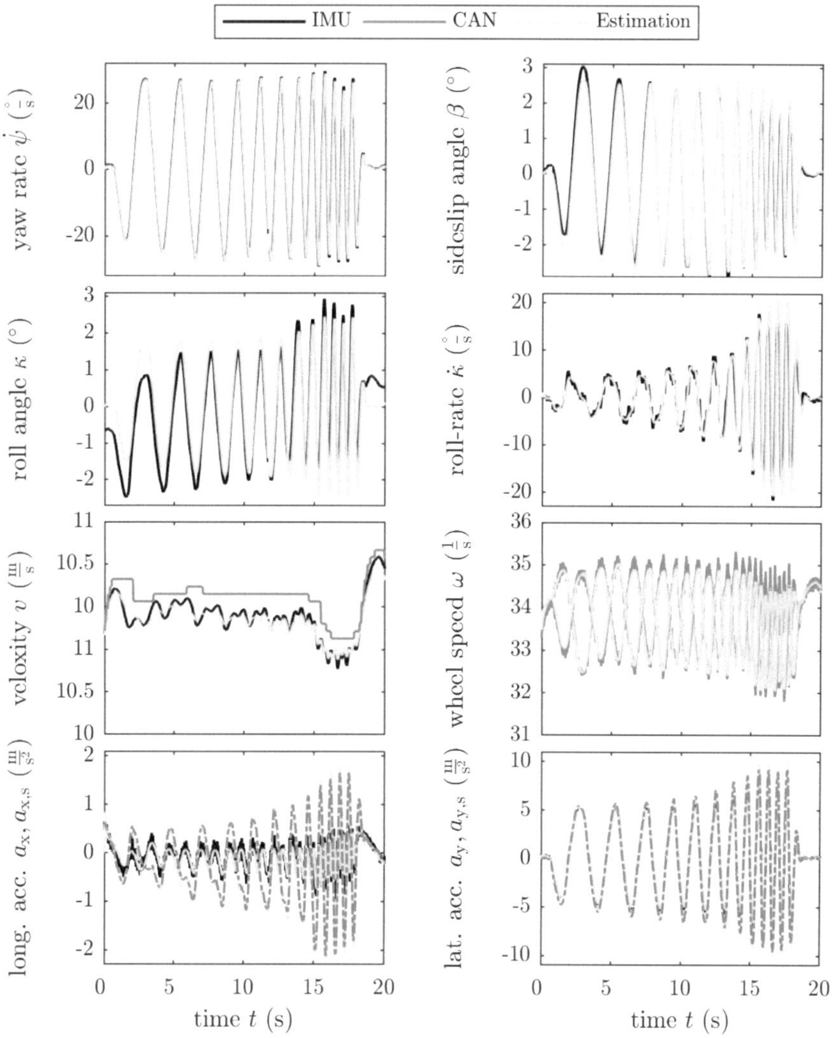

Figure 6. Vehicle state estimation during steering sweep maneuver.

A double-lane change maneuver with moderate lateral acceleration represents a typical maneuver on highways or freeways occurring by overtaking another slower moving vehicle. However, a double-lane change maneuver with high lateral acceleration represents obstacle avoidance maneuver. This kind of maneuver is displayed with a high lateral acceleration up to $a_y = 7\ \text{m/s}^2$ and with nearly constant velocity of $v = 37\,\text{km/h}$ in **Figure 7**. It can be stated that also an accurate estimation of all relevant states can be seen over the whole maneuver, and again particularly an improved velocity estimation compared to onboard measurement can be emphasized. The higher estimation performance of the velocity can be advantageous for some control applications, such as collision avoidance.

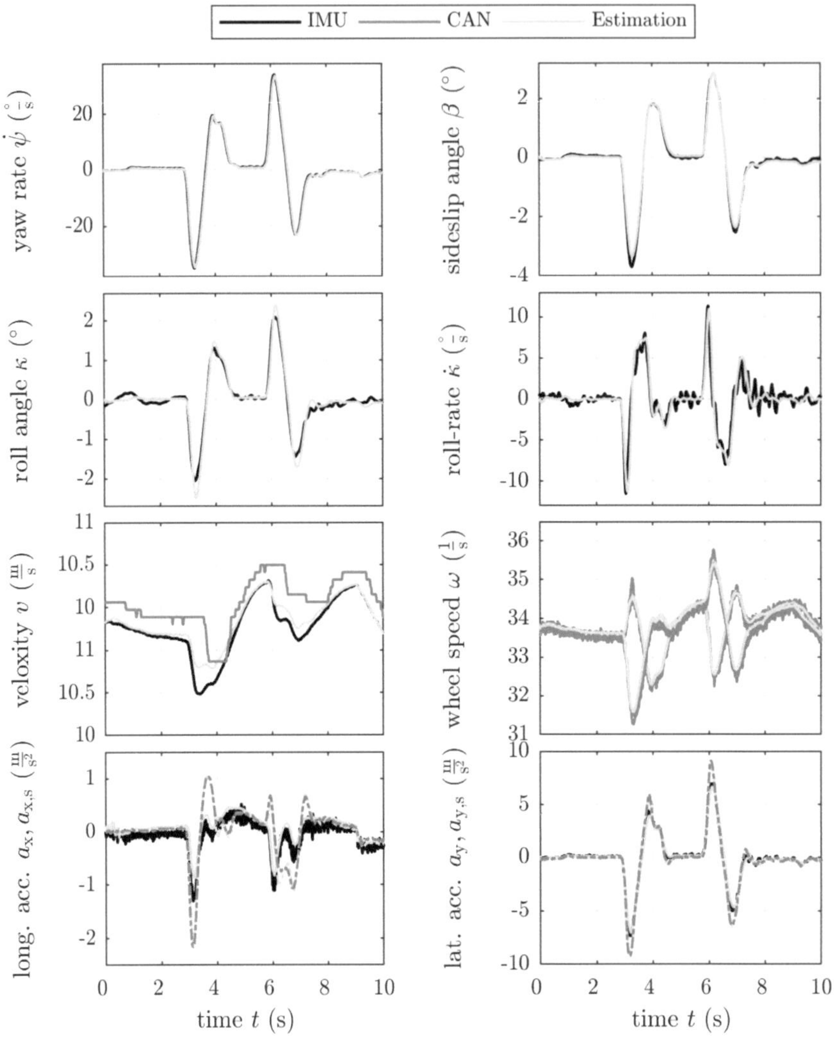

Figure 7. Vehicle state estimation during double-lane change maneuver.

4.4. Dead-time compensation

A major problem in control systems is dead time. Dead time may lead either to poor control results or to unstable control. In case of stability control systems and therefore for robust car steering, the real-time information of sideslip angle and yaw rate is very important. As pointed previously, an accurate estimation of this vehicle states can be realized using UKF with nonlinear two-track model even if vehicles move at its stability limits. However, the estimated states are not the true vehicle states at this particular time. They are delayed due to the dead-time-shifted measurement update, which occurs particularly of the communication on the CAN network. In dependence on the length of dead time, different arrangements exist for dead-time compensation. In addition, to consider the dead time in the controller designing, a further possibility is model-based dead-time compensation within state estimation. A simple method to predict the system's state and covariance during the dead time is to execute only the filter process update without doing the measurement update as defined in Section 3.2. Alternatively, to reduce computational cost, the prediction can be executed only considering the mean of the system states by neglecting further sigma points. During the prediction from time step k to $k + n_{t_d}$, there are no further information of the system input. Therefore, the last known system input is used for the prediction. From this follows a prediction error that decreases with less dynamics of the system input. A comparison of both possibilities and the quality of the dead-time compensation within the vehicle state estimation is shown in **Figure 8**. A step steering maneuver on dry asphalt at a velocity of approximately 50 km/h with maximum lateral acceleration of almost $a_y = 7\,\mathrm{m/s^2}$ is considered. The top diagram shows the yaw rate and the next diagram the sideslip angle, while measurements are displayed in black, the UKF's state estimation without dead-time compensation in grey, the dead-time compensated state estimation by using all sigma points in dashed light grey, and the dead-time compensated state estimation by using only the mean in dash-dotted grey. In the present case, the predominant dead time amounts about 30 ms that corresponds to $n_{t_d} = 3$. The dead time is determined by comparing redundant measurements that are obtained from the IMU of the ADMA and the vehicle's onboard CAN bus.

It is quite obvious that both methods for dead-time compensation do not really differ in the application of vehicle dynamics; thus, it is at an advantage due to reduced computational costs only to consider the mean of the state. Furthermore, the UKF results with dead-time-compensated

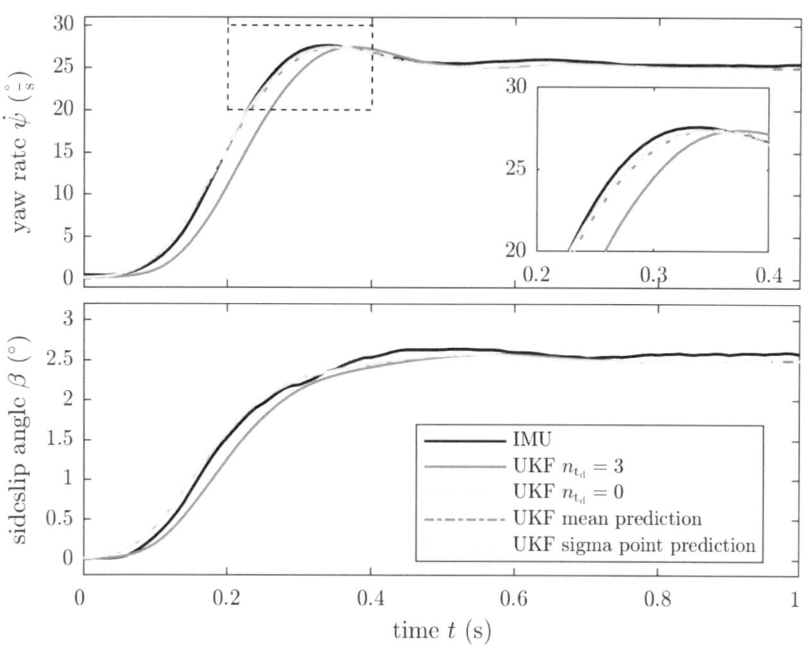

Figure 8. Dead-time-compensated vehicle state estimation during step steering maneuver.

states also do not differ from the time-shifted UKF prediction without dead time displayed in light grey. Therefore, it is possible to include dead-time compensation in the state estimation in a simple manner and to obtain precise estimation results. However, this method for dead-time compensation has the restriction that the system does not contain varying dead time and the dead time is well known.

All previously presented test maneuvers are executed on dry asphalt. Thus, the maximum friction coefficient between road and tire is well known. Under different road conditions, for example, wet, snow, and ice and without adaption of the friction coefficient, the accuracy of the state estimation decreases highly. Hence, for precise state estimation, it is essential to estimate the maximum friction coefficient as well.

4.5. Maximum friction coefficient estimation

However, not only for improved vehicle state estimation a simultaneous estimation of the maximum friction coefficient between road and tire is of paricular importance. To ensure a proper functionality of safety functions not only the knowledge of the driving situation but also of the driving environment, in particular the road condition, is required. This section focuses on the road condition classification using a joint unscented Kalman filter approach as described in Section 3.3. The extended process, covariance matrix $\boldsymbol{Q}_{\mathrm{p}}$, for the maximum friction coefficient estimation is empirically determined and set up to

$$\boldsymbol{Q}_{\mathrm{p}} = \mathrm{diag}\big(\mathrm{diag}(\boldsymbol{Q}), 10^{-8}\big). \tag{49}$$

Furthermore, the maximum friction coefficient is bounded according to Eq. (37) with a upper bound of 1.1 and a lower bound of 0.1. The upper bound corresponds to the best traction potential that may occur when the roads are dry and the tires are in good condition. The lower bound corresponds to the lowest traction potential that may occur when the roads are icy and the tires have a low tire profile.

In **Figure 9**, again a double-lane change maneuver on dry asphalt at a velocity of approximately 43 km/h with maximum lateral acceleration of almost $a_y = 9\ \mathrm{m/s^2}$ is considered. The top diagram shows the measured system input namely the steering angle. The following diagrams show the yaw rate, the lateral acceleration, and the sideslip angle, while the IMU's measurements are displayed in black and the onboard measurements in grey, and the UKF's estimations in light grey. The bottom diagram shows the estimated maximum friction coefficient, while the initial value is wrongly set to $\mu_{\mathrm{max}} = 0.4$. By using offline identification algorithms, the reference value for the maximum friction coefficient was determined at approximately 1. The light red lines show the state estimation without adaptation of friction coefficient. Obviously, an accurate sideslip and yaw-rate estimation can only be guaranteed with adaptation of the friction coefficient. However, an adaption of the friction coefficient is only possible during phases of sufficient excitation. At the beginning and at the end of the maneuver without steering, no adaption may take place.

http://dx.doi.org/10.5772/intechopen.71900

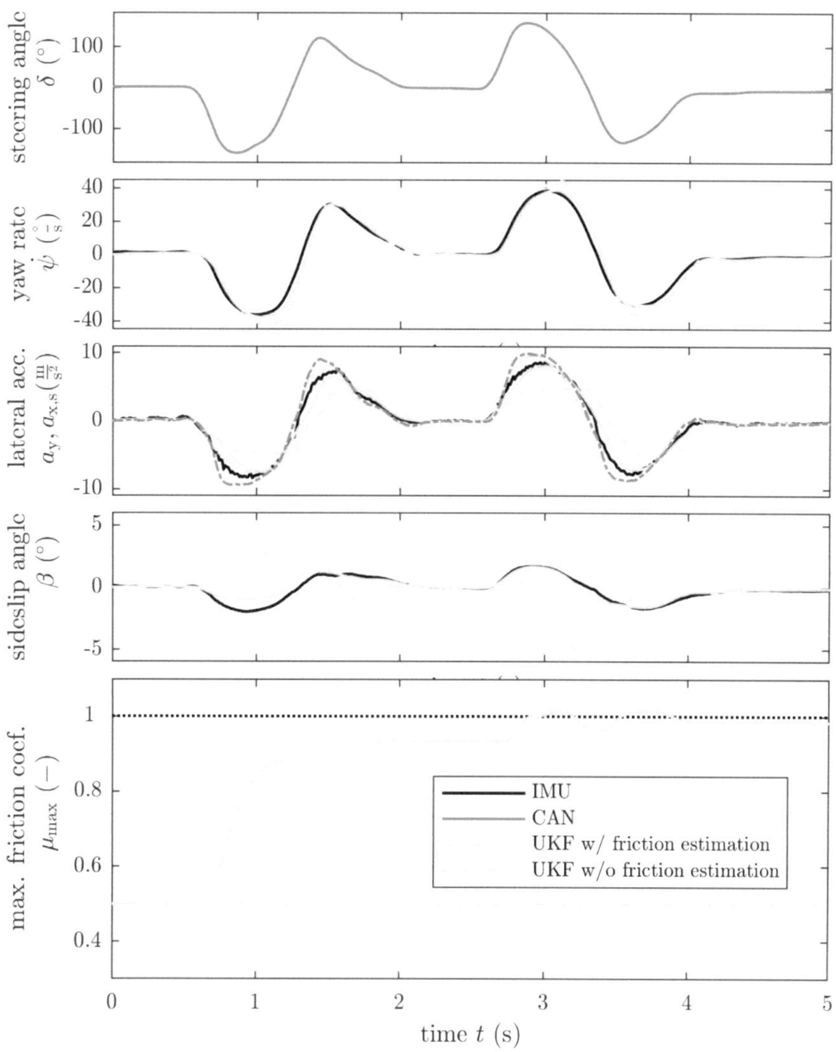

Figure 9. Vehicle state and maximum friction coefficient estimation during double-lane change maneuver on dry asphalt.

When sufficient excitation exists, the maximum friction coefficient can be estimated within few seconds.

In **Figure 10**, the estimation of the maximum friction coefficient on different roads is displayed. Beginning on dry asphalt, the test vehicle drives a sine steering maneuver and changes over to wet steel. Wet steel exhibits similar properties regarding traction potential as an icy road. Because it is much easier to build up a road composed of wet steel than of ice to carry out a test, in this test, the wet steel represents a road with low traction potential. Over the entire period of the sine maneuver, a sufficient excitation is existent. Hence, the unscented Kalman

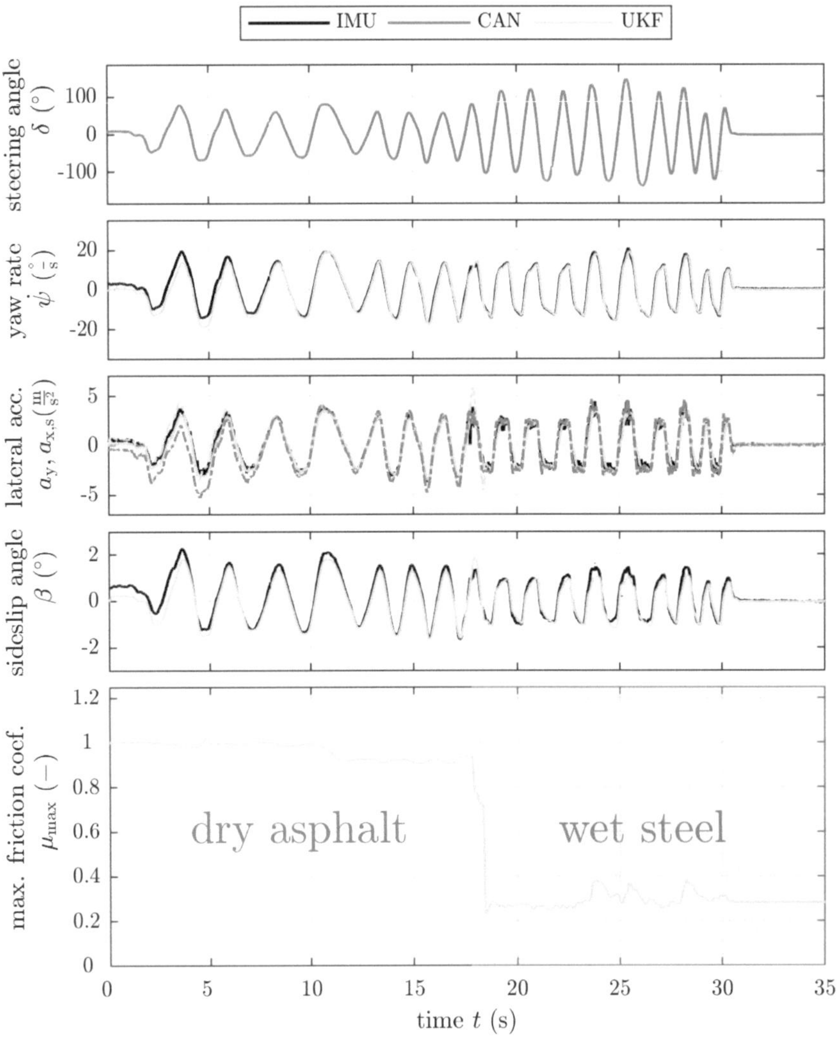

Figure 10. Vehicle state and maximum friction coefficient estimation during sine steering maneuver on dry asphalt and wet steel.

filter estimates while driving over the dry asphalt a maximum friction coefficient of approximately 1. After changing to wet steel, clearly, a lower friction level with a maximum friction coefficient of approximately 0.3 is detected. It can be spotted that while sufficient excitation is existent, different friction levels according to different road conditions can also be clearly recognized. The last 5 s of the maneuver is without steering and acceleration, and the estimated maximum friction coefficient remains constant. Therefore, due to missing excitation, it would not be possible to distinguish between dry asphalt and wet steel. Nonsufficient excitation is a major disadvantage of Kalman filter-based approaches for parameter estimation in general and thus also for friction estimation.

A promising approach to solving this problem is to use further source of information, for example, optical sensors and to do an information fusion, so that disadvantages of one information can be compensated by advantages of other information. Information fusion is the next step for improved friction estimation and current research.

5. Conclusion

In this chapter, state and parameter estimation in vehicle dynamics using the unscented Kalman filter is presented. Therefore, a detailed nonlinear process and measurement model of the vehicle are introduced, representing the vehicle's stability and the measurements taken with standard sensors. Dead times, due to CAN communication, are faced and compensated using model-based prediction. The validation of the introduced methods is realized by using a Volkswagen Golf GTE Plug-In Hybrid for high dynamic test maneuvers, e.g., double-lane change. The estimation results are compared with high-precision measurements using a Genesys ADMA inertial measurement unit. Accurate estimation even in situations with lateral acceleration above 7 m/s^2 can be achieved. Furthermore, real-time estimation, compensated for dead times can be realized using model-based prediction.

The parameter estimation is presented using the example of friction coefficient estimation utilizing the joint unscented Kalman filter. Thus, maneuvers with different excitation on different road surfaces are executed. Again, precise estimation in the presence of sufficient excitation can be shown.

Author details

Mark Wielitzka*, Alexander Busch, Matthias Dagen and Tobias Ortmaier

*Address all correspondence to: mark.wielitzka@imes.uni-hannover.de

Institute of Mechatronic Systems, Leibniz Universität Hannover, Germany

References

[1] Best MC, Gordon TJ, Dixon PJ. An extended adaptive Kalman filter for real-time state estimation of vehicle handling dynamics. Vehicle System Dynamics. 2000;**34**(1):57-75. DOI: 10.1076/0042-3114(200008)34:1;1-K;FT057

[2] Cheli F, Sabbioni E, Pesce M, Melzi S. A methodology for vehicle sideslip angle identification: Comparison with experimental data. Vehicle System Dynamics. 2007;**45**(6):549-563. DOI: 10.1080/00423110601059112

[3] Chen BC, Hsieh F-C. Sideslip angle estimation using extended Kalman filter. Vehicle System Dynamics. 2009;**46**(S1):353-364. DOI: 10.1080/00423110801958550

[4] Doumiati M, Victorino AC, Charara A, Lechner D. Onboard real-time estimation of vehicle lateral tire–road forces and sideslip angle. IEEE Transactions on Mechatronics. 2011;**16**(4):601-614. DOI: 10.1109/TMECH.2010.2048118

[5] Pacejka HB, Bakker E. The magic formula tire model. Vehicle System Dynamics. 1992;**21** (S1):1-18. DOI: 10.1080/00423119208969994

[6] Li L, Jia G, Ran X, Song J, Wu K. A variable structure extended Kalman filter for vehicle sideslip angle estimation on a low friction road. International Journal of Vehicle Mechanics and Mobility. 2014;**52**(2):280-308. DOI: 10.1080/00423114.2013.877148

[7] Antonov S, Fehn A, Kugi A. Unscented Kalman filter for vehicle state estimation. Vehicle System Dynamics. 2011;**49**(9):1497-1520. DOI: 10.1080/00423114.2010.527994

[8] Wielitzka M, Dagen M, Ortmaier T. State estimation of vehicle's lateral dynamics using unscented Kalman filter. Los Angeles, USA: IEEE 53rd Annual Conference on Decision and Control (CDC); 15-17 December 2014; 2015. pp. 5015-5020

[9] J Dakhlallah J, Glaser S, Mammar S, Sebsadji Y. Tire-Road Forces Estimation Using Extended Kalman Filter and Sideslip Angle Evaluation. Seattle, USA: 2008 American Control Conference (ACC); 11–13 June 2008; 2008. pp. 4597-4602

[10] Wenzel TA, Burnham KJ, Blundell MV, Williams RA. Dual extended Kalman filter for vehicle state and parameter estimation. Vehicle System Dynamics. 2006;**44**(2):153-171. DOI: 10.1080/00423110500385949

[11] Wielitzka M., Dagen M., Ortmaier T.: Joint unscented Kalman filter for state and parameter estimation in vehicle dynamics. Sydney, Australia: IEEE Conference On Control Applications (CCA); 21-23 September 2015. pp. 1945-1950

[12] Ahn C, Peng H, Tseng HE. Robust estimation of road friction coefficient using lateral and longitudinal vehicle dynamics. Vehicle System Dynamics. 2012;**50**(6):961-985. DOI: 10.1080/00423114.2012.659740

[13] Ahn C, Peng H, Tseng HE. Robust estimation of road friction. IEEE Transactions on Control Systems Technology. 2013;**521**(1):1-13. DOI: 10.1109/TCST.2011.2170838

[14] Wielitzka M, Dagen M, Ortmaier T. State and maximum friction coefficient estimation in vehicle dynamics using UKF. Seattle, USA: American Control Conference (ACC); 24-26 May 2017. pp. 4322-4327

[15] Wielitzka M, Dagen M, Ortmaier T. Nonlinear modeling and parameter identification of Vehicle's lateral dynamics. Tokio, Japan: 12thth International Symposium on Advanced Vehicle Control (AVEC); 22–16 September 2014

[16] Rajamani, Rajesh. Vehicle dynamics and control; Springer Science & Business Media; 2011

[17] Julier SJ. The Scaled Unscented Transformation. 2002: American Control Conference (ACC); 08-10 May 2002. USA: Anchorage; 2012. pp. 4555-4559

[18] Wan EA, Van der Merwe R. The unscented Kalman filter for nonlinear estimation. Alberta, Canada: Adaptive Systems for Signal Processing, Communications, and Control Symposium (AS-SPCC); 04 October 2000. p. 153-158

Chapter 4

Sensitivity-Based Adaptive SRUKF for State, Parameter, and Covariance Estimation on Mechatronic Systems

Mauro Hernán Riva, Mark Wielitzka and
Tobias Ortmaier

Additional information is available at the end of the chapter

http://dx.doi.org/10.5772/intechopen.72470

Abstract

Since the initial developments in the state-space theory in the 1950s and 1960s, the state estimation has become an extensively researched and applied discipline. All systems that can be modelled mathematically are candidates for state estimators. The state estimators reconstruct the states that represent internal conditions and status of a system at a specific instant of time using a mathematical model and the information received from the system sensors. Moreover, the estimator can be extended for system parameter estimation. The resulting Kalman filter (KF) derivatives for state and parameter estimation also require knowledge about the noise statistics of measurements and the uncertainties of the system model. These are often unknown, and an inaccurate parameterization may lead to decreased filter performance or even divergence. Additionally, insufficient system excitation can cause parameter estimation drifts. In this chapter, a sensitivity-based adaptive square-root unscented KF (SRUKF) is presented. This filter combines a SRUKF and the recursive prediction-error method to estimate system states, parameters and covariances online. Moreover, local sensitivity analysis is performed to prevent parameter estimation drifts, while the system is not sufficiently excited. The filter is evaluated on two testbeds based on an axis serial mechanism and compared with the joint state and parameter UKF.

Keywords: Unscented Kalman, filter, recursive prediction-error method, state estimation, parameter estimation, covariance estimation, sensitivity analysis

1. Introduction

State estimation is applicable to almost all areas of engineering and science. It is interesting to engineers for different reasons such as the control of a system using a state-feedback controller or monitoring the system states that are not measureable with sensors, or the sensors are too expensive or too difficult to install. The system states can be defined as variables, which provide a representation of internal conditions and status of a system at a specific instant of time. Applications that include a mathematical model of any system are candidates for state estimation. The estimations can be useful, for example, car assistance systems [1], predictive maintenance [2], structure health estimation [3], and many other applications (see [4] and references therein).

Different algorithms were proposed for online state estimation. A historical survey of the filtering algorithms can be found in [5]. The Kalman filter (KF) was presented in [6] and nowadays is the most widely applied algorithm for state estimation on linear systems. The KF is a linear optimal estimator [7]. This means that the KF is the best filter that uses a linear combination of the system measurements and states in order to estimate the last ones. The main operation of the KF is the propagation of the mean and covariance of the (Gaussian) random variables (RVs) through time. The KF assumes that the model and the noise statistics affecting the system are known. Otherwise, the estimates can degrade.

Different derivatives of the KF have been developed for nonlinear systems during the last decades. The extended Kalman filter (EKF) presented in [8] is the most commonly used estimator for nonlinear system. This filter linearizes the system and measurement equations at the current estimate. This may lead to poor performances for highly nonlinear or highly noisy systems [9]. To address the linearization errors of the EKF, the unscented Kalman filter (UKF) was presented in [10]. This filter uses the unscented transformation (UT) to pick a minimal set of points around the mean of the GRV. These points capture the true mean and covariance of the GRV, and they are then propagated through the true nonlinear function capturing the a posteriori mean and covariance more accurately.

The mathematical models usually describe the behaviour of the systems, and generally the structure and the parameters need to be determined. Once the structure is defined, system inputs and measurements can be used to identify the model parameters. This can be performed offline [11, 12]. However, the parameters usually may vary during operations. In order to monitor these variations online, the nonlinear extensions of the KF can be extended for parameter estimation [9].

The KF derivatives can only achieve good performances under a priori assumptions, for example, accurate system models, noise statistics knowledge, and proper initial conditions [7, 9, 13]. If one of these assumptions is not guaranteed, the KF derivative can potentially become unstable and the estimations can be diverged [14–16]. Moreover, tuning the performance of these filters implies primarily adjusting the process and measurement noise covariances to match the (unknown) real-system noise statistics. In the last decades, numerous methods were presented to estimate these unknown covariances. The autocovariance least-square method was presented in [17, 18], and it was extended (and simplified) in [19], and diagonal process and noise covariances were considered in [20]. This method estimates the noise covariances using least squares and it can only be used with KF. The method was extended for nonlinear or time-varying systems using an EKF in [21]. Online covariance estimation for EKF and square-root cubature Kalman filter (SRCuKF) was presented in [22]. These methods implement a combination of a KF derivative and a recursive prediction-error method (RPEM) to estimate covariances online. In [23], an adaptive UKF was presented to estimate only covariances online.

In this chapter, a sensitivity-based adaptive square-root unscented Kalman filter (SB-aSRUKF) is presented. This filter estimates system states, parameters and covariances online. Using local state sensitivity models (SMs), this filter prevents parameter and covariance estimation drifts, while the system is not sufficiently excited. Sensitivity analysis (SA) for the UKF is also presented. The performance of this filter is validated in simulations on two testbeds and compared with the joint UKF for parameter and state estimation.

Section 2 covers some algorithms for recursive estimation of states, parameters, and covariances. The SB-aSRUKF is the main topic of this chapter. This filter uses a KF derivative for state estimation. In Section 2.1, the KF for state estimation in linear dynamic systems is presented. The UKF, a nonlinear extension of the KF, is described in Section 2.2 and also extended for estimating system parameters. Section 2.3 covers parameter estimation using the RPEM. The UKF and the RPEM are combined in Section 2.4 to obtain the aSRUKF. In order to identify unknown parameters, the system inputs should be persistently exciting. Sensitivity models (SMs) are presented in this section and are used to evaluate the system excitation and prevent parameter estimation drifts while the system is not sufficiently excited.

Section 3 covers the testbed used for the filter evaluations. A planar one-link robot system is described in Section 3.1, and a pendulum robot (pedubot) is mathematically modelled in Section 3.2. The first testbed is used for the SM analysis, and the chaotic system is used to compare the filter performance with the joint SRUKF. The evaluation results of the SB-aSRUKF are presented in Section 4. The SMs are analysed with different system inputs on the first testbed in Section 4.1, and the filter performance for state and parameter estimation is compared with the joint SRUKF in Section 4.2. Section 5 completes the chapter with conclusions.

2. Recursive estimation

This section discusses some recursive approaches to estimate states, parameters and covariances of a general system. The KF as the optimal linear estimator for linear dynamic systems is presented. Nonlinear extensions of the KF are discussed, as well as an extension for parameter estimation. A recursive Gauss-Newton method for parameter estimation is also presented in this section. Finally, the last subsection discusses the SB-aSRUKF, which is the main topic of this chapter, and the SMs that are used for excitation monitoring.

2.1. Kalman filter (KF)

The KF is the most widely applied algorithm for state estimation on linear dynamic systems that are corrupted by stochastic noises (e.g. Gaussian noise). It uses a parametric mathematical model of the system and a series of (noisy) measurements from, for example, sensors to estimate the system states online [4]. In general, the state distribution of a system can be approximated by random variables (RVs). The main operation of the KF is the propagation of the mean and covariance of these (Gaussian) RVs through time. The KF is an optimal linear filter for these types of systems [7, 9]. It is a recursive algorithm, which enables new measurements to be processed as they arrive to correct and update the state and measurement estimates.

In general, a linear discrete-time system corrupted by additive noises can be written as follows:

$$\begin{aligned} x_k &= Ax_{k-1} + Bu_{k-1} + w_k, \\ y_k &= Cx_k + Du_k + v_k, \end{aligned} \tag{1}$$

where $x_k \in \mathbb{R}^{n_x}$ is the system state vector at discrete time k, and $u_k \in \mathbb{R}^{n_u}$ and $y_k \in \mathbb{R}^{n_y}$ correspond to the system input and measurement vectors, respectively. The matrices $A \in \mathbb{R}^{n_x \times n_x}$, $B \in \mathbb{R}^{n_x \times n_u}$, $C \in \mathbb{R}^{n_y \times n_x}$ and $D \in \mathbb{R}^{n_y \times n_u}$ are often called system, input, output and feedforward matrices, respectively, and describe the system behaviour. The random variable vectors w_k and v_k represent the process and measurement noises. These are considered white Gaussian, zero mean, and uncorrelated and have covariance matrices Q_k and R_k, respectively, as

$$\begin{aligned} w_k &\backsim N(0, Q_k), \\ v_k &\backsim N(0, R_k). \end{aligned} \tag{2}$$

The KF iterative nature can be separated in two main steps: the process update and the correction step. In the process update, based on the knowledge of the system dynamics, the state estimate $(\widehat{x}^+_{k-1})$[1] from the previous time step $(k-1)$ is used to calculate a new estimate at the current time (k). This step does not include any information of the system measurements and the resulting state estimate is called a priori estimate $(\widehat{x}^-_k)$. In the correction step, the a priori estimate is combined with the current system measurement (y_k) to improve the state estimate. This estimate is called the a posteriori state estimate $(\widehat{x}^+_k)$. The vectors $\widehat{x}^-_k$ and $\widehat{x}^+_k$ estimate both the same quantity, but the difference between them is that the last one takes the measurement (y_k) into account. A Kalman gain matrix (K_k) is calculated at every discrete step and weights the influence of the model and the measurements on the current state estimate. This gain is calculated using the system matrices and the process (Q_k) and measurement (R_k) covariances. More information about the KF equations and generalizations can be found in [4, 7, 9].

The KF is a linear optimal estimator, but it assumes that the system model and noise statistics are known. Otherwise, the filter estimates can degrade. Tuning the performance of the filter implies primarily adjusting the process and measurement covariance matrices to match the (unknown) real-system noise statistics. In practical implementations of the KF, the filter tuning is performed online, and empirical values are normally used. Extensive research has been done in this field to estimate the noise covariances from data (see [17–20] and references therein).

As mentioned before, the KF is the optimal linear estimator, which estimates states of a linear dynamic system using the inputs, measurements and a parametric mathematical model of the system. Even though many systems are close enough to linear and linear estimators give

[1] The hat $\widehat{}$ over a vector represents the estimate of the vector, for example, $\widehat{x}$ describes the estimate of the state vector x.

acceptable results, all systems are ultimately nonlinear. Extensions of the KF have been presented in the last decades to deal with nonlinear systems. Some examples are the EKF and the sigma-point Kalman filters (SPKFs).

2.2. Nonlinear filtering

The EKF and the UKF (a SPKF type) are derivatives of the KF for nonlinear systems. The EKF was originally proposed in [8] and is the most commonly applied state estimator for nonlinear systems. However, if the system nonlinearities are severe or the noises affecting the system are high, the EKF can be difficult to tune, often gives wrong estimates and can lead to filter divergence easily. This is because the EKF uses linearized system and measurement models at the current estimate and propagates the mean and covariance of the GRVs through these linearizations. The UKF was presented in [10] and addresses the deficiencies of the EKF linearization providing a direct and explicit mechanism for approximating and transforming the mean and covariance of the GRVs.

In general, a discrete-time state-space model of a nonlinear system can be described by

$$\begin{aligned} \boldsymbol{x}_k &= \boldsymbol{f}(\boldsymbol{x}_{k-1}, \boldsymbol{\theta}_{k-1}, \boldsymbol{u}_{k-1}) + \boldsymbol{w}_{k-1}, \\ \boldsymbol{y}_k &= \boldsymbol{h}(\boldsymbol{x}_k, \boldsymbol{\theta}_k, \boldsymbol{u}_k) + \boldsymbol{v}_k, \end{aligned} \tag{3}$$

where $\boldsymbol{\theta}_k \in \mathbb{R}^{n_p}$ is the (unknown) parameter vector and $\boldsymbol{f}$ and $\boldsymbol{h}$ are arbitrary vector-valued functions usually called system and measurement functions. As a KF derivative, the UKF aim is to minimize the covariance of the state estimation error to find an optimal estimation of the state true dynamic probability density function (pdf). The main component of this filter is the UT. This transformation uses a set of appropriately selected weighted points to parameterize the mean and covariance of the pdf. Two steps characterize also the UKF. In the process update, the sigma points are calculated and then propagated through the nonlinear system functions to recover the mean and covariance of the new a priori estimates. The estimated measurement $(\widehat{\boldsymbol{y}}_k)$ is calculated in the correction step and together with the actual measurement are used to correct the a priori estimate. This results in the a posteriori state estimate. While the UKF matches the true mean of $\boldsymbol{x}_k$ correctly up to the third order, the EKF only matches up to the first order. Both filters approximate the true covariance of $\boldsymbol{x}_k$ up to the third order. However, the UKF correctly approximates the signed of the terms to the fourth power and higher meaning that the resulting error should be smaller [7, 9].

The nonlinear extensions of the KF can also estimate the unknown parameters of a system. The UKF was extended for joint state and parameter estimation in [24]. In this case, the system state vector $\boldsymbol{x}_k$ was extended by including the unknown parameters $\boldsymbol{\theta}_k$ to obtain a joint state and parameter vector as

$$\tilde{\boldsymbol{x}}_k = \begin{pmatrix} \boldsymbol{x}_k \\ \boldsymbol{\theta}_k \end{pmatrix}, \tag{4}$$

remaining $\boldsymbol{\theta}_k = \boldsymbol{\theta}_{k-1}$ during the process update.

Square-root (SR) filtering increases mathematically the precision of the KF when hardware precision is not available. In [25], an SR version of the UKF was presented, which uses linear algebra techniques such as the QR decomposition and the Cholesky factor [26] to calculate the SR of the estimation error covariance. The SR form improves the numerical stability of the filter and guarantees positive semi-definiteness of this covariance. Additionally, the computational complexity for state and parameter estimation is reduced [25].

2.3. Recursive prediction-error method

In this section, the recursive prediction-error method (RPEM) is briefly discussed. This method is extensively analysed in [11, 12] and uses a parameterized predictor that estimates the system outputs at the current time step. The resulting predicted system output is then compared to the actual system measurement, and the predictor parameters are corrected such as that the prediction error is minimized.

The quadratic criterion function defined as

$$V_k(\boldsymbol{\theta}_k) = \frac{1}{2}\boldsymbol{e}_k^{\mathrm{T}}(\boldsymbol{\theta}_k)\boldsymbol{\Lambda}^{-1}\boldsymbol{e}_k(\boldsymbol{\theta}_k), \tag{5}$$

is minimized using the stochastic Gauss-Newton method in order to obtain the predictor parameters. The prediction error $\boldsymbol{e}_k(\boldsymbol{\theta}_k)$ at the discrete time k is described as

$$\boldsymbol{e}_k(\boldsymbol{\theta}_k) = \boldsymbol{y}_k - \widehat{\boldsymbol{y}}_k(\boldsymbol{\theta}_k), \tag{6}$$

where $\boldsymbol{y}_k$ corresponds to the actual system measurement, $\widehat{\boldsymbol{y}}_k(\boldsymbol{\theta}_k)$ refers to the parameterized predictor output using parameter set $\boldsymbol{\theta}_k$ and Λ is a user-defined weight factor.

The recursive solution that minimizes the quadratic criterion function in Eq. (5) is given by the following scheme:

$$\begin{aligned}
\boldsymbol{\Delta}_k &= \boldsymbol{\Delta}_{k-1} + (1-\lambda)\left(\boldsymbol{e}_k\boldsymbol{e}_k^{\mathrm{T}} - \boldsymbol{\Delta}_{k-1}\right),\\
\boldsymbol{S}_k &= \lambda\boldsymbol{\Delta}_k + \frac{\mathrm{d}\widehat{\boldsymbol{y}}_k}{\mathrm{d}\widehat{\boldsymbol{\theta}}_{k-1}}\boldsymbol{\Theta}_{k-1}\frac{\mathrm{d}\widehat{\boldsymbol{y}}_k^{\mathrm{T}}}{\mathrm{d}\widehat{\boldsymbol{\theta}}_{k-1}},\\
\boldsymbol{L}_k &= \boldsymbol{\Theta}_{k-1}\frac{\mathrm{d}\widehat{\boldsymbol{y}}_k^{\mathrm{T}}}{\mathrm{d}\widehat{\boldsymbol{\theta}}_{k-1}}\boldsymbol{S}_k^{-1},\\
\boldsymbol{\Theta}_k &= \left(\boldsymbol{I}^{n_\theta} - \boldsymbol{L}_k\frac{\mathrm{d}\widehat{\boldsymbol{y}}_k}{\mathrm{d}\widehat{\boldsymbol{\theta}}_{k-1}}\right)\boldsymbol{\Theta}_{k-1}\left(\boldsymbol{I}^{n_\theta} - \boldsymbol{L}_k\frac{\mathrm{d}\widehat{\boldsymbol{y}}_k^{\mathrm{T}}}{\mathrm{d}\widehat{\boldsymbol{\theta}}_{k-1}}\right)/\lambda + \boldsymbol{L}_k\boldsymbol{\Delta}_k\boldsymbol{L}_k^{\mathrm{T}},\\
\widehat{\boldsymbol{\theta}}_k &= \widehat{\boldsymbol{\theta}}_{k-1} + \boldsymbol{L}_k\boldsymbol{e}_k.
\end{aligned} \tag{7}$$

The user-defined parameter $0 < \lambda \leq 1$ is often called the forgetting factor. The matrix $\boldsymbol{\Delta}_k$ is calculated using the prediction error. This matrix is used to calculate $\boldsymbol{S}_k$, where the derivative of the output w.r.t. to the unknown parameter vector $\left(\frac{\mathrm{d}\widehat{\boldsymbol{y}}_k}{\mathrm{d}\widehat{\boldsymbol{\theta}}_{k-1}}\right)$ appears. The gain vector $\boldsymbol{L}_k$ is multiplied by the innovation error to update then the parameter estimation. It should be noted that besides the matrix $\widehat{\boldsymbol{y}}_k = \frac{\mathrm{d}\widehat{\boldsymbol{y}}_k}{\mathrm{d}\widehat{\boldsymbol{\theta}}_{k-1}}$, all parameters, vectors, and matrices are defined after an initialization. The matrix $\widehat{\boldsymbol{y}}_k$ can be calculated modifying a KF derivative.

The selection of the forgetting factor essentially defines the measurements that are relevant for the current estimation of the parameter predictor. The most common choice is to take a constant forgetting factor for systems that change gradually. Other criterions for selection of this factor and the convergence of the RPEM are discussed extensively in [11, 12].

2.4. Sensitivity-based adaptive square-root Kalman filter

This is the main section of this chapter. The earlier sections were written to provide the needed methods for this section, and the later sections are written to analyse and test the performance of the filter described in this section.

The aSRUKF is discussed in this section. This filter combines the SRUKF and the RPEM. While the KF derivative estimates the system states and measurements, the RPEM calculates the unknown parameters and covariances.

In this filter, the innovation error in Eq. (6) is calculated and minimized using the recursive scheme presented in Eq. (7) in order to estimate the unknown system parameters and covariances. Besides the matrix $\hat{y}_k$, all parameters, vectors, and matrices of the recursive scheme are defined. The derivative of the estimated measurement ($\hat{y}_k$) w.r.t. the vector ($\hat{\boldsymbol{\theta}}_{k-1}$) containing the unknown values of parameters and covariances needs to be calculated. This matrix is also called the output sensitivity and describes the influence of a variation of a parameter on the system output. The output sensitivity can be obtained using a KF derivative.

The equations of a SRUKF are then extended in order to calculate the output sensitivity. To simplify the reading flow, the following definitions are presented:

$$\begin{aligned}
&w_i^m = w_i^c = \frac{1}{2(n_x+\lambda_f)}, && i=1,\dots,m=2n_x,\ \ \lambda_f=\alpha^2(n_x+\kappa), && \\
&w_0^m = \frac{\lambda_f}{2(n_x+\lambda_f)}, && w_0^c = w_0^m + (1-\alpha^2+\beta), && \eta=\sqrt{n_x+\lambda_f}, \\
&\boldsymbol{m}_1 = 1_{1,n_x}, && \boldsymbol{m}_2 = 1_{1,l}, && l = 2n_x+1, \\
&\boldsymbol{w}^c = (w_0^c,\dots,w_m^c), && \boldsymbol{w}^m = (w_0^m,\dots,w_m^m), && \mathbf{W}_{\mathrm{kr}}^c = \boldsymbol{w}^c \otimes \boldsymbol{m}_1^{\mathrm{T}},
\end{aligned} \tag{8}$$

where $w_i^{m,c}$ are a set of scalar weights, α determines the spread of sigma points around the estimated state $\hat{x}_k$, β incorporates information about the noise distribution (e.g. $\beta = 2$ assumes that the system is affected by Gaussian noise), and κ is a scaling factor, which can be used to reduce the higher-order errors of the mean and covariance approximations [9]. The Kronecker product [27] is described by $\otimes$.

The process update step of the aSRUKF is presented in **Table 1**. After the filter initialization, the sigma points ($\boldsymbol{X}_{k-1}$) that describe the pdf of the state estimate are calculated using the UT. At the same time, the sigma-point derivatives ($\boldsymbol{\Phi}_{k-1}$) are also obtained. The sigma points and its derivatives are propagated through the system function and the system derivative function, respectively, to obtain the a priori state estimate ($\hat{x}_k^-$) and the a priori state estimate sensitivity ($\hat{\boldsymbol{\mathcal{X}}}_k^-$). Considering additive process noises, the SR factor of the estimation error covariance ($\boldsymbol{S}_{xx,k}^-$) is calculated using the QR decomposition (qr()) and the rank-one update to Cholesky

aSRUKF	Initialization	$\widehat{x}_0 = x_{\text{init}} \in \mathbb{R}^{n_x}, \quad \widehat{\theta}_0 = \theta_{\text{init}} \in \mathbb{R}^{n_\theta}, \quad \mathcal{S}_{yy.k} = \frac{dS_{yy,k}}{d\theta_k}, \quad \Theta_k = \Theta_{\text{init}} \in \mathbb{R}^{n_\theta \times n_\theta}$ $\mathcal{P}_{xx,k} = \frac{dP_{xx,k}}{d\theta_k}, \quad \mathcal{P}_{yy,k} = \frac{dP_{yy,k}}{d\theta_k}, \quad \mathcal{S}_{xx.k} = \frac{dS_{xx,k}}{d\theta_k}, \quad \Delta_k = 0 \in \mathbb{R}^{n_y \times n_y},$ $P_{xx,0} = S_{xx,0} S_{xx,0}^{\mathrm{T}} = P_{xx,\text{init}} \in \mathbb{R}^{n_x \times n_x},$ $\mathcal{P}_{xx,0} = \mathcal{S}_{xx,0} = 0 \in \mathbb{R}^{n_x \times n_x n_\theta}, \quad \mathcal{P}_{yy,0} = \mathcal{S}_{yy,0} = 0 \in \mathbb{R}^{n_y \times n_y n_\theta}$
SRUKF	Sigma points	$X_{k-1} = (X_{k-1,1}, \ldots, X_{k-1,l}) = \left(\widehat{x}_{k-1} \widehat{x}_{k-1} \otimes m_1 + \eta S_{xx,k-1} \widehat{x}_{k-1} \otimes m_1 - \eta S_{xx,k-1}\right)$
	Sigma-points propagation	$X_k^* = f\left(X_{k-1}, \widehat{\theta}_{k-1}, u_{k-1}\right)$
SM	Sigma-point derivatives	$\Phi_{k-1,j} = (\Phi_{k-1,1,j}, \ldots, \Phi_{k-1,l,j}) = \left(\frac{dX_{k,1}}{d\widehat{\theta}_{k,j}}, \ldots, \frac{dX_{k,m}}{d\widehat{\theta}_{k,j}}\right) = \left(\widehat{\mathcal{X}}_{k-1,j} \widehat{\mathcal{X}}_{k-1,j} \otimes m_1 + \eta \mathcal{S}_{xx,k-1,j} \widehat{\mathcal{X}}_{k-1,j} \otimes m_1 - \eta \mathcal{S}_{xx,k-1,j}\right)$ with $\widehat{\mathcal{X}}_{k-1} = \frac{d\hat{x}_{k-1}}{d\hat{\theta}_{k-1}}$
	Sensitivity propagation	$\Phi_{k,j,i}^* = \left.\frac{\partial f}{\partial x_k}\right\vert_{X_{k-1}, \hat{\theta}_{k-1}} \Phi_{k-1,j,i} + \left.\frac{\partial f}{\partial \theta_{k,j}}\right\vert_{X_{k-1}, \hat{\theta}_{k-1}}$
SRUKF	A priori state estimate	$\widehat{x}_k^- = X_k^* (w^m)^{\mathrm{T}}$
	SR estimation error covariance	$S_{xx,k}^- = \mathrm{qr}\left(\left(\sqrt{w_1^c}\left(X_{1:2n_x,k}^* - \widehat{x}_k^-\right) \quad S_{Q,k}\right)\right),$ $S_{xx,k}^- = \mathrm{cholupdate}\left(S_{xx,k}^-, \sqrt{\lvert w_0^c \rvert}\left(X_{0,k}^* - \widehat{x}_k^-\right), \mathrm{sign}\left(w_0^c\right)\right)$
SM	A priori state sensitivity	$\widehat{\mathcal{X}}_k^- = \Phi_k^* \left(I^{n_\theta} \otimes (w^m)^{\mathrm{T}}\right)$
	Derivative of estimation error covariance	$\mathcal{S}_{xx,k}^- = \mathrm{treshapeM}\left(\left(I^{n_\theta} \otimes A_{\mathrm{ls}, S_{xx,k}^-}^{\dagger}\right)\left(\mathcal{P}_{xx,k}^-\right)_s\right),$ with $A_{\mathrm{ls}, S_{xx,k}^-} = \left(S_{xx,k}^- \otimes I^{n_x} + \left(I^{n_x} \otimes \mathcal{S}_{xx,k}^-\right) \mathcal{N}_{(n_x)}\right)) \mathcal{E}_{(n_x)}$ and $\mathcal{P}_{xx,k}^- = \left(\Phi_k^* - \widehat{\mathcal{X}}_k^- \otimes m_2\right)\left(I^{n_\theta} \otimes \left(\left(W_{\mathrm{kr}}^c\right)^{\mathrm{T}} \odot \left(X_k^* - \widehat{x}_k^- \otimes m_2\right)^{\mathrm{T}}\right)\right)$ $\quad + W_{\mathrm{kr}}^c \odot \left(X_k^* - \widehat{x}_k^- \otimes m_2\right)\left(\mathcal{T}_{(l)} I^{n_\theta} \otimes \left(\Phi_k^* - \widehat{\mathcal{X}}_k^- \otimes m_2\right)^{\mathrm{T}}\right) + \left.\frac{dQ_k}{d\theta_k}\right\vert \widehat{\theta}_{k-1}$ in which $\mathcal{T}_{(l)} = \left(I^0, T_1, \ldots, T_{n_\theta}\right)$, and $T = \left(0_1^{l \times l}, \ldots, 0_{n_\theta}^{l \times l}, I^l\right)$. Further details of the construction of the matrix $\mathcal{N}_{(n_x)}$ and the elimination matrix $\mathcal{E}_{(n_x)}$ can be found in [28]

Table 1. aSRUKF: filter initialization, sigma-points calculation and filter process update step. The Kronecker product is described by $\otimes$ and $\odot$ defines the element-wise multiplication. The $(\,)_s$ operator stacks the matrix columns to form a column vector.

factorization (cholupdate() [2]) in which the signum function (sign()) is used to determine the sign of the first calculated weight factor. If the weight factor results negative, a Cholesky downdate takes place; otherwise, a Cholesky update occurs. The next step calculates the derivative of the SR factor of the estimation error covariance ($\mathcal{S}_{xx,k}^-$). In this step, the function treshapeM() is used. This function converts a vector of dimension $((n_x(n_x+1)/2)n_\theta \times 1)$ into a $(n_x \times n_x n_\theta)$ matrix with n_θ lower triangular blocks of size $(n_x \times n_x)$. Additionally, the operator $()_s$ is utilized to stack the matrix columns to form a column vector. Further information about this step can be found in [28].

[2] Matlab does not allow the use of cholupdate() in real-time applications; using coder.extrinsic('cholupdate'), it is possible to use the function in Simulink but the application does not run in real time. The cholupdate() line can be replaced with $\mathrm{chol}\left(\left(S_{xx,k}^-\right)^{\mathrm{T}} S_{xx,k}^- + w_0^c\left(X_{0,k}^* - \widehat{x}_k^-\right)\right)$.

The correction step is shown in **Table 2**. A new set of sigma points ($\boldsymbol{X}_k^-$) and its derivatives ($\boldsymbol{\Phi}_k$) can be generated using steps (a) and (b) from **Table 2**. If the nonlinearity is not severe, these steps can be skipped. This saves computational effort but sacrifices filter performance. The (new) sigma points and its derivatives are then propagated through the measurement function and its derivative, respectively. The resulting points are used to calculate the estimated measurements ($\boldsymbol{Y}_k$) as well as the output sensitivities ($\widehat{\boldsymbol{\mathcal{Y}}}_k$). These are used within the RPEM to estimate the system parameters and covariances.

The SR factor of the innovation error covariance ($\boldsymbol{S}_{yy,k}$), the cross-covariance ($\boldsymbol{P}_{xy,k}$) together with its derivatives matrices ($\boldsymbol{\mathcal{S}}_{yy,k}$, $\boldsymbol{\mathcal{P}}_{xy,k}$) are calculated in order to obtain the Kalman gain matrix ($\boldsymbol{K}_k$) and its sensitivity ($\boldsymbol{\mathcal{K}}_k$). The aSRUKF treats also the measurement noises as additive. The a posteriori state estimation ($\widehat{\boldsymbol{x}}_k^+$), the a posteriori state sensitivity ($\widehat{\boldsymbol{\mathcal{X}}}_k^+$) together with the SR factor of the estimation error covariance ($\boldsymbol{S}_{xx,k}$) and its sensitivity ($\boldsymbol{\mathcal{S}}_{xx,k}$) close the loop of the aSRUKF.

Local sensitivity analysis can be utilized to determine if a system input or a system modification can excite the system parameters in order to identify them. The a posteriori state sensitivity from **Table 2 (d)** can be used to determine the influence of parameters to the system states. This sensitivity results from the correction step of the aSRUKF. As long as the sensitivity $\widehat{\boldsymbol{\mathcal{X}}}_k^+$ remains below a user-defined threshold, the parameter update from **Table 2 (e)** can be skipped to prevent parameter estimation drifts. A time window (N_w) can be used to calculate $\max\left(\left\|\widehat{\boldsymbol{\mathcal{X}}}_{k-N_w}^+\right\|_2, \ldots, \left\|\widehat{\boldsymbol{\mathcal{X}}}_k^+\right\|_2\right)$ to normalize the sensitivity values. A threshold vector $\boldsymbol{t}_v$ is then defined with values between 0 and 1. The update procedure can be described as

$$\begin{aligned} &\textbf{for } p = 1 \textbf{ to } n_\theta \textbf{ do} \\ &\quad sa = 0 \\ &\quad \textbf{for } n = 1 \textbf{ to } n_x \textbf{ do} \\ &\qquad \mathrm{sa} = \mathrm{sa} + \frac{dx_{k,n}}{d\theta_{k,p}} \\ &\textbf{if } \mathrm{sa} > \boldsymbol{t}_v(p) \textbf{ then} \\ &\mathrm{update\_values} = \mathrm{True} \end{aligned} \tag{9}$$

The variable sa represents the sensitivity sum w.r.t. a system parameter $\theta_{k,p}$ over all system states ($x_{k,1}, \ldots, x_{k,n_x}$). The threshold vector $\boldsymbol{t}_v$ should be selected with caution. Too high values prevent parameter estimation drifts but can increase the convergence time of the filter. Moreover, the parameter excitation can be significantly reduced and the resulting estimation can be biased. The performance of the SB-aSRUKF is evaluated in Section 4.

The local state sensitivity can be also calculated as follows (cf. [29]):

$$\frac{\mathrm{d}\boldsymbol{x}_k}{\mathrm{d}\theta_{k,j}} = \left.\frac{\partial f}{\partial \boldsymbol{x}_k}\right|_{\widehat{\boldsymbol{x}}_{k-1},\widehat{\boldsymbol{\theta}}_{k-1}} \frac{\mathrm{d}\boldsymbol{x}_{k-1}}{\mathrm{d}\theta_{k,j}} + \left.\frac{\partial f}{\partial \theta_{k,j}}\right|_{\widehat{\boldsymbol{x}}_{k-1},\widehat{\boldsymbol{\theta}}_{k-1}} \tag{10}$$

This sensitivity computation is compared in Section 4 with a posterior state sensitivity obtained using the SB-aSRUKF.

SRUKF	Sigma points	$X_k^- = (X_{k,1}, \ldots, X_{k,l}) = \left(\widehat{x}_k^{\,-} \widehat{x}_k^- \otimes m_1 + \eta S_{xx,k}^- \widehat{x}_k^- \otimes m_1 - \eta S_{xx,k-1}^- \right)$	(a)
	Output sigma points	$Y_k = h\left(X_k^-, \widehat{\theta}_{k-1}, u_k\right)$	
	Estimated measurement	$\widehat{y}_k = Y_k (w^m)^{\mathrm{T}}$	
SM	Sigma-points derivative	$\Phi_{k,j} = (\Phi_{k,1,j}, \ldots, \Phi_{k,l,j}) == \left(\widehat{\mathcal{X}}_{k,j} \widehat{\mathcal{X}}_{k,j} \otimes m_1 + \eta \mathcal{S}_{xx,k,j}^- \widehat{\mathcal{X}}_{k,j} \otimes m_1 - \eta \mathcal{S}_{xx,k,j}^- \right)$	(b)
	Output sigma-points derivative	$\Psi_{k,j,i} = \frac{\partial h}{\partial x_k}\Big\vert_{X_k^-, \widehat{\theta}_{k-1}} \Phi_{k,j,i} + \frac{\partial h}{\partial \theta_{k,j}}\Big\vert_{X_k^-, \widehat{\theta}_{k-1}}$	
	Output sensitivity	$\widehat{y}_k = \Psi_k \left(I^{n_\theta} \otimes (w^m)^{\mathrm{T}} \right)$ with $\Psi_k = (\Psi_{k,1}, \ldots, \Psi_{k,n_\theta})$	(c)
SRUKF	SR innovation error covariance	$S_{yy,k} = \mathrm{qr}\left(\left(\sqrt{w_1^c}(Y_{1:2n_x,k} - \widehat{y}_k) \quad S_{R,k} \right)\right),$ $S_{yy,k} = \mathrm{cholupdate}\left(S_{yy,k}, \sqrt{\lvert w_0^c \rvert}(Y_{0,k} - \widehat{y}_k), \mathrm{sign}(w_0^c) \right)$	
	Cross-covariance matrix	$P_{xy,k} = W_{\mathrm{kr}}^c \odot \left((Y_k - \widehat{y}_k \otimes m_2)(X_k - \widehat{x}_k^- \otimes m_2)^{\mathrm{T}} \right)$	
SM	Derivative of innovation error covariance	$\mathcal{S}_{yy,k} = \mathrm{treshapeM}\left(\left(I^{n_\theta} \otimes A_{\mathrm{ls}, S_{yy,k}}^{\dagger} \right) (\mathcal{P}_{yy,k})_s \right)$, with $A_{\mathrm{ls}, S_{yy,k}} = \left(S_{yy,k} \otimes I^{n_y} + (I^{n_y} \otimes \mathcal{S}_{yy,k}) \mathcal{N}_{(n_y)} \right)) \mathcal{E}_{(n_y)}$ and $\mathcal{P}_{yy,k} = (\Psi_k - \widehat{y}_k \otimes m_2) \left(I^{n_\theta} \otimes \left((W_{\mathrm{kr}}^c)^{\mathrm{T}} \odot (Y_k - \widehat{y}_k \otimes m_2)^{\mathrm{T}} \right) \right)$ $+ W_{\mathrm{kr}}^c \odot (Y_k - \widehat{y}_k \otimes m_2) \left(\mathcal{T}_{(l)} I^{n_\theta} \otimes (\Psi_k - \widehat{y}_k \otimes m_2)^{\mathrm{T}} \right)$	
	Derivative of Cross-covariance matrix	$\mathcal{P}_{xy,k} = \left(\Phi_k - \widehat{\mathcal{X}}_k^- \otimes m_2 \right) \left(I^{n_\theta} \otimes \left((W_{\mathrm{kr}}^c)^{\mathrm{T}} \odot (Y_k - \widehat{y}_k \otimes m_2)^{\mathrm{T}} \right) \right)$ $+ W_{\mathrm{kr}}^c \odot (X_k - \widehat{x}_k^- \otimes m_2) \left(\mathcal{T}_{(l)} I^{n_\theta} \otimes (\Psi_k - \widehat{y}_k \otimes m_2)^{\mathrm{T}} \right)$	
SRUKF	Kalman gain	$K_k = \left(P_{xy,k} / S_{yy}^{\mathrm{T}} \right) / S_{yy,k}$	
	A posteriori state estimate	$\widehat{x}_k^+ = \widehat{x}_k^- + K_k (y_k - \widehat{y}_k)$	
	SR estimation error covariance	**for** $z = 1$ **to** n_y **do** $S_{xx,k} = \mathrm{cholupdate}\left(S_{xx,k}^-, U_k(:, z), '-' \right)$ with $U_k = K_k S_{yy,k}$	
SM	Kalman gain derivative	$\mathcal{K}_k = \mathcal{P}_{xy,k} / (I^{n_\theta} \otimes P_{yy,k}) - (P_{xy,k}/P_{yy,k})(\mathcal{P}_{yy,k}/(I^{n_\theta} \otimes P_{yy,k}))$	
	A posteriori state sensitivity	$\widehat{\mathcal{X}}_k^+ = \widehat{\mathcal{X}}_k^- - K_k \widehat{y}_k + \mathcal{K}_k (I^{n_\theta} \otimes (y_k - \widehat{y}_k))$	(d)
	Derivative of estimation error covariance update	$\mathcal{S}_{xx,k} = \mathrm{treshapeM}\left(\left(I^{n_\theta} \otimes A_{\mathrm{ls}, S_{xx,k}^-}^{\dagger} \right) (\mathcal{P}_{xx,k})_s \right)$, with $\mathcal{P}_{xx,k} = \mathcal{P}_{xx,k}^- - P_{xy,k} \mathcal{T}_{(n_y)} (I^{n_\theta} \otimes \mathcal{K}_k^{\mathrm{T}}) - \mathcal{P}_{xy,k} (I^{n_\theta} \otimes K_k^{\mathrm{T}})$	
RPEM	Parameter and covariance estimation	$\Delta_k = \Delta_{k-1} + (1 - \lambda)(e_k e^{\mathrm{T}} - \Delta_{k-1}),$ $S_k = \lambda \Delta_k + \widehat{y}_k \Theta_{k-1} \widehat{y}_k^{\mathrm{T}},$ $L_k = \Theta_{k-1} \widehat{y}_k^{\mathrm{T}} S_k^{-1},$ $\Theta_k = (I^{n_\theta} - L_k \widehat{y}_k) \Theta_{k-1} \left(I^{n_\theta} - L_k \widehat{y}_k^{\mathrm{T}} \right) / \lambda + L_k \Delta_k L_k^{\mathrm{T}},$ $\widehat{\theta}_k = \widehat{\theta}_{k-1} + L_k e_k.$	(e)

Table 2. aSRUKF: filter correction step and the RPEM for parameter and covariance estimation.

http://dx.doi.org/10.5772/intechopen.72470

The RPEM can be combined with different KF derivatives to estimate system parameters and covariances. An EKF and a SRCuKF were used to calculate the output sensitivity in [22], which is then used to estimate the unknown values. More information about the aSRUKF can be found in [28, 30].

3. Testbeds

In this section, two testbeds are presented and modelled. These modelled systems are used in Section 4 to test the performance of the SB-aSRUKF. The planar one-link robot system is presented and extended with a second arm to form a pendulum robot (pendubot). The pendubot is a chaotic and inherently unstable system.

3.1. Planar one-link robot system

This section describes the planar one-link robot system shown in **Figure 1**. It consists of a long rectangular aluminium link driven by a DC motor via a shaft and a one-state toothed gear.

The angular position is measured with an incremental rotary encoder and the motor torque is determined by measuring the motor current. To simplify the motor model, it is assumed that the motor current is directly proportional to the armature current and that the motor torque is proportional to this current by a constant factor. Additionally, the link acceleration is measured using a micro-electro-mechanical sensor (MEMS) accelerometer attached to the link. The motor position is controlled by a classical cascade structure with P- and P-feedback controllers for position and speed.

The corresponding continuous state-space representation of the planar one-link robot system, where the link angular position and acceleration are measured, can be written as follows:

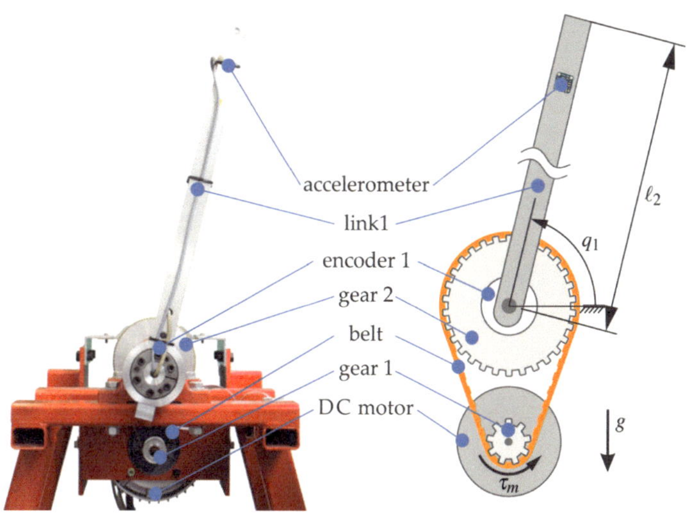

Figure 1. Planar one-link robot system: structure and functionality.

$$\dot{x} = Ax + b(x,u),$$
$$y = c(x,u).$$

The system states are the link angular position ($x_1 = q_1$) and the link speed ($x_2 = \dot{q}_1$). The input u corresponds to the motor torque ($u = \tau_m$). The measurements are the link angular position ($y_1 = q_1$) and acceleration ($y_2 = \ddot{q}_1$). The matrix A and the vector-valued functions b and c are then described as

$$A = \begin{pmatrix} 0 & 1 \\ 0 & -\frac{\mu_v}{J_{tot}} \end{pmatrix},$$

$$b(x,u) = \begin{pmatrix} 0 \\ -\frac{\mu_d}{J_{tot}} \arctan(k\,x_2) - \frac{m_a l_2}{2\,J_{tot}} \sin(x_1) + \frac{\tau_m}{J_{tot}} \end{pmatrix},$$

$$c(x,u) = \begin{pmatrix} x_1 \\ -\frac{\mu_v}{J_{tot}} x_2 - \frac{\mu_d}{J_{tot}} \arctan(k\,x_2) - \frac{m_a l_2\,g}{2\,J_{tot}} \sin(x_1) + \frac{\tau_m}{J_{tot}} \end{pmatrix},$$

where J_{tot} represents the total inertia including motor, shaft and link inertias. The link friction is modelled as a dry Coulomb (μ_d and k) and viscous friction (μ_v). The parameters m_a, l_2, and g are the link mass, length, and the gravity of Earth coefficient, respectively.

3.2. Pendubot

This section describes the pendulum robot (pendubot) that is presented in **Figure 2**. The pendubot is built adding an under-actuated link to the planar one-link robot system described in Section 3.1. The actuated joint (q_1) is located at the shoulder of the first link (arm) and the elbow joint (q_2) allows the second link (forearm) to swing free. This joint contains only a second incremental rotatory encoder that measures the angle between the two links.

The system states result as $x_1 = q_1$, $x_2 = \dot{q}_1$, $x_3 = q_2$, and $x_4 = \dot{q}_2$, where q_i and $\dot{q}_i$ are the corresponding position and speed of the i–joint, respectively. The state-space representation of the pendubot can be written as

$$\dot{x} = Ax + b(x,u),$$
$$y = (x_1, x_3, \dot{x}_2)^{\mathrm{T}},$$

in which

$$A = \begin{pmatrix} 0 & 1 & 0 & 0 \\ 0 & 0 & 0 & 0 \\ 0 & 0 & 0 & 1 \\ 0 & 0 & 0 & 0 \end{pmatrix},$$

$$b(x,u) = \begin{pmatrix} 0 & \dot{x}_2 & 0 & \dot{x}_4 \end{pmatrix}^{\mathrm{T}},$$

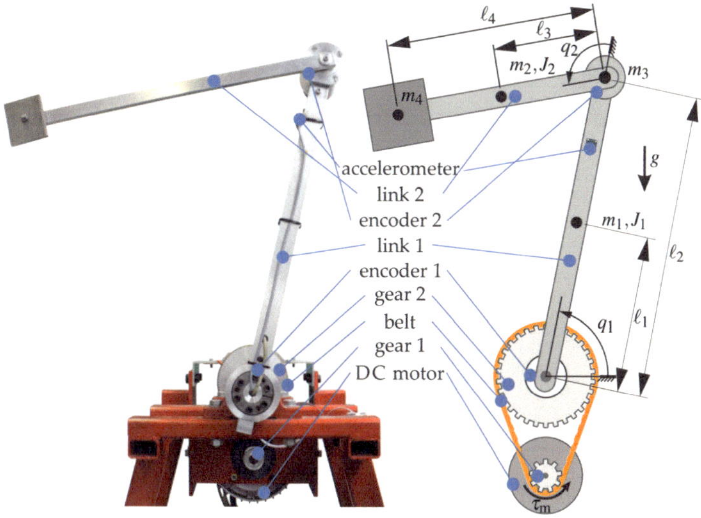

Figure 2. Pendubot: structure and functionality.

where

$$\begin{pmatrix} \dot{x}_2 \\ \dot{x}_4 \end{pmatrix} = \ddot{\boldsymbol{q}} = \begin{pmatrix} \ddot{q}_1 \\ \ddot{q}_2 \end{pmatrix} = \boldsymbol{D}(\boldsymbol{q})^{-1} \begin{pmatrix} \tau_{\mathrm{m}} - \mu_{\mathrm{v}_1} \dot{q}_1 - \mathrm{d}(\dot{q}_1) \\ \mu_{\mathrm{v}_2} \dot{q}_2 \end{pmatrix} - \boldsymbol{D}(\boldsymbol{q})^{-1} \boldsymbol{C}(\boldsymbol{q}, \dot{\boldsymbol{q}}) \dot{\boldsymbol{q}} - \boldsymbol{D}(\boldsymbol{q})^{-1} \boldsymbol{g}(\boldsymbol{q}).$$

The viscous and Coulomb frictions are described with the parameters μ_{v_1} and μ_{v_2} and the function $\mathrm{d}(\dot{q}_1) = \mu_{\mathrm{d}} \arctan(k\,\dot{q}_1)$. The matrices $\boldsymbol{D}(\boldsymbol{q})$ and $\boldsymbol{C}(\boldsymbol{q}, \dot{\boldsymbol{q}})$ and the vector $\boldsymbol{g}(\boldsymbol{q})$ are the inertial and the Coriolis matrices and the gravity vector, respectively. They are defined as follows

$$\boldsymbol{D}(\boldsymbol{q}) = \begin{pmatrix} \vartheta_1 + \vartheta_2 + 2\vartheta_3 \cos(q_2) & \vartheta_2 + \vartheta_3 \cos(q_2) \\ \vartheta_2 + \vartheta_3 \cos(q_2) & \vartheta_2 \end{pmatrix},$$

$$\boldsymbol{C}(\boldsymbol{q}, \dot{\boldsymbol{q}}) = \vartheta_3 \sin(q_2) \begin{pmatrix} -\dot{q}_2 & -\dot{q}_1 - \dot{q}_2 \\ \dot{q}_1 & 0 \end{pmatrix},$$

$$\boldsymbol{g}(\boldsymbol{q}) = \begin{pmatrix} \vartheta_4\, g \cos(q_1) + \vartheta_5\, g \cos(q_1 + q_2) \\ \vartheta_5\, g \cos(q_1 + q_2) \end{pmatrix},$$

where the ϑ_i parameters are defined as

$$\vartheta_1 = m_1 l_1^2 + (m_2 + m_3 + m_4) l_2^2 + J_1,$$

$$\vartheta_2 = m_2 l_3^2 + m_4 l_4^2 + J_2,$$

$$\vartheta_3 = (m_2 l_3 + m_4 l_4)\, l_2,$$

$$\vartheta_4 = m_1 l_1 + (m_2 + m_3 + m_4) l_2,$$
$$\vartheta_5 = m_2 l_3 + m_4 l_4.$$

The parameters J_1 and J_2 correspond to the moments of inertia of link 1 and link 2 about their centroids. J_1 includes motor, gear and shaft inertias. The m_i and l_i parameters are defined in **Figure 2**.

4. Experiments

In this section, the SB-aSRUKF is tested on the planar one-link robot system and on the pendubot. Both testbed models were discretized using first-order explicit Euler with a sample time of 0.2 ms. System states, parameters and covariances were estimated online. The SB-aSRUKF is also compared with the joint state and parameter SRUKF in this section. Sensitivity analysis is also discussed.

4.1. State sensitivity analysis and parameter and covariance estimation

Sensitivity analysis (SA) was performed on simulation using the planar one-link robot system. The system parameters were identified offline on the real testbed using Prediction-Error Method. The parameters defined as

$$\hat{\boldsymbol{\theta}}_{\text{true}} = \left(\hat{J}_{\text{tot}}\ \hat{\mu}_{\text{v}}\ \hat{\mu}_{\text{d}}\ \widehat{m_{\text{a}} l_2}\ \hat{k} \right)^{\text{T}} = \left(5.59{\cdot}10^{-2}\text{kg m}^2 \quad 0.05\ \text{N m}\tfrac{\text{s}}{\text{rad}} \quad 0.27\ \text{N m} \quad 0.11\ \text{kg m} \quad 10\tfrac{\text{s}}{\text{rad}} \right)^{\text{T}},$$

were used for the simulation. Noise distributions with the following covariance matrices

$$Q = \text{diag}\left(\left(10^{-11} \quad 5{\cdot}10^{-5} \right) \right)^{3},$$
$$R = \text{diag}\left(\left(5{\cdot}10^{-7} \quad 5 \right) \right),$$

were added to the simulation to incorporate model and measurement uncertainties. An s-curve profile was considered as a desired link angular position.

The following system states, parameters and covariances were estimated:

$$\begin{aligned}
x_1 &= q_1 && \rightarrow \text{link position,}\\
x_2 &= \dot{q}_1 && \rightarrow \text{link speed,}\\
\theta_{\text{a}} &= q_{11} && \rightarrow \text{process covariance value (related to the link ang.pos.),}\\
\theta_1 &= \mu_{\text{v}} && \rightarrow \text{viscous friction coefficient,}\\
\theta_2 &= \mu_{\text{d}} && \rightarrow \text{Coulomb friction coefficient,}\\
\theta_3 &= J_{\text{tot}}^{-1} && \rightarrow \text{inverse inertia.}
\end{aligned}$$

[3] The function $\text{diag}(v)$ transforms the $v \in \mathbb{R}^n$ vector into a $(n \times n)$ square matrix with the elements of v on the diagonal of the matrix.

The remaining tuning factors for SB-aSRUKF were set as

$$Q = q_{22} = 5{\cdot}10^{-5}, \quad R = \operatorname{diag}\left(\left(5{\cdot}10^{-7} \quad 2{\cdot}10^{1}\right)\right),$$

$$\alpha = 0.08 \quad \beta = 2 \quad \kappa = 3 - n_x \quad \lambda = 0.999,$$

$$P_{xx,\,\mathrm{init}} = \boldsymbol{I}^2, \quad \boldsymbol{\Theta}_{\mathrm{init}} = \operatorname{diag}\left(\left(5{\cdot}10^{-13} \quad 10^{-1} \quad 10^{-6} \quad 10^{-4} \quad 10^{-8}\right)\right).$$

The values of $\boldsymbol{\Theta}_{\mathrm{init}}$ tune the parameter and covariance estimation, and the index order is the same as the above-defined θ_i values. This means that the first value tunes the estimation of θ_{a} (process covariance value), the second value tunes θ_1 (viscous friction coefficient) and so on.

The filter initial system states were set to zero and the initial system parameters were set as the true values multiply by a random factor between $0 < \theta_{\mathrm{factor},\,i} \leq 10$ as

$$\boldsymbol{x}_{\mathrm{init}} = \boldsymbol{0},$$

$$\boldsymbol{\theta}_{\mathrm{init}} = \left(2{\cdot}10^{-8}\mathrm{rad}^2 \quad 1.5\left(1/\widehat{J}_{\mathrm{tot}}\right) \quad 3\widehat{\mu}_{\mathrm{v}} \quad 8\widehat{\mu}_{\mathrm{d}} \; 0.25\left(\widehat{m}_{\mathrm{a}}\widehat{l}_2\right)\right)^{\mathrm{T}}.$$

In order to test the sensitivity-based section of the filter, the link angular position was held at $q_1 = \pi/2$ after ca. 11 s for about 4.5 s. At the same time, the system parameters $\widehat{\mu}_{\mathrm{v}}$, $\widehat{\mu}_{\mathrm{d}}$, and $\widehat{m}_{\mathrm{a}}$ were quadrupled.

Figure 3 compares the a posteriori state sensitivity calculated using the SB-aSRUKF and the state sensitivity using Eq. (10). The first diagram shows the estimated and true link angular position of the planar one-link robot system. The following diagrams present the normalized SA related to the link angular speed ($\widehat{\dot{q}}_1$) and corresponding to the inverse inertia, viscous and Coulomb friction coefficients, and the link mass and length parameter. While the state sensitivity calculated using Eq. (10) was affected directly by input noises, the a posteriori state sensitivity provided an almost noise-free estimation. While the SAJ_{tot}^{-1} maxima were related with the acceleration (speed-up and brake phases), the SAμ_{v} maxima coincided with the link maximal speed. The SA$m_{\mathrm{a}}l_2$ was only zero while the arm was crossing the 0 rad position and the SAμ_{d} was the sensitivity value more affected by the system noise. This is caused because the maximal change rate of arctan() occurs when the argument is near zero. When the link speed is zero, the added noise varies near this value and its effect is amplified by arctan().

Figure 4 shows the state, parameter and covariance estimation of the planar one-link robot system. The aSRUKF was used in two configurations: SB-aSRUKF utilized SA to monitor the system excitation while aSRUKF did not. After the initialization, the two estimators needed almost the same time to converge to the offline identified values. The parameters estimated using the SB-aSRUKF did not diverge while the link position was held. Because two of the estimated parameters using the aSRUKF diverged, this filter needed more time to converge after the stop phase. The two filters were able to estimate the link mass and length parameter during the stop phase. While the viscous and the Coulomb friction coefficients and the inverse inertia estimated with the SB-aSRUKF remained constant during the stop phase, the aSRUKF was able to estimate the Coulomb friction with bias (before μ_{v} diverged). Because of the added noise, the parameter was excited and could be identified. This can be seen in the fourth

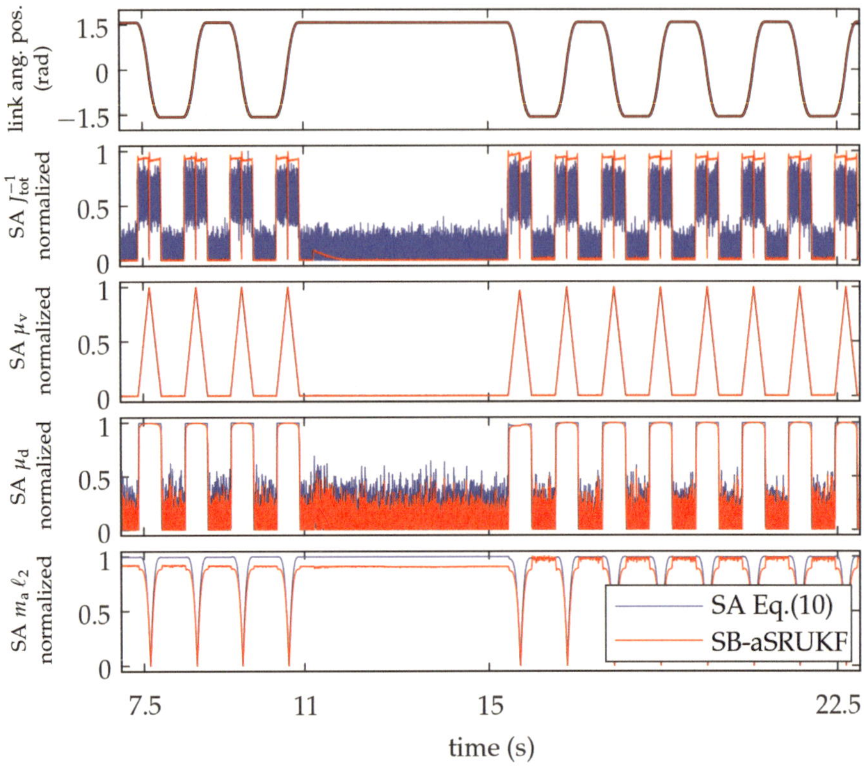

Figure 3. Sensitivity analysis (SA): comparison between the a posteriori state sensitivity obtained using the SB-aSRUKF and the resulting using Eq. (10). The desired link position was set as an s-curve between $-\pi/2$ and $\pi/2$. The link position was held at $\pi/2$ after ca. 11 s for about 4.5 s. The parameter sensitivities are related to the link angular speed ($\hat{x}_2 = \hat{\dot{q}}_1$).

diagram of **Figure 3**. These SA values remained under the threshold used on the SB-aSRUKF and were filtered. The parameter estimation stayed then constant. It should be noted that the diagram corresponding to the viscous friction coefficient is zoomed to present the parameter change, and the oscillations of the aSRUKF are cut. These reached up more than 50 times the parameter true value.

The SB-aSRUKF was able to estimate the parameters and covariance of the proposed testbed. The online estimations converged to the offline identified values without bias. The sensitivity-based approach used as a system excitation monitor prevented parameter estimation drifts and did not modify the convergence time of the filter.

4.2. Comparison between SB-aSRUKF and joint state and parameter SRUKF

The SB-aSRUKF and the joint SRUKF were compared on the pendubot for state and parameter estimation. The SB-aSRUKF identified also covariances.

The system parameters were identified offline and used for the simulation as

$$\hat{\theta}_{\text{true}} = \left(\hat{\vartheta}_1\ \hat{\vartheta}_2\ \hat{\vartheta}_3\ \hat{\vartheta}_4\ \hat{\vartheta}_5\right)^{\text{T}} = \left(0.339\,\text{kg}\,\text{m}^2 \quad 0.0667\,\text{kg}\,\text{m}^2 \quad 0.0812\,\text{kg}\,\text{m}^2 \quad 0.717\,\text{kg}\,\text{m} \quad 0.146\,\text{kg}\,\text{m}\right)^{\text{T}},$$

$$\left(\hat{\mu}_{\text{v1}}\ \hat{\mu}_{\text{d}}\ \hat{\mu}_{\text{v2}}\ \hat{k}\right) = \left(0.09\,\text{N}\,\text{m}\frac{\text{s}}{\text{rad}} \quad 0.226\,\text{N}\,\text{m} \quad 0.003\,\text{N}\,\text{m}\frac{\text{s}}{\text{rad}} \quad 10\frac{\text{s}}{\text{rad}}\right).$$

http://dx.doi.org/10.5772/intechopen.72470

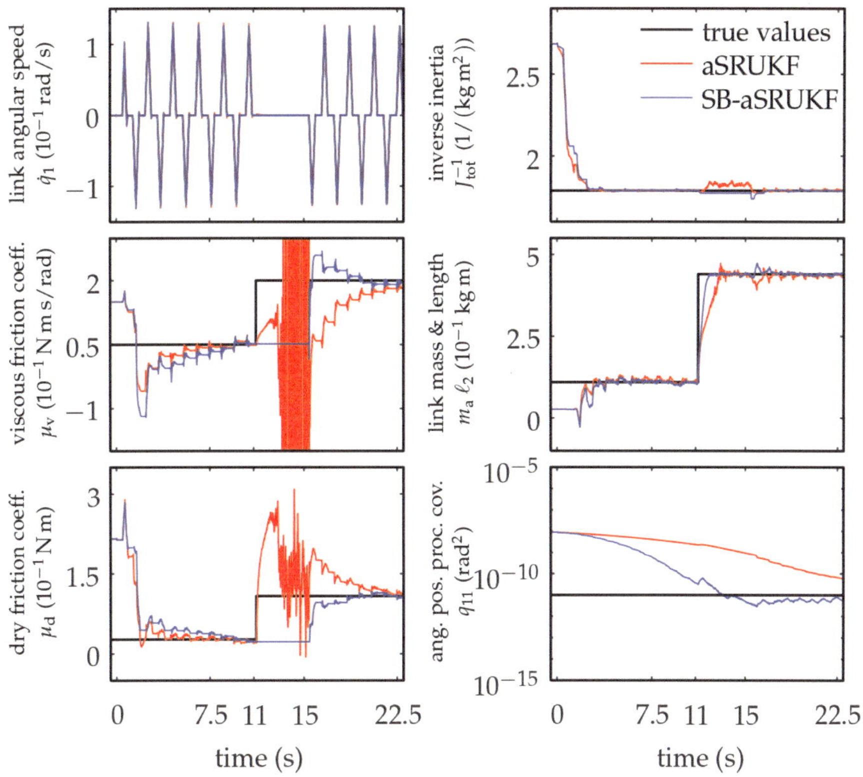

Figure 4. Planar one-link robot system: parameter and covariance estimation. The SB-aSRUKF uses SA to monitor the system excitation. The initial parameter $\boldsymbol{\theta}_{\text{init}}$ was randomly selected. The link position was held after ca. 11 s for about 4.5 s, and simultaneously the system parameters $\widehat{\mu}_v$, $\widehat{\mu}_d$, and $\widehat{m}_a$ were quadrupled.

An s-curve profile was selected as the desired position of the first link. The following states and parameters were estimated online:

$$
\begin{aligned}
x_1 &= q_1 && \rightarrow \text{link 1 position,}\\
x_2 &= \dot{q}_1 && \rightarrow \text{link 1 speed,}\\
x_3 &= q_2 && \rightarrow \text{link 2 position,}\\
x_4 &= \dot{q}_2 && \rightarrow \text{link 2 speed,}\\
\theta_a &= q_{11} && \rightarrow \text{process covariance (related to the link 1 ang.pos.),}\\
\theta_b &= q_{33} && \rightarrow \text{process covariance (related to the link 2 ang.pos.),}\\
\vartheta_1, \dots, \vartheta_5 & && \rightarrow \text{pedubot minimal set of parameters.}
\end{aligned}
$$

The values θ_a and θ_b, which correspond to the process covariance values, were only estimated using the SB-aSRUKF. The viscous and Coulomb friction coefficients were identified offline and remained constant for both filters.

To simulate model and measurement uncertainties, noise distributions with the following covariance matrices were added to the system for the simulation:

$$
\begin{aligned}
\boldsymbol{Q} &= \text{diag}\left(\begin{pmatrix} 2\cdot 10^{-10} & 1.5\cdot 10^{-7} & 2\cdot 10^{-10} & 1.5\cdot 10^{-7} \end{pmatrix}\right),\\
\boldsymbol{R} &= \text{diag}\left(\begin{pmatrix} 5\cdot 10^{-7} & 5\cdot 10^{-7} & 1 \end{pmatrix}\right).
\end{aligned}
$$

The tuning parameters for the joint SRUKF were chosen as

$$Q_{0_1} = \mathrm{diag}\big(\big(2{\cdot}10^{-10} \quad 1.5{\cdot}10^{-7} \quad 2{\cdot}10^{-10} \quad 1.5{\cdot}10^{-7}$$

$$10^{-7} \quad 10^{-7} \quad 10^{-7} \quad 5{\cdot}10^{-10} \quad 10^{-10}\big)\big),$$

$$R_{0_1} = \mathrm{diag}\big(\big(5{\cdot}10^{-7} \quad 5{\cdot}10^{-7} \quad 10\big)\big),$$

$$\alpha = 0.85, \quad \beta = 2, \quad \kappa = 3 - n_x - n_p,$$

$$P_{0_1} = \mathrm{diag}\big(\big(1 \quad 1 \quad 1 \quad 1 \quad 10^{-3} \quad 10^{-5} \quad 10^{-5} \quad 5 \quad 1\big)\big),$$

and the parameters for the SB-aSRUKF were set as

$$Q_{0_2} = \mathrm{diag}\big(\big(1.5{\cdot}10^{-7} \quad 1.5{\cdot}10^{-7}\big)\big), \quad P_{xx,0_2} = I^4,$$

$$R_{0_2} = \mathrm{diag}\big(\big(5{\cdot}10^{-7} \quad 5{\cdot}10^{-7} \quad 10\big)\big),$$

$$\alpha = 0.85, \quad \beta = 2, \quad \kappa = 3 - n_x, \quad \lambda = 0.999,$$

$$\Theta_{\mathrm{init}} = \mathrm{diag}\big(\big(10^{-25} \quad 10^{-25} \quad 10^{-7} \quad 10^{-7} \quad 10^{-7} \quad 5{\cdot}10^{-10} \quad 10^{-10}\big)\big).$$

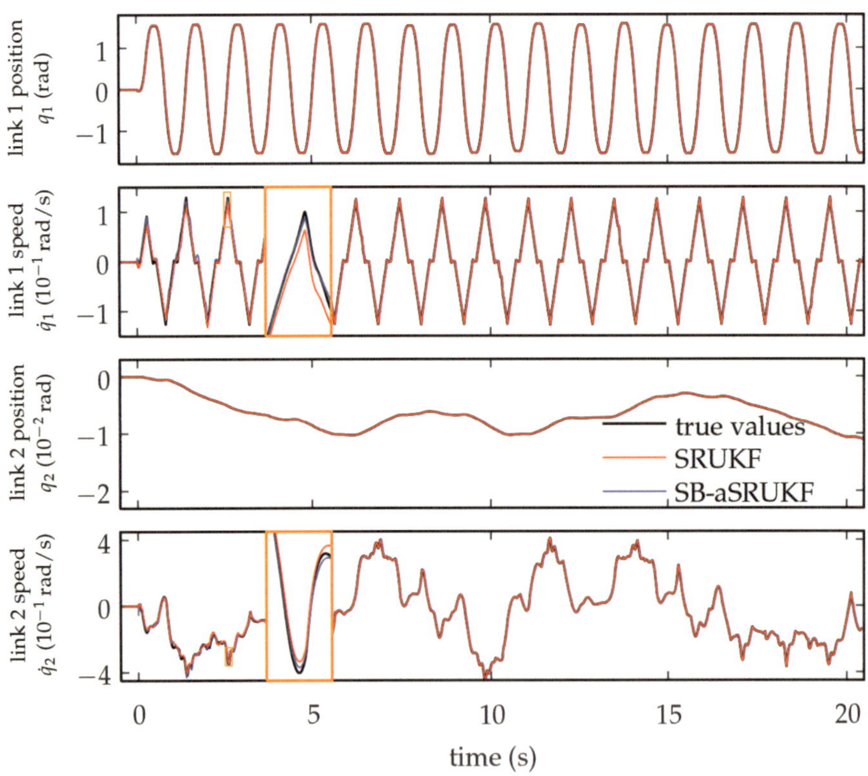

Figure 5. Pendubot: state estimation using the SB-aSRUKF and joint SRUKF. Both filters followed the dynamic of the true system states without any significant bias.

The filter initial system states were set to zero and the initial system parameters were set as the true values multiply by a random factor between $0 < \theta_{\text{factor},i} \leq 5$ as

$$\boldsymbol{x}_{\text{init}} = \mathbf{0},$$
$$\boldsymbol{\theta}_{\text{init}_1} = \left(1.5\,\widehat{\vartheta}_1\ 1.3\,\widehat{\vartheta}_2\ 1.5\,\widehat{\vartheta}_3\ 2\,\widehat{\vartheta}_4\ 2\,\widehat{\vartheta}_5\right)^{\mathrm{T}},$$
$$\boldsymbol{\theta}_{\text{init}_2} = \left(2{\cdot}10^{-10}\text{rad}^2\quad 2{\cdot}10^{-10}\text{rad}^2\quad (\boldsymbol{\theta}_{\text{init}_1})^{\mathrm{T}}\right)^{\mathrm{T}}.$$

The first four values of $\boldsymbol{P}_{0_1}$ tune the initial state estimation, while the last ones the parameter estimation. The first two values of $\boldsymbol{\Theta}_{\text{init}}$ tune the estimation of the covariance values θ_{a} and θ_{b} while the last values follow the index order of ϑ_i defined in Section 3. It should be noted that the settings related to the state and parameters estimation were equally tuned for both filters.

The state estimation of the pendubot is presented in **Figure 5**. The SB-aSRUKF was slightly faster to reach the true system states (cf. diagrams 1 and 4) and after ca. 5 s both filters followed the dynamic of the true system states without any significant bias.

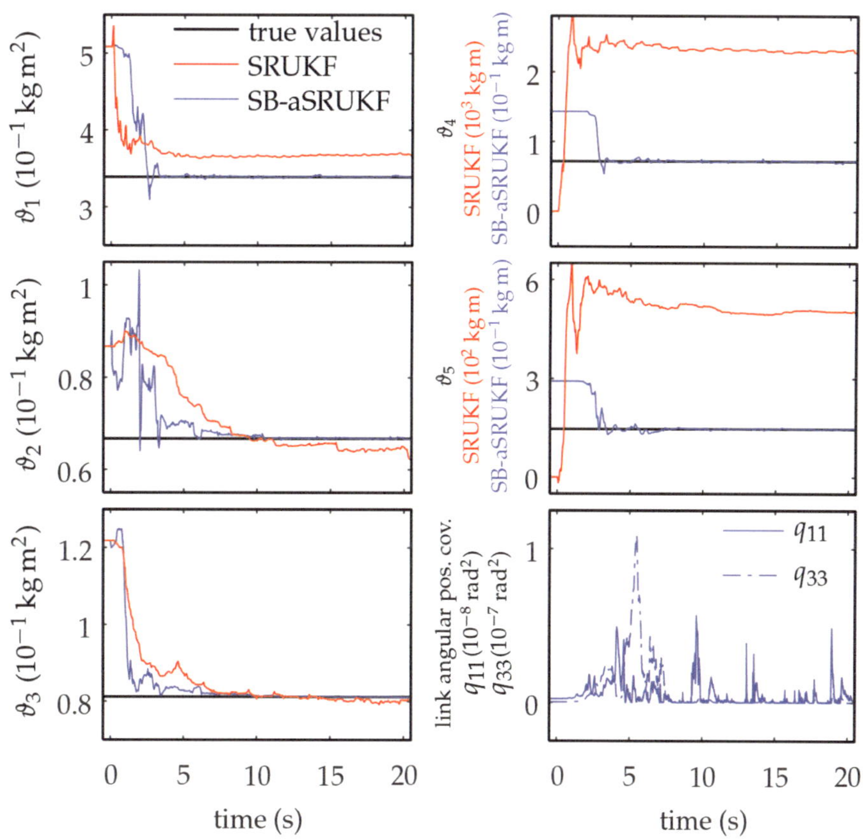

Figure 6. Pendubot: parameter and covariance estimation using the SB-aSRUKF and joint SRUKF. The SB-aSRUKF was configured to estimate the system parameters ϑ_1 to ϑ_5 and the process covariances corresponding to the link positions. The joint SRUKF estimates only the system parameters. It should be noted that the diagram scales for parameters ϑ_4 and ϑ_5 are different between the filters.

Figure 6 shows the parameter estimation of the pendubot for both filters. Using the same tuning parameter set, while the SB-aSRUKF estimated the ϑ_1 to ϑ_5 parameters without bias, the joint SRUKF estimated ϑ_1 to ϑ_3 with slight bias, and it was not able to estimate ϑ_4 and ϑ_5. These two parameters did not converge to the true system values. It should be noted that the diagram scales corresponding to parameters ϑ_4 and ϑ_5 are different between the filters. The parameter initialization and the tuning settings for the two filters were the same. The SB-aSRUKF outperforms the joint SRUKF for the parameter estimation of the pendubot.

5. Conclusions

In this chapter, some approaches for state, parameter and covariance estimation were discussed. Moreover, a sensitivity-based adaptive square-root unscented Kalman filter (SB-aSRUKF) was discussed and its performance was analysed. This filter estimates system states, parameters and covariances online. Additionally, sensitivity models were presented and used as system excitation monitor to prevent parameter and covariance estimation drifts while the system excitation was not sufficient.

Two testbeds based on an axis serial mechanism were modelled and used for testing the performance of the filter. Two experiments were performed in simulation on these two testbeds: a state sensitivity analysis and a comparison between the joint state and parameter square-root unscented Kalman filter (SRUKF) and the SB-aSRUKF. Simulation results show that the SB-aSRUKF outperforms the joint SRUKF with the same tuning configuration. While the joint SRUKF did not estimate two of the five parameters correctly, the SB-aSRUKF identified all the parameters. Moreover, the estimation of covariances reduces the parameter estimation bias. Configuring the right excitation thresholds for the system excitation monitor in Eq. (9) prevented parameter estimation drifts, while the input was not persistently exciting and did not only increased but also in some cases reduced the convergence time of the filter.

Author details

Mauro Hernán Riva*, Mark Wielitzka and Tobias Ortmaier

*Address all correspondence to: mauro.riva@imes.uni-hannover.de

Institute of Mechatronic Systems, Leibniz Universität Hanover, Hanover, Germany

References

[1] Wielitzka M, Dagen M, Ortmaier T. State estimation of vehicle's lateral dynamics using unscented Kalman filter. 53rd IEEE Conference on Decision and Control (CDC); Los Angeles, USA. 2014:5015-5020. DOI: 10.1109/CDC.2014.7040172

[2] Ding SX. Model-based Fault Diagnosis Techniques: Design Schemes, Algorithms, and Tools. 1st ed. Berlin, Germany: Springer; 2008. p. 493

[3] Simon D, Simon DL. Kalman filtering with inequality constraints for turbofan engine health estimation. IEE Proceedings - Control Theory and Applications. 2006;**153**(3):371-378. DOI: 10.1049/ip-cta:20050074

[4] Grewal MS, Andrews AP. Kalman Filtering: Theory and Practice with MATLAB. 4th ed. Hoboken, New Jersey, USA: Wiley-IEEE Press; 2014. p. 640

[5] Sorenson HW. Least-squares estimation: from Gauss to Kalman. IEEE Spectrum. 1970;**7**(7):63-68. DOI: 10.1109/MSPEC.1970.5213471

[6] Kalman RE. A new approach to linear filtering and prediction problems. Journal of Basic Engineering. 1960;**82**(1):35-45. DOI: 10.1115/1.3662552

[7] Anderson BDO, Moore JB. Optimal Filtering. 1st ed. Mineola, NY, USA: Dover Publications; 2006. p. 368

[8] Schmidt SF. Applications of state space methods to navigation problems. C. T. Leondes (Ed.) Advanced in Control Systems. 1966;**3**:293-340

[9] Simon D. Optimal State Estimation: Kalman, H Infinity, and Nonlinear Approaches. 1st ed. Hoboken, New Jersey, USA: Wiley-Interscience; 2006. p. 526

[10] Julier SJ, Uhlmann JK. New extension of the Kalman filter to nonlinear systems. In: Signal Processing, Sensor Fusion, and Target Recognition VI. Orlando, FL, USA: International Society for Optics and Photonics; 1997. pp. 182-194. DOI: 10.1117/12.280797

[11] Ljung L, Soderstrom T. Theory and Practice of Recursive Identification. 1st ed. Cambridge, Massachusetts, USA: The MIT Press; 1983. p. 551

[12] Ljung L. System Identification: Theory for the User. 2nd ed. Prentice Hall: Upper Saddle River, NJ, USA; 1999. p. 672

[13] Saab SS, Nasr GE. Sensitivity of discrete-time Kalman filter to statistical modeling errors. Optimal Control Applications and Methods. 1999;**20**(5):249-259. DOI: 10.1002/(SICI)1099-1514(199909/10)20:5<249::AID-OCA659>3.0.CO;2-2

[14] Fitzgerald R. Divergence of the Kalman filter. IEEE Transactions on Automatic Control. 1971;**16**(6):736-747. DOI: 10.1109/TAC.1971.1099836

[15] Price C. An analysis of the divergence problem in the Kalman filter. IEEE Transactions on Automatic Control. 1968;**13**(6):699-702. DOI: 10.1109/TAC.1968.1099031

[16] Ljung L. Asymptotic behavior of the extended Kalman filter as a parameter estimator for linear systems. IEEE Transactions on Automatic Control. 1979;**24**(1):36-50. DOI: 10.1109/TAC.1979.1101943

[17] Odelson B. Estimating Disturbance Covariances From Data For Improved Control Performance [dissertation]. Madison, WI, USA: University of Wisconsin-Madison; 2003. p. 309

[18] Odelson BJ, Rajamani MR, Rawlings JB. A new autocovariance least-squares method for estimating noise covariances. Automatica. 2006;**42**(6):303-308. DOI: 10.1016/j.automatica.2005.09.006

[19] Rajamani MR, Rawlings JB. Estimation of the disturbance structure from data using semidefinite programming and optimal weighting. Automatica. 2009;**45**(1):142-148. DOI: 10.1016/j.automatica.2008.05.032

[20] Riva MH, Díaz Díaz J, Dagen M, Ortmaier T. Estimation of covariances for Kalman filter tuning using autocovariance method with Landweber iteration. In: IASTED International Symposium on Intelligent Systems and Control; Marina del Rey, CA, USA. 2013. DOI: 10.2316/P.2013.807-009

[21] Rajamani MR. Data-based Techniques to Improve State Estimation in Model Predictive Control [dissertation]. Madison, WI, USA: University of Wisconsin-Madison; 2007. p. 262

[22] Riva MH, Beckmann D, Dagen M, Ortmaier T. Online parameter and process covariance estimation using adaptive EKF and SRCuKF approaches. In: 2015 IEEE Conference on Control Applications (CCA); Sydney, Australia. 2015. p. 1203-1210. DOI: 10.1109/CCA.2015.7320776

[23] Han J, Song Q, He Y. Adaptive unscented Kalman filter and its applications in nonlinear control. In: Moreno VM, Pigazo A, editors. Kalman Filter Recent Advances and Applications. Croatia: InTech; 2009. pp. 1-24. DOI: 10.5772/6799

[24] Wan EA, Van der Merwe R. The unscented Kalman filter for nonlinear estimation. In: IEEE 2000 Adaptive Systems for Signal Processing, Communications, and Control Symposium; Lake Louise, Canada. 2000. p. 153-158. DOI: 10.1109/ASSPCC.2000.882463

[25] Van der Merwe R, Wan EA. The square-root unscented Kalman filter for state and parameter-estimation. In: IEEE International Conference on Acoustics, Speech, and Signal Processing (ICASSP 01); Salt Lake City, USA. 2001. p. 3461-3464. DOI: 10.1109/ICASSP.2001.940586

[26] Press WH, Flannery BP, Teukolsky SA, Vetterling WT. Numerical Recipes in C: The Art of Scientific Computing. 2nd ed. Cambridge, New York, USA: Cambridge University Press; 1992. p. 994

[27] Henderson HV, Searle SR. The vec-permutation matrix, the vec operator and Kronecker products: A review. Linear and Multilinear Algebra. 1981;**9**(4):271-288. DOI: 10.1080/03081088108817379

[28] Riva MH, Dagen M, Ortmaier T. Adaptive Unscented Kalman Filter for online state, parameter, and process covariance estimation. In: 2016 American Control Conference (ACC); Boston, MA, USA. 2016. p. 4513-4519. DOI: 10.1109/ACC.2016.7526063

[29] Eykhoff P. System Identification Parameter and State Estimation. 1st ed. Hoboken, New Jersey, USA: Wiley-Interscience; 1974. p. 555

[30] Riva MH, Dagen M, Ortmaier T. Comparison of covariance estimation using autocovariance LS method and adaptive SRUKF. In: 2017 American Control Conference (ACC); Seattle, WA, USA. 2017. p. 5780-5786. DOI: 10.23919/ACC.2017.7963856

Chapter 5

Kalman Filters for Parameter Estimation of Nonstationary Signals

Sarita Nanda

Additional information is available at the end of the chapter

http://dx.doi.org/10.5772/intechopen.71874

Abstract

An adaptive Taylor-Kalman filter with PSO tuning for tracking nonstationary signal parameters in a noisy environment with primary focus on time-varying power signals has been presented in this piece of work. In order to deal with the dynamic envelope of the power signal, second-order Taylor expansion has been used such that the Taylor coefficients are updated with the PSO-tuned Taylor-Kalman Filter algorithm. In addition to this, for fast convergence, a self-adaptive particle swarm optimization technique has been used for obtaining the optimal values of model and measurement error covariances of the Kalman filter. The proposed algorithm is linear and therefore has less computational burden, which is easier to be implemented on a hardware platform like DSP processor or FPGA. The proposed PSO-tuned Taylor-Kalman filter exhibits robust tracking capabilities even under changing signal dynamics, immune to critical noise conditions, harmonic contaminations, and also reveals excellent convergence properties.

Keywords: nonstationary signals, power signal frequency and phasor estimation, hybrid Kalman approach, PSO tuning

1. Introduction

Signal parameter estimation, which dates back to the late 19th century, describes the various methods employed to track amplitude, phase, and frequency-like parameters of a signal. Among all the signal parameters, frequency is the primary concern, as it is a nonlinear function in the received data sequence, and once that is measured accurately, tracking of other parameters like phase, amplitude, and damping factor of a signal can be relatively easier. Most real-world signals are nonstationary in nature, i.e., they have a time-varying frequency behavior. Some of the popular sources of nonstationary signals include speech, audio, sounds of mammals, machine vibrations, electrical power networks, and a variety of biomedical signals like electromyogram (EMG), electroencephalogram (EEG), phonocardiogram (PCG), and vibroarthrogram (VAG)).

These signals are rich in information and when analyzed properly provides with information that could be used to improve many aspects of our lives. Hence, the information of interest of the signal can be extracted, which includes the estimation of parameters like amplitude, phase, frequency, and damping factor directly from the discrete measurement in the presence of noise both in stationary and nonstationary environments. Precise and smooth operation of the power generation and distribution system is very much required in the present day scenario. With the increasing demand for power, the number and type of load are having deteriorating effects on the power quality. Power quality is defined as the ability of the electrical grid to deliver clean and stable power to the consumer. Between generation and supply, the power being delivered encounters large number of transformers and several lengths of overhead lines and underground cables. Phenomena like lightning strikes, system faults, load switching, and other such intentional or unintentional events are the main cause of electromagnetic disturbances, which results in voltage or current waveform distortions to propagate in the entire power system. Recently, the increase in the number of power electronic loads in the system causes nonlinear loading effect on the power system signal, leading to degradation of power quality.

Recently, harmonic estimation has become a challenging and critical issue for electrical engineers. Estimating harmonics and other faults is important for maintaining power quality. Research works carried out recently sheds light on various techniques for estimating harmonics. FFT [1]-based techniques are the conventional ones, and they suffer from some pitfalls such as aliasing and picket fence effects, which lead to inaccurate estimation results. There are some other methods suffering from these three problems, and this is because of existing high frequency components measured in the signal; however, truncation of the sequence of sampled data, when only a fraction of the sequence of a cycle exists in the analyzed waveform, can boost leakage problem of the DFT method. So, the need of new algorithms that process the data, sample-by-sample and not in a window as in FFT and DFT, is of paramount importance. Another very robust algorithm for the purpose of estimating sinusoidal signals with unknown noise content is the Kalman filter (KF) [2, 3].

However, when cases related to system dynamics, like sudden changes in frequency, amplitude and phase of a signal, arise, KF exhibits serious drawbacks. Study of several literature shows that single methods employed for the purpose of signal estimation are not efficient on their own, so hybrid methods based on the combination of different need to be formulated. The major contribution of this chapter is the accurate tracking of nonstationary power signal parameters, i.e., phasor, frequency, and harmonics. The power signal is modeled using Taylor series, and the coefficients of the Taylor series are updated using the Kalman Filter [4, 5], which are again utilized to estimate the time varying amplitude, phase, and frequency of the test signal. Moreover, a self-adaptive particle swarm optimization approach is deployed to choose the optimum values of the Kalman filter parameters like model and measurement error covariances, which in turn enables the filter to attain convergence in a faster rate.

2. Literature review

Work on harmonic and parameter estimation has been going since the introduction of AC power generation. Over the course of time, several methods have been proposed to fulfill this particular requirement, but so far, the existing methodologies have exhibited significant drawbacks. Here are some of the research works that have been performed in the last 5 years.

For the estimation of harmonics and interharmonics, a technique using simple techniques like least mean square [4] and a two-stage ADALINE network has been studied [5]. The method utilized here provided a better accuracy even when power frequency deviation and interharmonic components are present in the measured signal. As the conventional ADALINE is unable to detect interharmonics, a two-stage ADALINE is used. The architecture is classified in two parts—the front stage that extracts the frequency value and the back stage that computes amplitude and phase. Here, the adaptive algorithm used in the filter is the RLS algorithm. The method yielded more accurate results in protection and monitoring applications.

Sliding window tracking (SWT) [6] accurately tracks the frequency and amplitude of a signal by processing only three (or more) recent data points. It works for a signal with any nonzero moving average and noise. Teager-Kaiser algorithm (TKA) is a well-known four-point method for online tracking of frequency and amplitude. TKA takes into assumption that the signal is purely harmonic, so any moving average in the signal can totally destroy the accuracy of TKA, whereas SWT uses a pair of windowed regular harmonics to estimate the frequency and amplitude thus eliminating the effect of moving average. In order to start the online tracking of frequency, SWT requires TKA to provide the first estimate of the frequency. The accuracies of SWT and TKA are compared using Hilbert-Huang transform, which is used to extract accurate time-varying frequencies and amplitudes by processing the whole data set without assuming the signal to be harmonic. Tracking accuracy increases when window length is equal to or greater than one quarter of the signal period. If the chosen window length is too long, then the estimated frequency is an average over the window length. The method requires constant frequency and amplitude to accurately track the parameters, and this shows that the dynamic response of the method is very poor and the accuracy deteriorates, if there is no change in the parameter values.

A real-time approach for the estimation of power system frequency based on Newton-type algorithm and least squares method has been used in this paper [7]. The adopted optimization technique has been based on a two-stage mathematical model. A Newton type algorithm has been used to model the first stage for estimating the line to neutral voltage-phase angle and its variation. The second stage has been modeled using LS minimization technique that extracts the power system frequency by processing the information in the phase angles estimated using NTA. The method also studies the modulating effect of time-varying frequency on the online estimation of the phase angle.

Taylor series expansion and Fourier algorithm have been used for frequency estimation [8]. To model the changing envelope of a power signal within an observation, a second-order Taylor series has been used, and the parameters of the model have been estimated using Fourier algorithm. Comparing with the traditional Fourier algorithm, this method introduces more computational load.

A modified ADALINE structure has been used in the paper [9] for online tracking of harmonics. Self-synchronized ADALINE network for power system harmonics estimation relies on the Levenberg Gradient Descent method for updating the system parameters. A faster response and better noise immunity are provided by conventional methods. A high computational load is the only drawback that exists in the proposed approach.

Ensemble Kalman Filter has been used in the proposed method [10] for filtering and estimating signal harmonics and interharmonics. To avoid the problem of singularity and for the computational feasibility of state covariance P, the state covariance P is replaced by a sample covariance C for the computation of Kalman gain.

The proposed method [11] is adopted for real-time estimation of phasor and harmonics. The technique reduces the turnaround time on two different off-the-shelf research and development DSP platforms. The proposed method has been found to be superior to that of ADALINE and RDFT techniques under the presence of noise sub-harmonics and frequency variations. The proposed technique has a computational efficiency that is higher than that of ADALINE and RDFT techniques.

The proposed algorithm in [12] is simple, computational efficient and makes the correction of the signal that enables to reach the mean square error. It provides a new kind of step adaptation for LMS algorithm. Two LMS algorithms have been utilized by this method. The first one has a fixed-step size, and the weight coefficient generated from the first algorithm is used to update the step size of the second algorithm, which has initial step size of 0.001.

An adaptive linear network (ADALINE) [13–15] for harmonic and interharmonic estimation (Martin) allows the computing of root mean square voltage and total harmonic distortion indices. Classification and detection of sags, swells, outages, and harmonics-interharmonics have been done using the indices computed before. Classification of spikes, notching, flicker, and oscillatory transients has been achieved by using a feed forward neural network through pattern recognition using horizontal and vertical histograms of a specific voltage waveform. The method used in [16] uses noneven item interpolation FFT based on triangular self-convolution window. Variances of frequency estimation are proportional to the energy of the adopted window. By choosing suitable values of length of FFT, sampling frequency, and the shape of the adopted window, the variances of frequency estimation have been determined.

3. PSO-tuned Taylor-Kalman filter

To improve the performance of Kalman Filter in this aspect, a hybrid adaptive filter has been proposed in this thesis work that consists of the combination of Taylor series, Kalman Filter, and self-adaptive PSO. Taylor series is used to model the changing envelope of the sinusoidal signal. The sinusoidal signal is expressed in its trigonometric components, which in turn are expanded by using Taylor series. The Taylor coefficients are stored in the state vector that is further used to estimate the signal and its amplitude, frequency, and phase. In each iteration, the state vector is updated in order to get a better estimate than the previous, and the process continues until convergence is reached. There are two parameters on which the performance of the KF depends—the model and measurement error covariances. In the traditional approach, the values for these parameters are chosen by trial and error that makes the algorithm time consuming and prone to errors. Self-adaptive PSO is used here to select the optimal values of the error covariances in order to achieve fast convergence.

3.1. Signal modeling using Taylor expansion

Let the discrete signal be represented as:

$$v(i) = A(i).\cos\left[i\omega(i) + \phi(i)\right] + \kappa(i) \tag{1}$$

where $A(i), \omega(i), and\ \phi(i)$ are "the amplitude", "angular frequency," and "phase" of the sinusoid, respectively. $\omega(i) = 2\pi f(i)$ and $f(i)$ is the fundamental frequency of the signal, while $\kappa(i)$ is an additive white noise with unknown variance $\sigma_g{}^2$ Now let us represent $\theta(i) = 2\pi fidt + \phi(i)$. The rate of change of phase angle is equal to frequency. So the signal frequency can be represented as [3]:

$$f = \frac{1}{2\pi}\frac{d}{dt}(\theta(i)) = f_0 + \frac{1}{2\pi}\frac{d}{dt}(\phi(i)) \tag{2}$$

Eq. (1) can be expressed according to trigonometric function as:

$$v(i) = Q(i)\cos(2\pi f(i)) - R(i)\sin(2\pi f(i)) \tag{3}$$

where $Q(i) = A(i)\cos\phi(i)\, and\, R(i) = A(i)\sin\phi(i)$.

The coefficient functions $Q(i)\, and\, R(i)$ express the envelope of the time varying sinusoid and can be expanded using Taylor series [17, 18] as shown:

$$Q(i) \cong m_0 + m_1 i + m_2 i^2 + \dots \quad \text{and} \quad R(i) \cong n_0 + n_1 i + n_2 i^2 + \dots \tag{4}$$

where

$m_0 = Q(0), m_1 = \frac{dQ(i)}{dt}$ at k = 0; $m_2 = \frac{d^2Q(i)}{dt^2}$ and $m_3 = \frac{d^3Q(i)}{dt^3}$ at k = 0.

$n_0 = R(0), n_1 = \frac{dR(i)}{dt}$ at k = 0, $n_2 = \frac{d^2R(i)}{dt^2}$ and $n_3 = \frac{d^3R(i)}{dt^3}$ at k = 0.

Now we can obtain the amplitude and phase angle of the described given sinusoid using Eq. (3) and (4) as follows at k = 0:

$$\hat{a} = \sqrt{m_0^2 + n_0^2} \tag{5}$$

and

$$\hat{\phi} = \arctan(n_0/m_0) \tag{6}$$

where $m_0 = A(0).\cos\phi(0)$

$$n_0 = A(0).\sin\phi(0) \tag{7}$$

Similarly for estimating the frequency of the given sinusoid, consider Eq.(4) at k = 0, the first derivative will be:

$$m_1 = \frac{d}{dt}\left(A(0).\cos\phi(0)\right) \tag{8}$$

$$n_1 = \frac{d}{dt}\left(A(0).\sin\phi(0)\right) \tag{9}$$

By substituting Eq. (7) in Eq. (8) and (9) and by neglecting

$$\frac{d}{dt}\left(A(0)\right) : \frac{d}{dt}\left(\phi(0)\right) = \frac{m_0 n_1 - n_0 m_1}{{m_0}^2 + {n_0}^2} \tag{10}$$

Now from Eq. (2) and Eq. (10), we get the formula for computing the frequency:

$$\widehat{f} = f_0. + \frac{1}{2\pi}\left(\frac{m_0 n_1 - n_0 m_1}{{m_0}^2 + {n_0}^2}\right) \tag{11}$$

3.2. Updation of Taylor coefficients using the PSO-tuned Kalman filtering algorithm

Let us consider the following discrete signal:

$$Y_i = a.\sin\left[i\omega T_S + \phi\right] + n_k \tag{12}$$

where $a, T_s, \omega, and \phi$ are the amplitude, sampling time, angular frequency, and phase of the signal, respectively, and n_k represents measurement noise with a covariance R.

We can represent the state space Eq. (10) of the discrete signal as:

$$\overset{\wedge -}{\chi}(i) = f_i \chi_i + \eta_i \tag{13}$$

$$\chi_i(1) = m_0;\ \chi_i(2) = m_1;\ \chi_i(3) = m_2; \chi_i(4) = n_0;\ \chi_i(5) = n_1;\ \chi_i(6) = m_2 \tag{14}$$

And the state transition matrix is given by:

$$f_i = \begin{bmatrix} 1 & 0 & 0 & 0 & 0 & 0 \\ 0 & 1 & 0 & 0 & 0 & 0 \\ 0 & 0 & 1 & 0 & 0 & 0 \\ 0 & 0 & 0 & 1 & 0 & 0 \\ 0 & 0 & 0 & 0 & 1 & 0 \\ 0 & 0 & 0 & 0 & 0 & 1 \end{bmatrix} \tag{15}$$

http://dx.doi.org/10.5772/intechopen.71874

The stochastic model of the signal is obtained as

$$\begin{bmatrix} \chi\ \mathbf{1(i)} \\ \chi\ \mathbf{2(i)} \\ \chi\ \mathbf{3(i)} \\ \chi\ \mathbf{4(i)} \\ \chi\ \mathbf{5(i)} \\ \chi\ \mathbf{6(i)} \end{bmatrix} = \begin{bmatrix} 1\ 0\ 0\ 0\ 0\ 0 \\ 0\ 1\ 0\ 0\ 0\ 0 \\ 0\ 0\ 1\ 0\ 0\ 0 \\ 0\ 0\ 0\ 1\ 0\ 0 \\ 0\ 0\ 0\ 0\ 1\ 0 \\ 0\ 0\ 0\ 0\ 0\ 1 \end{bmatrix} \begin{bmatrix} \chi\ \mathbf{1(i)} \\ \chi\ \mathbf{2(i)} \\ \chi\ \mathbf{3(i)} \\ \chi\ \mathbf{4(i)} \\ \chi\ \mathbf{5(i)} \\ \chi\ \mathbf{6(i)} \end{bmatrix} \tag{16}$$

The measurement model of the signal expressed in Eq. (12) can be calculated as:

$$S_i = h_i \chi_i + v_i \tag{17}$$

where the observation matrix can be calculated as:

$$H_i = \begin{bmatrix} \sin\left(2\pi f_0 idt\right) idt \sin\left(2\pi f_0 idt\right) \\ \cos\left(2\pi f_0 idt\right) idt \cos\left(2\pi f_0 idt\right) \end{bmatrix} \tag{18}$$

The error signal can be obtained as

$$E_i = S - H_i \widehat{\chi}(i) \tag{19}$$

Using Eq. (19) the updated state estimate can be obtained from the following equation

$$\hat{\chi}(i) = \hat{\chi}^{-}(i-1) + K(i)\left(S_i - H\hat{\chi}(i)\right) \tag{20}$$

where the Kalman gain K(i) is given as:

$$K(i) = \widehat{P}(i-1) H_i{}^T \left(H\widehat{P}(i-1)_i H_i^T + r\right)^{-1} \tag{21}$$

where $\widehat{P}(i)$ is the covariance matrix given by

$$\widehat{P}(i) = \widehat{P}(i-1) - K_i H_i \widehat{P}(i-1) \tag{22}$$

and

$$\widehat{P}(i+1) = \widehat{P}(i) + q \tag{23}$$

where q is the model noise covariance matrix given by

$$q = \begin{bmatrix} q_a & 0 & 0 & 0 & 0 & 0 \\ 0 & q_b & 0 & 0 & 0 & 0 \\ 0 & 0 & q_c & 0 & 0 & 0 \\ 0 & 0 & 0 & q_d & 0 & 0 \\ 0 & 0 & 0 & 0 & q_e & 0 \\ 0 & 0 & 0 & 0 & 0 & q_f \end{bmatrix} \quad (24)$$

r is the measurement noise covariance which is fine tuned by using the error between the desired and estimated signals for

$$r(i) = \lambda_g(r + (E(i))\hat{}2) \quad (25)$$

where λ_g is the forgetting factor in the range (0.9–1). Finally using the EKF time updated equations the $X(i)$ matrix is computed which determines the values of the Taylor series coefficients m_0, m_1, n_0, and n_1.

3.3. Particle swarm optimization-based tuning of the Kalman filter

For fast convergence, optimum values of q and r are selected by applying a self-adaptive PSO [19]. For this purpose, a cost function is formulated, which passed in the PSO algorithm to get the optimum value for q and r. Here, the cost function is:

$$F = \frac{1}{L}\sum_{i=1}^{L} E_i^2 \quad (26)$$

Particle swarm optimization is used to minimize the value of Eq. (26). Each particle is characterized by two attributes:

i. pbest or Personal best: it holds the best value of position with respect to the previous positions of the particular particle.

ii. gbest or Global best: it holds the best value of position in the entire search space.

The PSO algorithm either minimizes or maximizes the value of gbest. Let x_{ij} and V_{ij} be the position and velocity of the ith particle in the jth dimension at k^{th} instance of time. The personal best value can be determined from Eq. (27).

$$pbest_i(k+1) = \begin{cases} pbest_i(k), if\ F(x_i(k+1)) > F(pbest_i(k)) \\ x_i(k+1), if\ F(x_i(k+1)) < F(pbest_i(k)) \end{cases} \quad (27)$$

where F indicates the cost function. The value of global best is obtained as:

$$gbest(k) = \min\{C..F(pbest_0(k)), C..F(pbest_1(k)), C..F(pbest_2(k)), \ldots., C..F(pbest_s(k)),\} \quad (28)$$

For each particle, the updated velocity and position at time $(k+1)$ are given by

$$V_i(k+1) = K\{\alpha v_i(k) + b_1 rnd_1 (pbest_i(k) - x_i(k)) + b_2 rnd_2 (gbest_i(k) - x_i(k))\} \tag{29}$$

$$x_i(k+1) = x_i(k) + V_i(k+1) \tag{30}$$

where α is the inertia weight factor, b_1 and b_2 are the acceleration constants, rnd_1 and rnd_2 are random numbers in the range [0, 1], K is a constriction factor given by:

$$K = \frac{2}{|2 - b - \sqrt{b^2 - 4b}|}, \quad \text{and} \quad b = b_1 + b_2; b > 4 \tag{31}$$

The performance of the PSO algorithm is significantly affected by the three factors w, c_1, and c_2. In this approach, a detection function defined as: $\varphi(k) = |(gbest_i - x_i(k))/(pbest_i - x_i(k))|$. The values of the three factors are adjusted dynamically using the following equations

$$\alpha(k) = \frac{\alpha_{initial} - \alpha_{final}}{1 + e^{\varphi(k)(k - ((1 + \ln(\varphi(k)))L_{\max}/\mu)}} + \alpha_{final} \tag{32}$$

$$c_1(k) = c_1 \varphi^{-1}(k) \tag{33}$$

$$c_2(k) = c_2 \varphi(k) \tag{34}$$

where $w_{initial}$ and w_{final} lie in the range $(0 < w < 2)$, $L_{\max}$ is the final evolutionary generation, and k is the current evolutionary generation.

3.4. PSO-based Taylor-Kalman filter structure

The adaptive filter structure with the proposed adaptive algorithm is shown in the **Figure 1**. This particular structure is modeled for only the fundamental component of the signal to be estimated. For a signal with Nth order harmonics, the same structure can be extended to meet the requirements. The signal is modeled using Taylor series up to the second order, so the filter

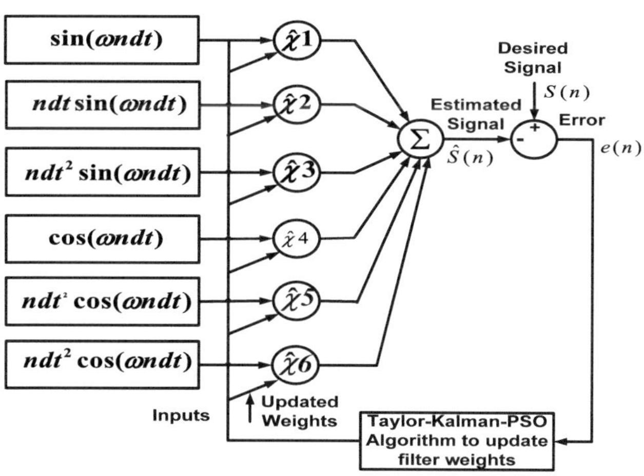

Figure 1. Filter structure of the proposed approach.

has six weights for the six inputs that are used for the purpose of estimation. The performance of the algorithm is judged on the basis of speed of convergence, which is verified from the simulation results in Section 4.

4. Simulation and results

The performance of the proposed algorithm for power system signals has been shown with the help of three computer simulated examples.

4.1. Tracking of a nonstationary signal with simultaneous change in amplitude, phase, and frequency

A nonstationary test signal as shown in Eq. (35) is generated in MATLAB. The simulation is done over 1000 samples of the signal. To make the signal nonstationary, a double step is introduced in the signal by changing the value of amplitude from 500 to 700 samples. This is done to simulate voltage surge occurrences in real time, where the amplitude increases from that of its desired value for some period of time. Similar disturbances also change the values of frequency and phase which is also simulated to test the tracking accuracy of the proposed algorithm. The results in **Figures 2–5** reveal that the accuracy of the proposed algorithm is very high and tracking is achieved within one cycle of the signal.

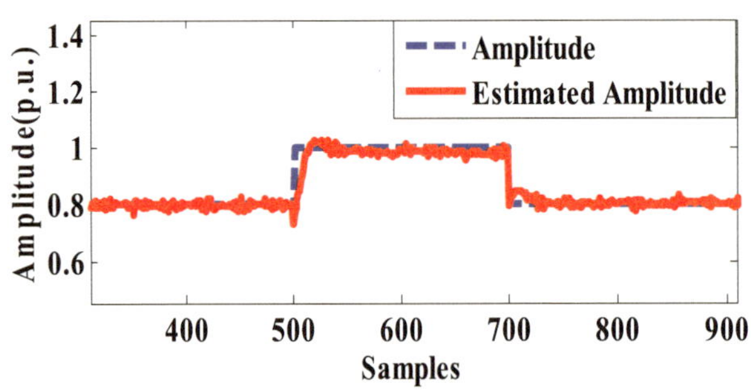

Figure 2. Estimated amplitude under 30 dB noise.

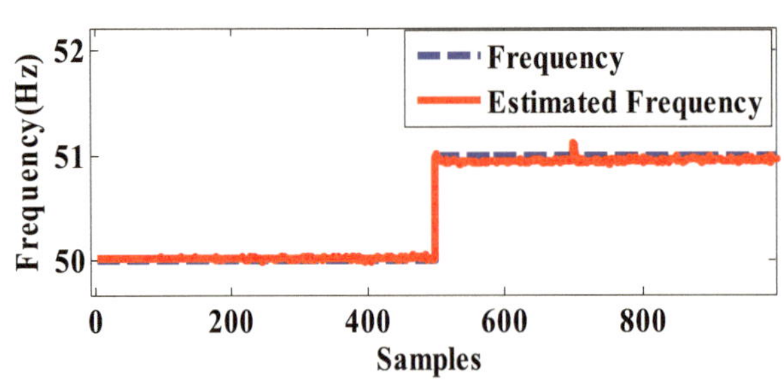

Figure 3. Estimated frequency under 30 dB noise.

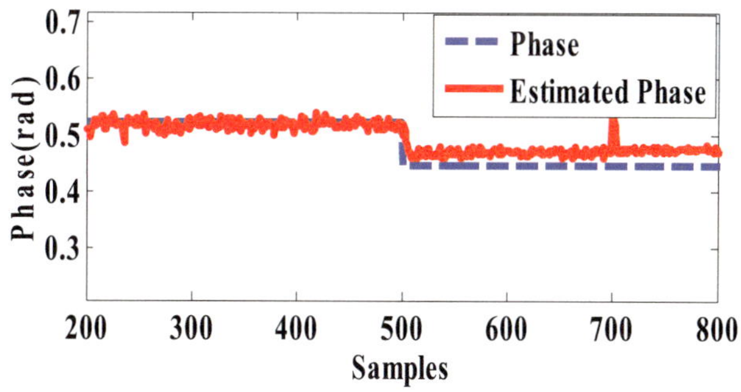

Figure 4. Estimated phase under 30 dB noise.

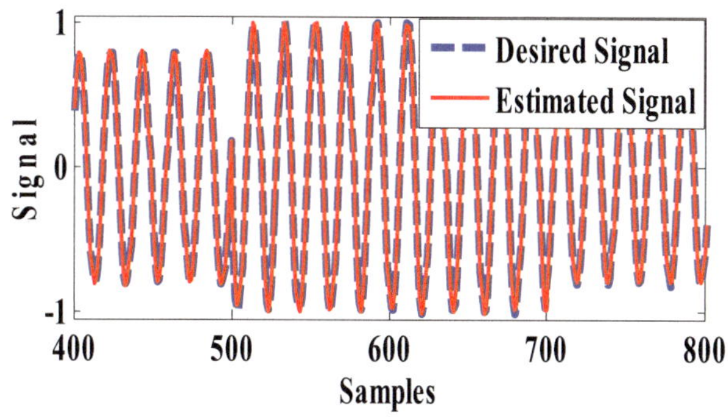

Figure 5. Estimated signal under 30 dB noise.

$$y(i) = a(i)\sin\left(i\omega(t)dt + \phi(i)\right) + n(i) \tag{35}$$

where,

$$a(i) = \begin{cases} 0.8 \ \ p.u, i < 500\, samples \\ 1 \ \ p.u, 500 < i < 700\, samples \\ 0.8 \ \ p.u, i > 700\, samples \end{cases}$$

is the signal amplitude.

$$f(i) = \begin{cases} 50 \ \ Hz, i < 500\, samples \\ 51 \ \ Hz, i > 500\, samples \end{cases}$$

The sampling frequency $f_s = 2\text{kHz}\ \phi(i) = \begin{Bmatrix} 0.5\,\text{rad} \\ 0.45\,\text{rad} \end{Bmatrix}$

$n(i)$ is the noise sequence with power level 30 dB. This signal fed into the algorithm and simulated in MATLAB 2013a environment.

Parameter	Samples	Actual value	Estimated value								
			LMS			EKF			Proposed method		
			20 dB	30 dB	40 dB	20 dB	30 dB	40 dB	20 dB	30 dB	40 dB
Amplitude	0–500	0.8	0.7	0.772	0.727	0.76	0.76	0.777	0.795	0.798	0.799
	500–700	1	0.885	0.946	0.919	0.89	0.88	0.93	0.896	0.942	0.997
	700–1000	0.8	0.7	0.772	0.727	0.78	0.79	0.787	0.789	0.794	0.797
Frequency	0–500	50	50.05	50.02	50.02	50.03	50.02	50.03	50.00	50.01	50.00
	500–1000	51	50.38	50.55	50.63	50.43	50.76	51.02	50.94	50.99	51
Phase	0–500	0.5	0.53	0.513	0.521	0.53	0.53	0.53	0.511	0.513	0.533
	500–1000	0.45	0.465	0.445	0.47	0.43	0.48	0.47	0.44	0.45	0.45

Table 1. Estimated values of amplitude, frequency, and phase under different noise conditions.

Table 1 contains the estimated values of the amplitude, frequency, and phase of the signal under different noise conditions. In this simulation, three different noise conditions have been considered. The simulation is carried out in a dynamic noise range from high noise (20 dB) to low noise (40 dB) conditions to test the performance of the proposed algorithm under noise. The analysis of the performance under noisy conditions shows that the proposed algorithm is able to track the desired signal very closely even under heavy noise conditions.

4.2. Performance of the proposed algorithm in harmonic estimation

In this case, the ability of the proposed algorithm is tested with respect to the tracking of harmonics. The number of harmonic components present in the system is not constant, and it can vary from few to a large number. It is not possible for any method to track infinite number of harmonics but can handle a substantial quantity. In the real-time scenario, harmonics occur as odd multiples of the fundamental frequency, so the simulation is carried out with a system generated signal containing harmonics up to the 19th order.

$$\begin{aligned} y(i) = {} & a1(i)\sin\left(i\omega(i)dt + \phi_1(i)\right) + 0.8(i)\sin\left(i3\omega(i)dt + 0.4\right) \\ & +0.6(i)\sin\left(i5\omega(i)dt + 0.3\right) + 0.5(i)\sin\left(i7\omega(i)dt + 0.25\right) \\ & +0.4(i)\sin\left(i11\omega(i)dt + 0.2\right) + 0.2(i)\sin\left(i19\omega(i)dt + 0.1\right) \end{aligned} \tag{36}$$

The signal parameters are taken as:

$$\begin{aligned} a(i) &= 0.1\sin\left(2\pi idt\right) + 0.05\sin\left(10\pi idt\right) \\ a1(i) &= 1 + a(i) \\ f(i) &= 50Hz \\ f_s(i) &= 4kHz \\ \phi_1 &= 0.5 - 0.2\sin\left(2\pi\ 5\ idtd + 0.3\right) \end{aligned}$$

The amplitude, frequency, and phase are estimated, and results are shown in **Figures 6–11**.

http://dx.doi.org/10.5772/intechopen.71874

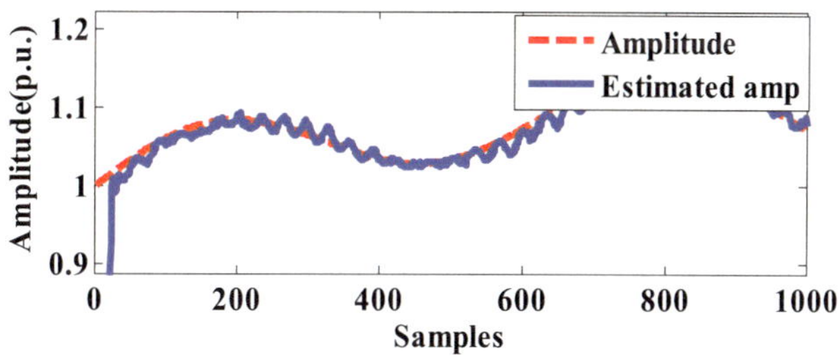

Figure 6. Estimated fundamental amplitude for case 4.2.

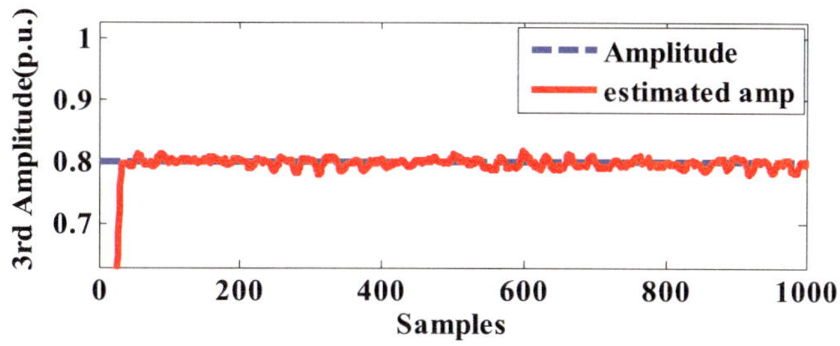

Figure 7. Estimated third amplitude for case 4.2.

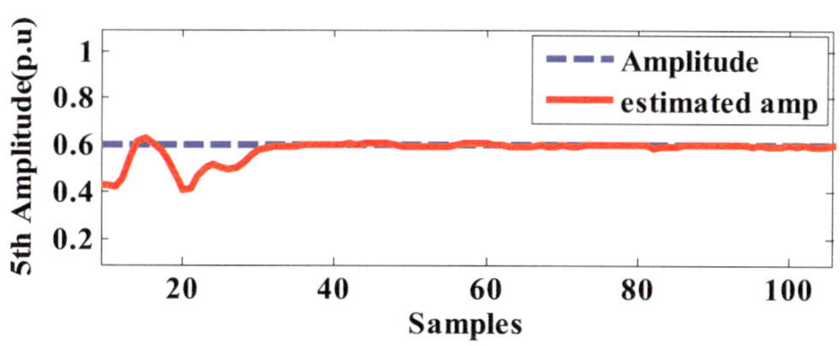

Figure 8. Estimated fifth amplitude for case 4.2.

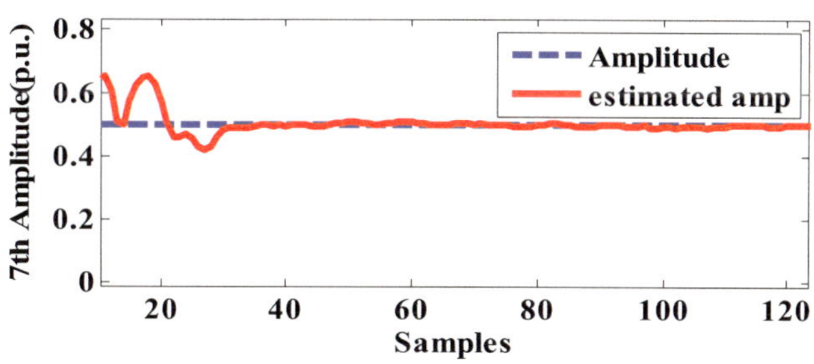

Figure 9. Estimated seventh amplitude for case 4.2.

Table 2 shows the comparison of the absolute errors in amplitude, frequency, and phase estimation for different harmonic components for EKF, LMS, RLS, and the proposed method. The values show that the higher order (>5th order) components exhibit higher error values for

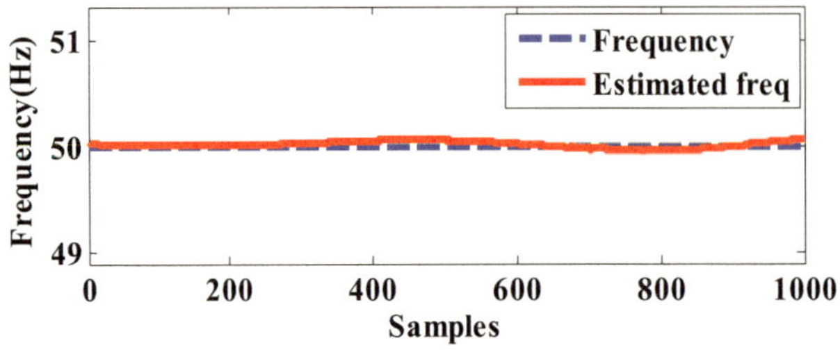

Figure 10. Estimated frequency of fundamental for case 4.2.

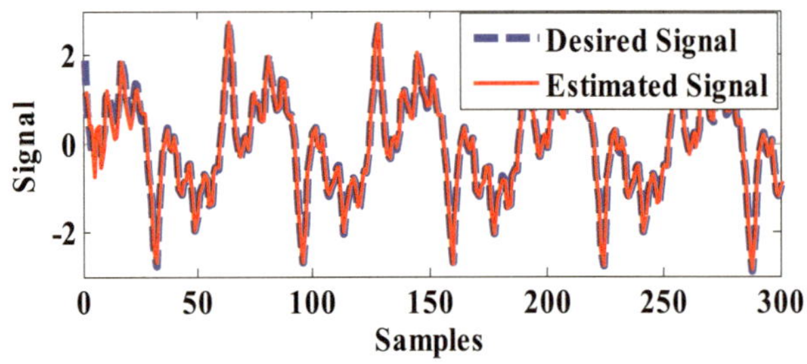

Figure 11. Estimated signal for case 4.2.

Parameter	Component	Absolute error			
		EKF	LMS	RLS	Proposed
Amplitude (harmonic order)	A_1	0.01	0.03	0.023	0.007
	A_3	0.013	0.032	0.04	0.003
	A_5	0.02	0.047	0.056	0.009
	A_7	0.04	0.058	0.0856	0.017
	A_{11}	0.025	0.0623	0.021	0.017
	A_{19}	0.03	0.0875	0.075	0.025
Frequency	Fundamental	0.065	0.045	0.0201	0.058
Phase	Φ_1	0.0029	0.0087	0.0047	0.0023
	Φ_3	0.006	0.04	0.032	0.0005
	Φ_5	0.024	0.045	0.054	0.007
	Φ_7	0.005	0.076	0.072	0.002
	Φ_{11}	0.03	0.085	0.088	0.0019
	Φ_{19}	0.067	0.083	0.0765	0.04

Table 2. Comparison of absolute errors.

all the methods, but the comparison shows that among all the methods compared, the proposed method has the least values of error. This comparison sheds light on the superiority of the proposed method over the other methods.

4.3. Estimation of a power signal in the presence of DC component

When a fault occurs it not only distorts the signal by changing the voltage and current waveforms but some DC component that decays over time also gets added to the signal. DC components are nonperiodic in nature and this simulation shows that the proposed algorithm efficiently tracks nonperiodic components in the signal which is clearly evident from **Figures 12–15**. A nonstationary test signal with a decaying DC component as shown in Eq.(37) is considered:

$$y(i) = a(i)\sin\left(i\omega(t)dt + \phi(i)\right) + n(i) \tag{37}$$

where,

$$a(i) = A\ \exp\left(-i/300\right)\ p.u.$$
$$A = 1p.u.$$

is the signal amplitude. $f(i) = 50Hz$

The sampling frequency $f_s = 2kHz$ and $\phi(i) = 0.52\ rad\ n(i)$ is the 30 dB noise.

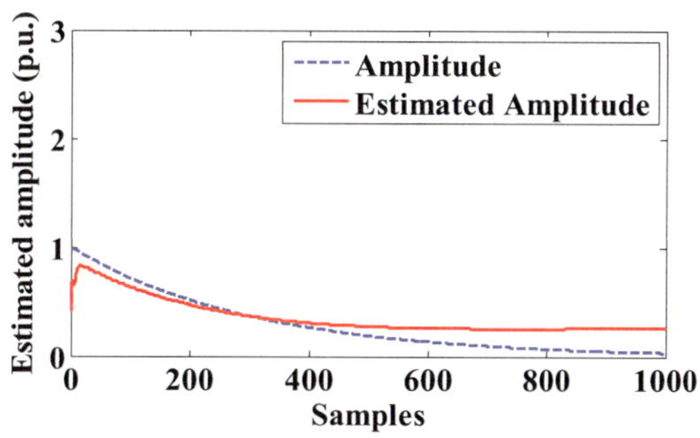

Figure 12. Estimated amplitude for signal with decaying DC component.

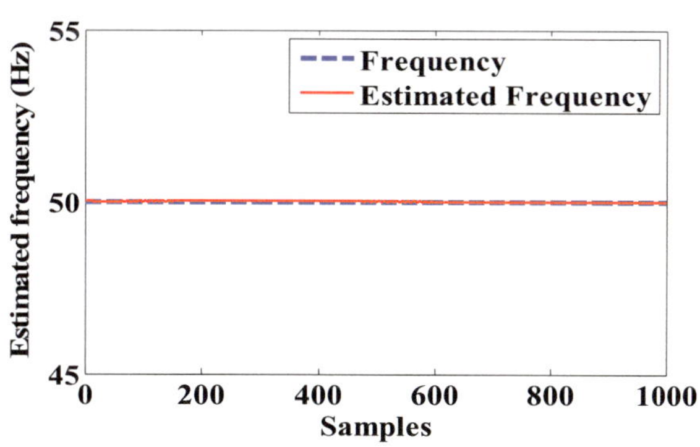

Figure 13. Estimated frequency for signal with decaying DC component.

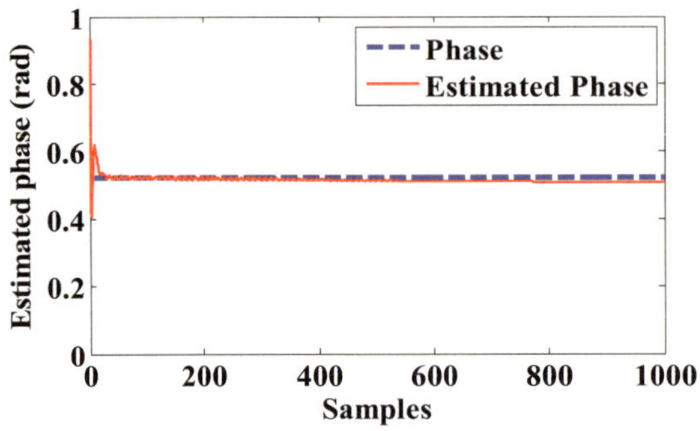

Figure 14. Estimated phase for signal with decaying DC component.

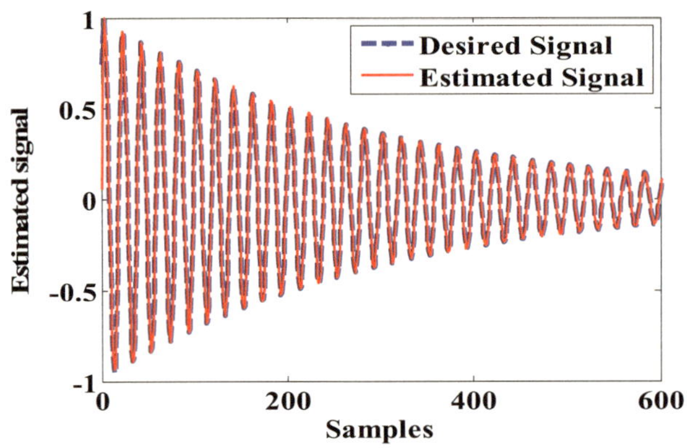

Figure 15. Estimated signal with decaying DC component.

5. Conclusion

The traditional Kalman filter has been extended to Taylor-Kalman filter which resulted in filters that are able to have flat magnitude and phase responses. These filters exhibit excellent tracking abilities and accurately estimate the amplitude, frequency and phase of a time varying power signal without any distortion. The further combination of the Taylor-Kalman filter with self-adaptive PSO makes the performance of the proposed method superior to the traditional approach. The methods can be individually used for the purpose of signal and parameter estimation, but individually, they suffer from some drawbacks. By combining the three methods into one hybrid method, the pitfalls of each are compensated by the other and hence much better results are obtained.

Further, the hardware implementation of the proposed method can be attempted for real-time applications [20–23]. The hardware implementation of the proposed method can be embedded within an integrated circuit that will result in a system on chip that can be installed at power distribution centers, from where power gets distributed to the consumers, thus equipping them with a tool for detecting anomalies in power quality before power is dispatched to the utility network. The objective of developing such a technology is to create a compact and versatile tool. It is a small contribution toward the development of smart grid technology.

Author details

Sarita Nanda

Address all correspondence to: sarita22579@rediffmail.com

School of Electronics, KIIT University, India

References

[1] Lin HC, Lee CS. Enhanced FFT-based parameter algorithm for simultaneous multiple harmonics analysis. Proceedings of the Institute of Electrical and Electronics Engineers: Generation, Transmission and Distribution Proceedings. May 2001;**148**(3):209-214. DOI: 10.1049/ip-gtd:20010278

[2] Chen CI, Chang GW, Hong RC, Li HM. Extended real model of Kalman filter for time-varying harmonics estimation. IEEE Transactions on Power Delivery. Jan 2010;**25**(1):17-26. DOI: 10.1109/TPWRD.2009.2035217

[3] Nanda S, Hasan S, Pujari SS, Dash PK. A fast hybrid adaptive filter for parameter estimation of non-stationary sinusoid under noise. IEEE Conference PCTIC. 2015. DOI: 10.1109/PCITC.2015.7438146

[4] Pradhan K, Routray A, Basak A. Power system frequency estimation using least mean square technique. IEEE Transactions on Power Delivery. Jul. 2005;**20**(3):1812-1816. DOI: 10.1109/TPWRD.2004.843453

[5] Chen C-I, Chang GW. A two stage ADALINE for harmonic and inter harmonic measurement. IEEE Transactions on Industrial Electronics. 2009;**56**(6):2220-2228. DOI: 10.1109/TIE.2009.2017093

[6] Pai PF. On-line tracking of instantaneous frequency and amplitude of dynamical system response. Mechanical Systems and Signal Processing. May 2010;**24**(4):1007-1024. DOI: org/10.1016/j.ymssp.2009.07.014

[7] Sadinezhad I, Agelidis VG. Slow sampling online optimization approach to estimate power system frequency. IEEE Transactions on Smart Grid. June 2011;**2**(2). DOI: 10.1109/TSG.2011.2114374

[8] Ren J, Kezunovic M. An improved fourier method for power system frequency estimation. North American Power Symposium (NAPS), 4–6 Aug, 2011, Boston, MA. pp. 1-6. DOI: 10.1109/NAPS.2011.6024845

[9] Arghya S, Roy Choudhury S, Sengupta S. A self-synchronized ADALINE network for on-line tracking of power system harmonics. Elsevier Measurement. May 2011;**44**(4):784-790. DOI: 10.1016/j.measurement.2011.01.009

[10] Ray PK, Subudhi B. Ensemble-Kalman-filter-based power system harmonic estimation. IEEE Transactions on Instrumentation and Measurement. December 2012;**61**(12). DOI: 10.1109/TIM.2012.2205515

[11] Sadinezhad I, Agelidis VG. Real-time power system phasors and harmonics estimation using a new decoupled recursive-least-squares technique for DSP implementation. IEEE Transactions on Industrial Electronics. 2013;**60**(6):2295-2308. DOI: 10.1109/TIE.2012.2192895

[12] Marchesan G, Cardoso Jr G, Mariotto L, Morais AP, Oliveira AL. An adaptive step-size least mean square algorithm for electric power systems frequency estimation in protective relays. 12th IET International Conference on IEEE Developments in Power System Protection (DPSP 2014). DOI: 10.1049/cp.2014.0118

[13] Valtierra-Rodriguez M, Romero-Troncoso R d J, Osornio-Rios RA, Garcia-Perez A. Detection and classification of single and combined power quality disturbances using neural networks. IEEE Transactions on Industrial Electronics. May 2014;**61**(5). DOI: 10.1109/TIE.2013.2272276

[14] Nanda S, Biswal M, Dash PK. Estimation of time varying signal parameters using an improved Adaline learning algorithm. AEU International Journal of Electronics and Communications. 2014;**68**(2):115-129. DOI: 10.1016/j.aeue.2013.07.014

[15] Nanda S, Dash PK, Chakraborty T, Hasan S. A quadratic polynomial signal model and fuzzy adaptive filter for frequency and parameter estimation of nonstationary power signals. Measurement. June 2016;**87**:274-293. DOI: 10.1016/j.measurement.2016.03.026

[16] He W, Guo S, Teng Z, Li F, Yang Y. Frequency estimation of distorted and noisy signals in power systems by FFT-based approach. IEEE Transactions on Power Systems. March 2014; **29**(2). DOI: 10.1109/TPWRS.2013.2283273

[17] Lee DG, Kang SH, Nam SR. Modified dynamic Phasor estimation algorithm for the transient signals of distributed generators. IEEE Transactions on Smart Grid. March.2013;**4**(1): 419-424. DOI: 10.1109/TSG.2012.2233772

[18] de la O Serna JA, Maldonado JR. Taylor Kalman Fourier filters for instantaneous oscillating Phasor and harmonic estimates. IEEE Transactions on Instrumentation and Measurement. April 2012;**61**(4):941-951. DOI: 10.1109/TIM.2011.2178677

[19] Kingshuk B, Nanda S. An adaptive filtering technique with self-adaptive PSO for estimation of non-stationary signals. 2016 International Conference on Communication and Signal Processing (ICCSP). IEEE, 2016. DOI: 10.1109/ICCSP.2016.7754311

[20] Cardenas A, Guzman C, Agbossou K. Development of a FPGA based real-time power analysis and control for distributed generation Interface. IEEE Transactions on Power Apparatus and Systems. August 2012;**27**(3):1343-1353. DOI: 10.1109/TPWRS.2012.2186468

[21] Jindapetch N, Chewae S, Phukpattaranont P. FPGA implementations of an ADALINE adaptive filter for power-line noise cancellation in surface electromyography signals. Measurement. 2012;**45**(3):405-414. DOI: 10.1016/j.measurement.2011.11.004

[22] Sarita N, Dash PK. A gauss-Newton ADALINE for dynamic Phasor estimation of power signals and its FPGA implementation. IEEE Transactions on Instrumentation and Measurement, (Early access). 2016. DOI: 10.1109/TIM.2016.2620841

[23] Sarita N, Dash PK. Field programmable gate array implementation of fuzzy variable step size adaptive linear element for adaptive frequency estimation. IET Signal Processing, (Available online. DOI: 10.1049/iet-spr.2016.0574

Chapter 6

Kalman Filter Models for the Prediction of Individualised Thermal Work Strain

Jia Guo, Ying Chen, Weiping Priscilla Fan, Si Hui Maureen Lee, Junxian Ong, Poh Ling Tan, Yu Li Lydia Law, Kai Wei Jason Lee and Kok-Yong Seng

Additional information is available at the end of the chapter

http://dx.doi.org/10.5772/intechopen.71205

Abstract

It is important to monitor and assess the physiological strain of individuals working in hot environments to avoid heat illness and performance degradation. The body core temperature (Tc) is a reliable indicator of thermal work strain. However, measuring Tc is invasive and often inconvenient and impractical for real-time monitoring of workers in high heat strain environments. Seeking a better solution, the main aim of the present study was to investigate the Kalman filter method to enable the estimation of heat strain from non-invasive measurements (heart rate (HR) and chest skin temperature (ST)) obtained 'online' via wearable body sensors. In particular, we developed two Kalman filter models. First, an extended Kalman filter (EFK) was implemented in a cubic state space modelling framework (HR versus Tc) with a stage-wise, autoregressive exogenous model (incorporating HR and ST) as the time update model. Under the second model, the online Kalman filter (OFK) approach builds up the time update equation depending only on the initial value of Tc and the latest value of the exogenous variables. Both models were trained and validated using data from laboratory- and outfield-based heat strain profiling studies in which subjects performed a high intensity military foot march. While both the EKF and OKF models provided satisfactory estimates of Tc, the results showed an overall superior performance of the OKF model (overall root mean square error, RMSE = 0.31°C) compared to the EKF model (RMSE = 0.45°C).

Keywords: heat strain, body core temperature, wearable body sensors extended Kalman filter, online Kalman filter

1. Introduction

Physically demanding tasks, environmental heat and humidity and various clothing requirements combine to create heat stress for workers. The associated physiological responses to that stress, e.g. increased body core temperature (Tc), heart rate (HR) and sweating, are collectively known as physiological strain. Physiological strain rises with the heat stress, and if not controlled, may diminish the quality and productivity of job performance. Left unchecked, high levels of heat strain may also result in increased accident rates and an increased risk of heat-related disorders including unconsciousness and cardiac arrest. Heat casualties are a concern to the military, first responders and industrial workers [1–3].

High Tc is one of the most reliable predictor of heat-related disorders and the ability to accurately monitor this variable could help mitigate the risk of heat injuries [4]. However, the measurement of Tc in an ambulatory setting is not straightforward. Traditional methods of Tc measurement typically require probes (e.g. rectal and oesophageal) but these are impractical for an ambulatory setting. While ingestible thermometer capsules (e.g. Philips Respironics, Murrysville, PA) have been used with success in laboratory and field settings, these instruments are relatively expensive, are unsuitable for individuals with food and drug administration contraindications, and while still in the stomach or upper intestine can suffer acute inaccuracies when cold fluids are ingested. This means that in many situations, the continuous ambulatory monitoring of Tc is still impractical. Alternative Tc surrogate methods, which seek some non-invasive core temperature correlate (e.g. surface heat flux), can be difficult to use consistently across different environments and lose precision when predicting for individuals [5].

Wearable activity trackers have emerged as an increasingly popular method for individuals to assess their daily physical activity and energy expenditure through sensing of physiological data, e.g. HR and surface skin temperature (ST) [6]. One means of overcoming Tc measurement problem is to estimate Tc based on other more readily available data obtained from such body worn sensors. From physiology, both HR and ST are closely related to work and heat stress. Serial HR measurements contain information about heat production [7] and heat transfer since HR is related to skin perfusion [8]. Similarly, because heat can be conducted from deep tissues to skin, an increase in Tc can lead to an elevation of ST over time [9]. Previous studies have also shown the promise of using HR and ST to estimate heat strain [10, 11].

Tapping on the wide availability of physiological measurements from increasingly ubiquitous wearable activity trackers and the physiological basis of associations between Tc with HR and ST, we applied the Kalman filter (KF) technique to track individual-specific Tc over time using time series observations of HR and ST. KF-based methods utilise a prediction-correction scheme to dynamically track and adjust both the system states (Tc for our application) and its uncertainty to agree with measurements (HR and ST) as they are made [12]. The system model expressed as a function of the state variable is used to iterate the distribution of Tc forward in time to produce a prediction, which is then corrected to both adjust the prediction and collapse its uncertainty.

The pursuit of reliable KF models to predict Tc is a subject of active investigation. Buller and co-authors have used the KF technique to estimate Tc by capturing the linear or quadratic relationship between time-varying HR and Tc [13–15]. Their results have indicated that 95% of all predictions fell within ±0.48–0.63°C for different study cohorts. However, the developmental datasets contained only a limited amount of data at high Tc (≥39°C) and thus most of these statistics are based on the lower Tc values, which may limit the model's ability to reliably predict hyperthermic body temperatures. Further, the validity of the Tc estimates in human subjects with differing demographics and working in a predominantly hot and humid climate was unclear. We implemented an extended Kalman filter (EKF) model using a non-linear (cubic) state space model (ST versus Tc) with a stage-wise, autoregressive exogenous model (incorporating HR) as the time update model [11]. We showed that the EKF model predicted Tc more precisely [root mean square error (RMSE) was 0.29°C] compared to KF models that relied only on HR as an explanatory variable (RMSD = 0.33°C). However, our model was developed using only laboratory data as developmental data and thus lacked assessment against data measured in the field settings.

While practical, the aforementioned KF models require previous estimates of Tc for continuous prediction of this latent variable. One major inherent limitation of such models is that when the forecast horizon increases, errors in the prediction would accumulate, which would progressively increase the prediction uncertainty even with the Kalman gains. This may give rise to grave clinical consequences since large prediction errors at high core temperature zones (for an individual who works continuously) could delay the application of cooling measures on heat casualties.

The main aim of this paper was to develop and investigate the potential of using online Kalman filter (OKF) models to improve the estimation of Tc over long time horizons as encountered during extended duration high intensity physical tasks, e.g. foot march. The OKF models comprised a time update equation that depends on the initial value of Tc and time-current value of the measurable exogenous variables such that the value of Tc at any time point is directly predicted. The second aim was to assess the comparative accuracy of Tc predictions by the EFK and OFK models vis-à-vis-observed Tc.

2. Methods

2.1. Data

Data for model development were derived from laboratory- and field-based heat strain profiling studies that involved different participants. The study protocols used in all studies were approved by the Institutional Review Board. All volunteers were briefed on the purpose, risks and benefits of the study and each gave their written informed consent prior to participation.

2.1.1. Study 1 (laboratory study)

A total of 29 male volunteers [mean (range); age = 30 (26–33years), bodyweight = 68.4 (48.9–87.6 kg), height = 1.71 (1.61–1.81 m), body mass index (BMI) = 23.7 (17.3–28.0 kg/m^2), body surface area (BSA) = 1.80 (1.52–2.07m^2)] performed a military 16 km foot march in a climatic chamber. During the trials, all participants donned a standard infantry full battle order (FBO), comprising camouflage uniform, combat boots, body armour, load bearing vest with standard accessories, Kevlar helmet, rifle replica and a backpack filled with additional accessories, for the foot march. All back packs used in the study were packed in the same configuration. The foot march was composed of three rounds of 4 km followed by one round each of 3 km and 1 km marches on the treadmill at 5.3 km/h and 0% gradient, with each exercise bout separated by 15 min seated rest. Water was provided *ad libitum* to all participants. Environmental conditions in the climatic chamber represented those present in hot-humid environments, with a mean dry bulb temperature of 32°C, relative humidity of 70%, solar radiation of 250 W/m^2 and wind speed of 1.5 m/s. The mean completion time of the full 16 km route march was 255 min.

2.1.2. *Study 2 (field study)*

A total of 43 male volunteers [age = 24 (18–33 years), bodyweight = 66.4 (49.9–89.3 kg), height = 1.72 (1.58–1.92 m), BMI = 22.4 (17.7–27.6 kg/m^2), BSA = 1.79 (1.54–2.09 m^2)], outfitted in FBO, performed a military 16 km foot march together as a group in the field. The foot march was conducted in the morning with cloudy skies (mean dry bulb temperature, relative humidity and wind speed during the trials were 27°C, 86% and 1.1 m/s, respectively). The foot march was composed of three rounds of 4 km followed by one round each of 3 km and 1 km marches on paved terrain, with each exercise bout separated by 15 min seated rest. All participants had *ad libitum* access to fluid from their water containers, which were refilled during each recess period. The total duration of the trials was approximately 285 min.

2.1.3. *Physiological measures*

For all heat profiling studies, Tc, HR and ST were recorded every 15 s using a chest belt physiological monitoring system (Equivital EQ02 LifeMonitor®, Hidalgo Ltd., Cambridge, UK) with an associated ingestible thermometer capsule (Philips Respironics, Murrysville, PA). Participants ingested one thermometer capsule at least 8 h prior to the foot march in order to ensure that the capsule had travelled far enough in the intestinal tract to avoid errors from ingested fluids. Each participant's real-time data were checked for accurate reporting of Tc, HR and ST prior to the trials. Tc data were not used if there were evident signs of fluid signatures (rapid decrease in Tc to below 32°C and slow recovery to normal body temperature).

For data modelling in the present study, Tc, HR and ST measured using the physiological monitoring system were reduced to 1 min intervals by taking the median of four 15 s samples for each 1 min epoch.

2.2. Assessment of model performance

Predictive performance of each model against data from study 1 and study 2 was assessed separately using in-sample and out-of-sample analyses. Conducting an in-sample analysis entailed using the model to estimate all observed Tc that formed the database for model training. Out-of-sample analysis: estimating observed Tc time series that was not part of the database for model training: was implemented using a four-fold cross-validation.

For cross-validation, the full dataset from study 1 and study 2 was randomly divided into four groups, each containing 25% of the participants (Tc measurements belonging to the same participant were kept in the same group). Four different subsets of three groups (i.e. 3 × 25%

of the studied profiles) were constituted to form four different index groups. Each remaining 25% of the studied profiles constituted a separate test group, generating four independent test groups. Then, a final model was separately identified using the four different index groups. To assess the predictive performance of the final model, the parameter estimates from each of the four subsets (i.e. index group) were used to predict the individual Tc time series in the respective test group.

Various evaluation criteria were used to assess the model performance. These were RMSE, Bland-Altman limits of agreement (LoA) [16] and percentage of prediction-data deviation (i.e. error) that were within ±0.1, 0.3 and 0.5°C [percentage of target attainment (PTA)].

The prediction error is computed using:

$$e_{t,i} = \widehat{Tc}_{t,i} - Tc_{t,i} \tag{1}$$

where $\widehat{Tc}_{t,i}$ denotes the predicted value of Tc at time t for the *i*th participant and $Tc_{t,i}$ is the measured (based on the thermometer capsule) value of Tc.

RMSE, a measure of the precision in the predicted Tc, is computed using:

$$\mathrm{RMSE} = \left(\frac{1}{N}\frac{1}{T}\sum_{i}^{N}\sum_{t}^{T} e_{t,i}^2\right)^{1/2} \tag{2}$$

where N and T denote the total number of participants in the relevant dataset and the total number of Tc measurements per participant, respectively.

LoA, which indicates the limits within which 95% of all prediction errors should fall assuming that the errors are normally distributed, is computed using:

$$\mathrm{LoA} = \mathrm{bias} + 1.96 \times (\mathrm{SD\ of\ } e_{t,i}) \tag{3}$$

where bias (mean error) $= \frac{1}{N}\frac{1}{T}\sum_{i}^{N}\sum_{t}^{T} e_{t,i}$ and SD is the standard deviation of the difference between the predicted and observed Tc.

3. Kalman filter models

In this section, we describe the KF approaches proposed by Buller and his co-authors [13–15], as well as the EKF [11] and the OKF models developed by our group. In the state-space models, Tc is not directly observed but considered as a latent state variable, while the other measurable physiological variables (e.g. HR, ST) are used as observable exogenous variables.

3.1. Kalman filter

The KF algorithm uses observed exogenous variables to estimate the latent or unobservable variable. The algorithm recursively operates on streams of noisy input variables to produce statistically optimal estimate of the state variable in a hypothesised state system. Without loss of generality, the system can be represented by a state-space model:

$$\text{Observation} : X_t = h(Y_t) + v_t, v_t \sim N(0, R), \tag{4}$$

$$\text{Time update} : Y_t = \phi_0 + \phi_1 Y_{t-1} + \theta_1 X_{t-1} + \theta_2 U_{t-1} + \epsilon_t, \epsilon_t \sim N(0, \sigma^2), \tag{5}$$

$$\text{Transition} : Y_t = g(Y_{t-1}, U_{t-1}) + \omega_t, \omega_t \sim N(0, Q), \tag{6}$$

where the functions $h(\cdot)$ and $g(\cdot)$ are differentiable for each state. The transition function is derived from the observation function and the time update equations. The innovations v_t, ϵ_t and ω_t are assumed to follow a Gaussian distribution with mean zero and constant variance. The partial derivatives of the Jacobian matrix can be derived as:

$$G_t = \left.\frac{\partial g}{\partial Y}\right|_{\widehat{y}_{t-1}, u_{t-1}} \tag{7}$$

$$H_t = \left.\frac{\partial h}{\partial Y}\right|_{\widehat{y}_t^*}. \tag{8}$$

The KF algorithm consists of two steps: predict and update. At any forecast origin t, we have.

Predict:

$$\widehat{y}_t^* = g(\widehat{y}_{t-1}, u_{t-1}) \tag{9}$$

$$P_t^* = G_t P_{t-1} G_t^T + Q \tag{10}$$

Update:

$$\widehat{y}_t = \widehat{y}_t^* + K_t \{x_t - h(\widehat{y}_t^*)\} \tag{11}$$

$$P_t = (1 - K_t H_t) P_t^* \tag{12}$$

where the Kalman Gain $K_t = P_t^* H_t^T (H_t P_t^* H_t^T + R)^{-1}$.

Buller et al. [13] proposed a KF model to predict Tc by tracking the observed exogenous HR time series. The KF model is represented as:

$$Tc_t = \varphi_0 + \varphi_1 Tc_{t-1} + \epsilon_t, \epsilon_t \sim N(0, \sigma_1^2) \tag{13}$$

$$HR_t = \alpha_1 + \alpha_2 Tc_t + v_t, \text{var}(v_t) = R. \tag{14}$$

To incorporate the nonlinear dependence between Tc and HR, Buller et al. [14, 15] further proposed a quadratic state space model, which was found to provide better fit in real data analysis:

$$Tc_t = \varphi_0 + \varphi_1 Tc_{t-1} + \epsilon_t, \epsilon_t \sim N(0, \sigma_1^2) \tag{15}$$

$$HR_t = \alpha_1 + \alpha_2 Tc_t + \alpha_3 Tc_t^2 + v_t, \text{var}(v_t) = R. \tag{16}$$

3.2. Extended Kalman filter

Our group extended the aforementioned work by proposing an EKF model in which both HR and ST are considered in the time update function and the nonlinear dependence is used in the time update function [11]. Moreover, work-rest regime-switching models were proposed to describe the different Tc dependency on HR and ST during the march (work) and the recess (rest) states. By permitting different formulations for the march and the rest time periods, we were able to harness the *a priori* knowledge of the work-rest cycles in the developmental data to enhance Tc estimates. Our EFK model is formulated as follows:

EKF:

$$\begin{aligned} \text{March (work)}: \mathrm{Tc_t} &= \varphi_0 + \varphi_1 \mathrm{Tc_{t-1}} + \varphi_2 \mathrm{HR_{t-1}} + \varphi_3 \mathrm{ST_{t-1}} + \epsilon_t, \epsilon_t \sim N(0, \sigma_1^2) \\ \mathrm{HR_t} &= \alpha_1 + \alpha_2 \mathrm{Tc_t} + \alpha_3 \mathrm{Tc_t^2} + \alpha_4 \mathrm{Tc_t^3} + v_t, \mathrm{var}(v_t) = R_1 \end{aligned} \tag{17}$$

$$\begin{aligned} \text{Recess (rest)}: \mathrm{Tc_t} &= \phi_0 + \phi_1 \mathrm{Tc_{t-1}} + \phi_2 \mathrm{HR_{t-1}} + \phi_3 \mathrm{ST_{t-1}} + \epsilon_t, \epsilon_t \sim N(0, \sigma_2^2) \\ \mathrm{HR_t} &= \beta_1 + \beta_2 \mathrm{Tc_t} + \beta_3 \mathrm{Tc_t^2} + \beta_4 \mathrm{Tc_t^3} + v_t, \mathrm{var}(v_t) = R_2 \end{aligned} \tag{18}$$

3.3. Online Kalman filter

The classical KF-type models depend on the previous forecasts of Tc, which may introduce significant uncertainty in the estimates when the forecast horizon increases and the prediction errors accumulate. To avoid concatenating forecast errors, we propose using a direct predictive model that relies on the dependence of Tc on its initial value and the latest information of the observed exogenous variables. We name this direct predictive model the online KF (OKF) model. Similar to the EFK model, the OKF model incorporated a regime-switching framework to better account for the varying dependence of Tc on the observed exogenous variables during work and rest periods. At each stage, the latest values of Tc, HR and ST are used to predict Tc:

OKF:

$$\text{March (work)}: \mathrm{Tc_t} = \varphi_{0t} + \varphi_{1t} \mathrm{Tc_0} + \varphi_{2t} \mathrm{HR_{t-1}} + \varphi_{3t} \mathrm{ST_{t-1}} + \epsilon_{1t}, \epsilon_{1t} \sim N(0, \sigma_{1t}^2) \tag{19}$$

$$\text{Recess (rest)}: \mathrm{Tc_t} = \phi_{0t} + \phi_{1t} \mathrm{Tc_0} + \phi_{2t} \mathrm{HR_{t-1}} + \phi_{3t} \mathrm{ST_{t-1}} + \epsilon_{2t}, \epsilon_{2t} \sim N(0, \sigma_{2t}^2) \tag{20}$$

The EKF and the OKF models were seeded with the actual starting Tc as measured by the ingestible thermometer capsule, with the assumption that initial Tc during real-life events could be either estimated or measured prior to the start of a physical activity.

4. Results

A total of 17,646 Tc-HR-ST data points were available for model development. The mean and range of Tc were 38.2 and [32.0, 40.1] °C, respectively. Approximately 5% of all Tc measurements were greater than or equal to 39.0°C.

4.1. Final model

For the sake of illustration, parameter estimates for the final EKF and OKF models trained using data from Study 1 (Laboratory Study) are reproduced in this paper. The EKF model is described in the equations below.

$$\begin{aligned} \text{March (work)} : Tc_t &= 0.36630 + 0.98368Tc_{t-1} + 0.00038HR_{t-1} \\ &+ 0.00586ST_{t-1} + \epsilon_t, \epsilon_t \sim N(0, 0.00051) \\ HR_t &= 6793.30385 - 673.08458Tc_t + 21.01836Tc_t^2 - 0.20822Tc_t^3 + v_t, \text{var}(v_t) = 280.36443 \end{aligned} \tag{21}$$

$$\begin{aligned} \text{Recess (rest)} : Tc_t &= 0.32403 + 0.98296Tc_{t-1} + 0.00060HR_{t-1} \\ &+ 0.00604ST_{t-1} + \epsilon_t, \epsilon_t \sim N(0, 0.00126) \\ HR_t &= 380042.09964 - 29325.80131Tc_t + 753.95823Tc_t^2 - 6.45618Tc_t^3 + v_t, \text{var}(v_t) = 292.55107. \end{aligned} \tag{22}$$

The transition functions are:

$$\begin{aligned} \text{March (work)} : Tc_t = 2.96438 + 0.72626Tc_{t-1} + 0.00804Tc_{t-1}^2 - 0.00008Tc_{t-1}^3 \\ + 0.00586ST_{t-1} + \epsilon_t, \epsilon_t \sim N(0, 0.00055), \end{aligned} \tag{23}$$

$$\begin{aligned} \text{Recess (rest)} : Tc_t = 228.97647 - 16.66092Tc_{t-1} + 0.45362Tc_{t-1}^2 \\ - 0.00388Tc_{t-1}^3 + 0.00604ST_{t-1} + \epsilon_t, \epsilon_t \sim N(0, 0.00136). \end{aligned} \tag{24}$$

The equations for the final OKF model are provided below, with different values for the four model parameters $[\varphi_{0t}, \varphi_{1t}, \varphi_{2t}, \varphi_{3t}]$ at different time points. The corresponding author may be contacted for values of these parameters.

$$\text{March (work)} : Tc_t = \varphi_{0t} + \varphi_{1t}Tc_0 + \varphi_{2t}HR_{t-1} + \varphi_{3t}ST_{t-1} + \epsilon_{1t}, \epsilon_{1t} \sim N(0, \sigma_{1t}^2) \tag{25}$$

$$\text{Recess (rest)} : Tc_t = \phi_{0t} + \phi_{1t}Tc_0 + \phi_{2t}HR_{t-1} + \phi_{3t}ST_{t-1} + \epsilon_{2t}, \epsilon_{2t} \sim N(0, \sigma_{2t}^2). \tag{26}$$

4.2. In-sample analysis

Figure 1 and **Table 1** summarise the performance of the final EKF model and the final OKF model on the study 1 data. **Figure 2** and **Table 2** summarise the performance of the final EKF model and the final OKF model on the study 2 data.

For both study 1 and study 2, the agreement between the observed and predicted Tc across the range of Tc was greater in the OKF model compared to the EKF model. For instance, under study 1, the LoA attained with the OKF model was [−0.49, 0.49]°C while that derived from the EKF model was [−0.70, 0.74]°C. For Study 2, the scatter plot of the observed versus predicted Tc departed from the line of identity markedly (observed Tc = 0.42 × predicted Tc + 22.15; units = °C) under the EKF model. By contrast, the scatter plot of the observed Tc versus the OKF model-predicted Tc for the same set of data was randomly distributed along the line of

http://dx.doi.org/10.5772/intechopen.71205

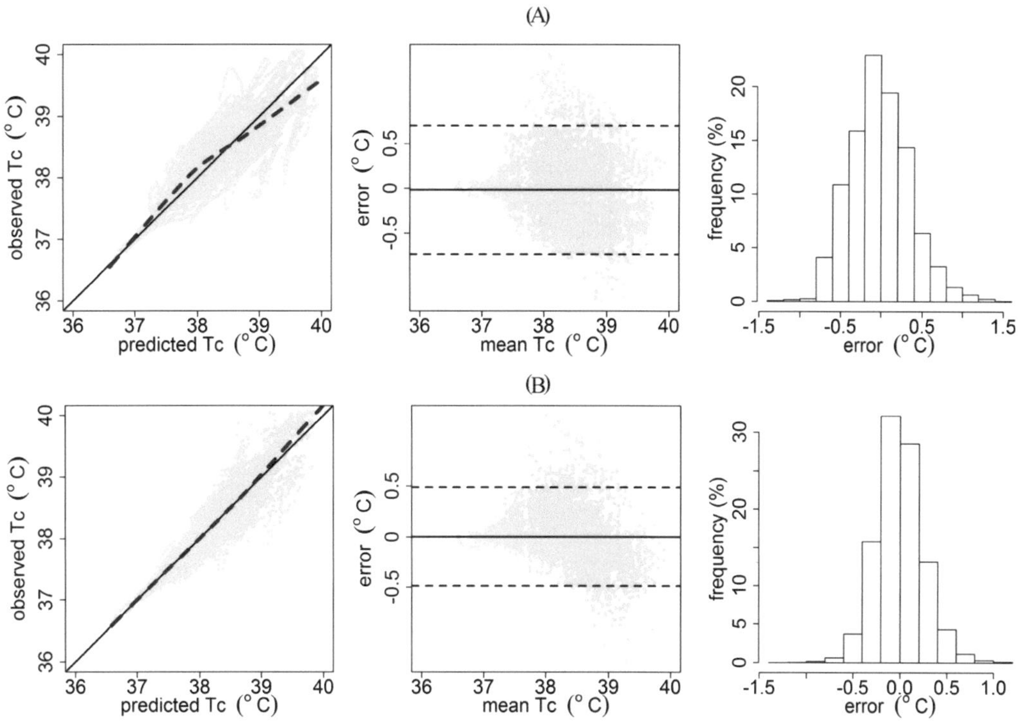

Figure 1. Diagnostic plots for assessment of the EKF (A) and OKF (B) models trained using study 1 data. For each model, the left side subplot shows the scatter plot of observed Tc versus predicted Tc together with the line of identity (black line) and the loess smooth plot (gray dashed line); the middle subplot shows the Bland-Altman plot showing bias (solid line) and ±1.96 × SD (dashed line); and the right side subplot shows the histogram of prediction error.

Model	RMSE (°C)	LoA (°C)	PTA ± 0.1°C (%)	PTA ± 0.3°C (%)	PTA ± 0.5°C (%)
EKF	0.37	0.02 ± 0.72	24	60	82
OKF	0.25	0.00 ± 0.49	40	78	95

Table 1. RMSE, LoA and PTA for the final EKF and OKF models trained using study 1 data.

identity. Combined across study 1 and study 2, the OKF model reduced the RMSE by 0.18°C. In addition, for both study 1 and study 2, the proportions of prediction errors within ±0.1, 0.3 and 0.5°C under the OKF model were also higher compared to those under the EKF model. In particular, the PTA ±0.3°C under the OKF model was 75%, which was about 25% higher compared to the PTA ±0.3°C under the EKF model. Collectively, the results indicated that the overall performance of the OKF model was superior to that of the EKF model based on the developmental data.

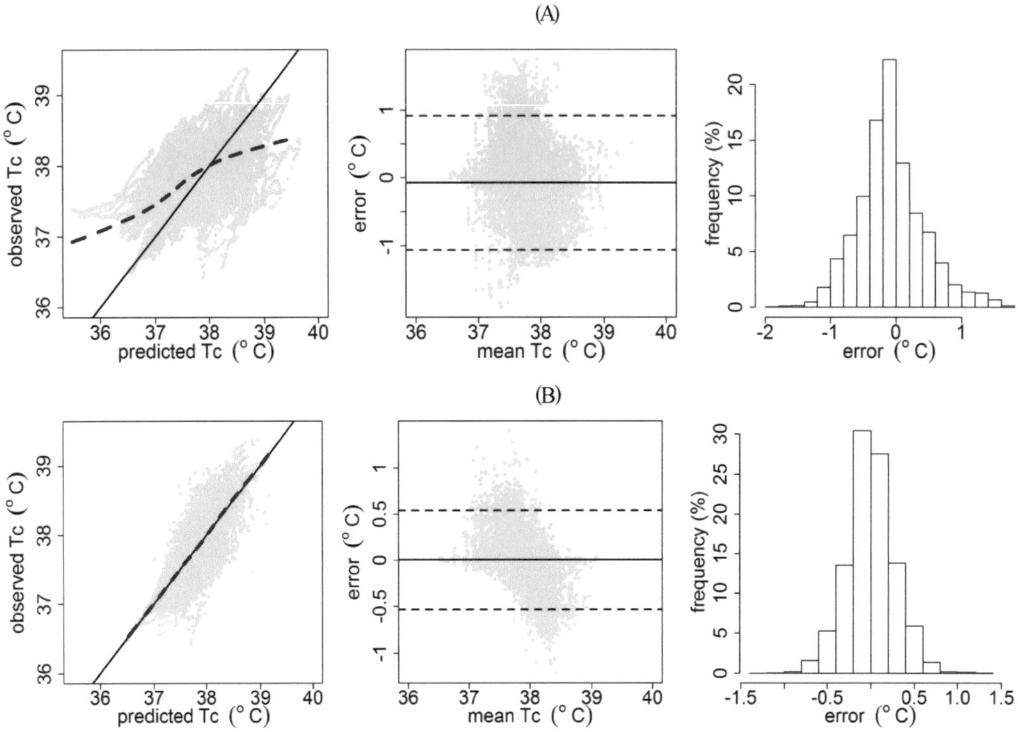

Figure 2. Diagnostic plots for assessment of the EKF (A) and OKF (B) models trained using study 2 data. For each model, the left side subplot shows the scatter plot of observed Tc versus predicted Tc together with the line of identity (black line) and the loess smooth plot (gray dashed line); the middle subplot shows the Bland–Altman plot showing bias (solid line) and ±1.96 × SD (dashed line); and the right side subplot shows the histogram of prediction error.

Model	RMSE (°C)	LoA (°C)	PTA ± 0.1°C (%)	PTA ± 0.3°C (%)	PTA ± 0.5°C (%)
EKF	0.51	0.07 ± 0.99	18	49	70
OKF	0.27	0.00 ± 0.54	33	75	92

Table 2. RMSE, LoA and PTA for the final EKF and OKF models trained using study 2 data.

4.3. Out-of-sample analysis

Tables 3–6 report the RMSE, LoA and PTA ±0.1, 0.3 and 0.5°C obtained in each of the four index sets under study 1 and study 2 based on the EKF and OKF approaches. Similar to the in-sample analysis, the comparison between the observed and predicted Tc showed a smaller RMSE and a greater agreement under the OKF model compared to the EKF model.

When averaged across all the index sets and both study 1 and study 2, the RMSE fell by 0.03°C and the PTA increased by 13% under the OKF model vis-a-vis the EKF model. In addition, the overall agreement between the observed and predicted Tc was closer under the OKF model.

Index Set	RMSE (°C)	LoA (°C)	PTA ± 0.1°C (%)	PTA ± 0.3°C (%)	PTA ± 0.5°C (%)
1	0.41	0.19 ± 0.71	33	62	72
2	0.42	0.04 ± 0.82	18	57	78
3	0.30	0.06 ± 0.58	34	69	87
4	0.37	−0.13 ± 0.68	20	53	87
Overall	0.38	0.04 ± 0.70	27	60	81

Table 3. RMSE, LoA and PTA for the four different study 1 index sets derived using the EKF model.

Index Set	RMSE (°C)	LoA (°C)	PTA ± 0.1°C (%)	PTA ± 0.3°C (%)	PTA ± 0.5°C (%)
1	0.23	−0.06 ± 0.43	40	80	98
2	0.45	0.08 ± 0.87	23	57	75
3	0.31	−0.08 ± 0.58	32	69	89
4	0.34	0.05 ± 0.65	29	74	91
Overall	0.33	0.00 ± 0.63	31	70	88

Table 4. RMSE, LoA and PTA for the four different Study 1 index sets derived using the OKF model.

Index Set	RMSE (°C)	LoA (°C)	PTA ± 0.1°C (%)	PTA ± 0.3°C (%)	PTA ± 0.5°C (%)
1	0.47	−0.02 ± 0.92	21	55	76
2	0.47	0.25 ± 0.78	17	48	70
3	0.43	−0.08 ± 0.83	23	57	80
4	0.58	0.07 ± 1.12	12	36	59
Overall	0.49	0.06 ± 0.91	18	49	71

Table 5. RMSE, LoA and PTA for the four different study 2 index sets derived using the EKF model.

Index Set	RMSE (°C)	LoA (°C)	PTA ± 0.1 °C (%)	PTA ± 0.3 °C (%)	PTA ± 0.5 °C (%)
1	0.53	0.00 ± 1.04	27	70	87
2	0.45	0.18 ± 0.81	26	61	83
3	0.39	−0.06 ± 0.75	31	68	87
4	0.56	−0.06 ± 1.09	24	65	84
Overall	0.48	0.01 ± 0.92	27	66	85

Table 6. RMSE, LoA and PTA for the four different study 2 index sets derived using the OKF model.

These trends were also evident at the index set level. Using index set 1 of study 1 dataset as an example, the RMSE under the EKF model was 0.41°C, which was larger compared to the OKF model's RMSE (0.23°C). As a further indication of the superior performance of the OKF model, the LoA under the OKF model was narrower compared to that under the EKF model [(−0.49, 0.37)°C versus (−0.52, 0.9) °C].

Figure 3 shows a comparison between the mean observed and EKF/OKF-predicted Tc time series for study 1 and study 2. The results showed that the mean Tc versus time profile generated by the OKF model largely matched that of the observed mean Tc time series. By contrast, mean Tc predictions produced from the EKF model were observed to deviate from the observed mean Tc and lie outside of the 95% confidence interval of the Tc measurements at various time periods during the foot march.

Figure 4 compares the mean error time series from the EKF and OKF models in study 1 and study 2. While the mean errors (prediction bias) were observed to be generally stable and contained to under approximately ±0.1°C across all time instances for the OKF model, those of the EKF model were comparatively larger in magnitude. In addition, the mean error from the

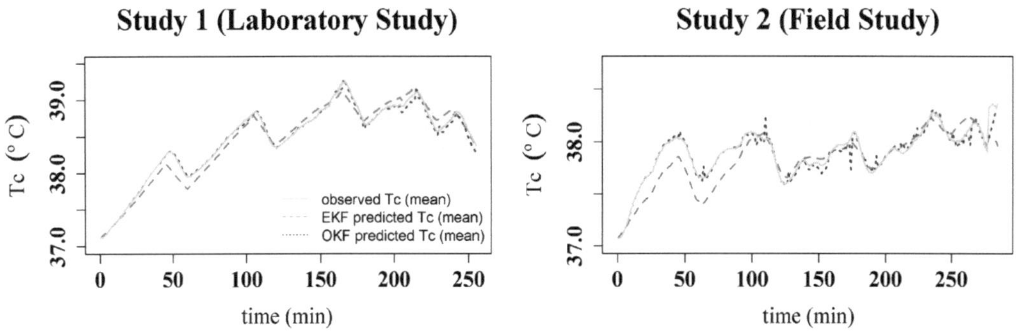

Figure 3. Comparison between the mean observed and predicted Tc time series for study 1 and study 2. For each study, the mean observed, EFK model-predicted and OKF model-predicted Tc time series are plotted in continuous dashed and dotted lines, respectively. The 95% confidence interval for the observed Tc time series is shown as a grey area.

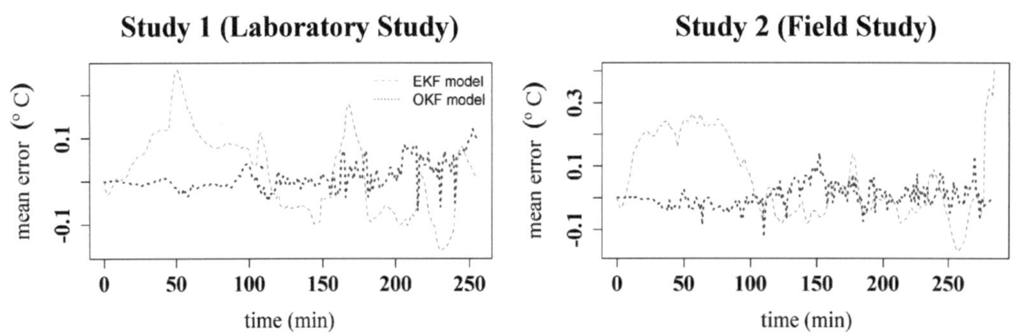

Figure 4. Comparison between the mean error time series generated from the EKF model and the OKF model for study 1 and study 2. For each study, the mean error time series produced by the EFK model and the OKF model are plotted in dashed and dotted lines, respectively.

EKF model was also observed to expand in magnitude with increasing time for both the laboratory and field datasets.

5. Discussion

In this study, the EKF and OKF models were validated against Tc measurements obtained from volunteers who participated in a high intensity foot march typically performed in the military. When pooled across study 1 and study 2, approximately 5% of all Tc measurements were equal to or greater than 39°C. This represented a respectable data volume for model assessment at the high thermal work zone. Using only measures of HR and ST, our results showed that the models estimated Tc with a small overall bias of 0.03°C, which was within the individual physiological variation of ±0.25°C [17]. In addition, the overall RMSE of the EKF and OKF models (0.31 and 0.45°C) were also comparable to those found in other comparisons of different measures of human core temperature (rectal probe versus oesophageal probe, rectal probe versus thermometer capsule and oesophageal probe versus thermometer capsule) [18].

The aforementioned observations notwithstanding, our results clearly indicated differences in the accuracy of the EKF and OKF approaches in Tc time series prediction during the studied high intensity foot march in both laboratory and outfield conditions. Classical Kalman filter strategies fundamentally rely on known model and noise information. Consequently, as depicted by results from the EKF approach, they cannot compensate for the effect of model-process mismatch and concatenating noise uncertainty. Our results showed that the OKF approach can estimate Tc continuously across time with less error than EKF model. Moreover, prediction bias arising from the OKF model appeared to be more stable in magnitude over time compared to that of the EKF model. This is significant in the practical settings because a progressively larger prediction error under a longer forecast horizon may lead to more false positives or false negatives for high thermal work strain. If the EKF model is deployed for tracking individualised heat strain, healthy workers with no imminent heat injury risk may be withdrawn from the physical activity prematurely (reducing work efficiency and processes) or actual heat casualties may not be identified, with the second scenario (false negatives) a more problematic one compared to the first (false positives). This makes the OKF method a more promising approach than the EKF method for predicting Tc based on real-time wearable sensor data in a continuous manner.

Technologies that reliably assess Tc in a non-invasive manner are expected to play a crucial role in supporting the development of tools, methods and techniques to enhance productivity, safety and well-being of military, first responders and industrial workers. During military training and operations, real-time monitoring of Tc can allow each soldier's thermo-physiological state to be assessed, which permits commanders to take effective measures to intervene and mitigate heat injuries. Monitoring of Tc of every firefighter in the fireground can provide objective information to either empower the trooper to stay in longer to finish a job or warn the trooper to exit the fireground sooner. In addition, the use of physiological monitoring, coupled with work physiology and ergonomics concepts, can foster the creation of innovative workforce management procedures allowing enhancements not only in productivity, but also in civilian workers' well-being and safety.

The main limitation of the present study is the usage of only Tc measurements from the military foot march for modelling. Such developmental data may limit the model's ability to reliably calculate Tc of human subjects in non-military tasks, e.g. first responders operating in uncompensable heat stress environments, civilian construction workers and professional sports athletes geared with light clothing. In the future, the strong influence of ST on Tc in our mathematical model will be verified in human subjects operating in clothing systems that either severely limit heat dissipation or facilitate sweat evaporation under less humid conditions. The current study did not assess the reliability of the model on repeated Tc measures derived on different trial occasions. Future work will include testing our Kalman filter model's reliability and precision on different test occasions based on repeated measures data from the same subjects. Last, while we showed that the OKF approach can estimate Tc with less error than the EKF model, appreciable variability in the Tc still remains unexplained by HR and ST. Future work will include the evaluation of breathing rate to improve Tc estimations since hyperthermia has been shown to increase ventilation [19].

6. Conclusions

In this paper, we have reported two different Kalman filter approaches for predicting real-time Tc trajectories of subjects engaged in a high intensity physical activity. In particular, we introduced the OKF model where the time update equation depends only on the initial value of Tc and time-current values of the exogenous variables. Both models leverage time-varying values of ST and HR to predict subject-specific Tc. Overall, Tc predictions from the OKF model matched the observed Tc better compared to those from the EFK models. Future work includes testing and qualification of our model against additional heat strain datasets including those derived from non-foot march tasks, and investigation of the influence of further exogenous observations, such as body acceleration, on Tc. While this approach may not be a complete replacement for direct Tc measurement, it offers a simple and promising new method to estimate subject-specific Tc in a non-invasive manner, and is accurate and practical enough for real-time monitoring of thermal work strain.

Acknowledgements

This study was funded by the Ministry of Defence in Singapore. We thank all volunteers who participated in this study. The authors are also grateful to Adam Chai and Leonard Chan for suggestions on model development.

Author details

Jia Guo[1], Ying Chen[1], Weiping Priscilla Fan[2], Si Hui Maureen Lee[2], Junxian Ong[2], Poh Ling Tan[2], Yu Li Lydia Law[2], Kai Wei Jason Lee[2,3] and Kok-Yong Seng[2,4]*

*Address all correspondence to: skokyong@dso.org.sg

1 Department of Statistics and Applied Probability, Faculty of Science, National University of Singapore, Republic of Singapore

2 Defence Medical and Environmental Research Institute, DSO National Laboratories, Republic of Singapore

3 Department of Physiology, Yong Loo Lin School of Medicine, National University of Singapore, Republic of Singapore

4 Department of Pharmacology, Yong Loo Lin School of Medicine, National University of Singapore, Republic of Singapore

References

[1] Update: Heat injuries, active component, U.S. Armed Forces, 2012. MSMR. 2013;**20**(3): 17-20

[2] Fahy RF, PR LB, Molis JL, editors. Firefighter Fatalities in the United States-2015. National Fire Protection Association. Quincy, MA; 2016. p. 33

[3] Cadarette BS, Levine L, Staab JE, Kolka MA, Correa M, Whipple M, Sawka MN. Heat strain imposed by toxic agent protective systems. Aviation, Space, and Environmental Medicine. 2001;**72**(1):32-37

[4] Montain SJ, Sawka MN, Cadarette BS, Quigley MD, McKay JM. Physiological tolerance to uncompensable heat stress: effects of exercise intensity, protective clothing, and climate. Journal of Applied Physiology. 1994;**77**(1):216-222

[5] Gunga HC, Werner A, Stahn A, Steinach M, Schlabs T, Koralewski E, Kunz D, Belavý DL, Felsenberg D, Sattler F, Koch J. The double sensor–A non-invasive device to continuously monitor core temperature in humans on earth and in space. Respiratory Physiology and Neurobiology. 2009;**169**(Suppl 1):S63-S68. DOI: 10.1016/j.resp.2009.04.005

[6] Coughlin SS, Stewart J. Use of consumer wearable devices to promote physical activity: A review of health intervention studies. Journal of Environment and Health Sciences. 2016;**2**(6). DOI: 10.15436/2378-6841.16.1123

[7] Fick AV. On liquid diffusion. Philosophical Magazine Series 4. 1855;**10**(63):30-39

[8] Richmond VL, Davey S, Griggs K, Havenith G. Prediction of core body temperature from multiple variables. The Annals of Occupational Hygiene. 2015;**59**(9):1168-1178. DOI: 10.1093/annhyg/mev054

[9] Yamakage M, Namiki A. Deep temperature monitoring using a zero-heat-flow method. Journal of Anesthesia. 2003;**17**(2):108-115

[10] Yokota M, Moran DS, Berglund LG, Stephenson LA, Kolka MA. Noninvasive warning indicator of the 'Red Zone' of potential thermal injury and performance impairment: A pilot study. In: International Conference of Environmental Ergonomics; Lund University, Sweden. 2005. p. 514-517

[11] Seng KY, Chen Y, Chai AKM, Wang T, Fun DCY, Teo YS, Tan PMS, Ang WH, Lee JKW. Tracking body core temperature in military thermal environments: An extended Kalman filter approach. In: 2016 IEEE 13th International Conference on Wearable and Implantable Body Sensor Networks (BSN). San Francisco: IEEE; 2016. p. 296-299. DOI: 10.1109/BSN.2016.7516277

[12] Kalman RE. A new approach to linear filtering and prediction problems. ASME Journal of Basic Engineering. 1960;**82**(Series D):35-45

[13] Buller MJ, Tharion WJ, Hoyt RW, Jenkins OC. Estimation of human internal temperature from wearable physiological sensors. In: Proceedings of the 22nd Conference on Innovative Applications of Artificial Intelligence (IAAI); Atlanta, GA. July 11–15: IAAI;2010: 1763-1768

[14] Buller MJ, Tharion WJ, Cheuvront SN, Montain SJ, Kenefick RW, Castellani J, Latzka WA, Roberts WS, Richter M, Jenkins OC, Hoyt RW. Estimation of human core temperature from sequential heart rate observations. Physiological Measurement. 2013;**34**(7):781-798. DOI: 10.1088/0967-3334/34/7/781

[15] Buller MJ, Tharion WJ, Duhamel CM, Yokota M. Real-time core body temperature estimation from heart rate for first responders wearing different levels of personal protective equipment. Ergonomics. 2015;**58**(11):1830-1841. DOI: 10.1080/00140139.2015.1036792

[16] Bland JM, Altman DG. Statistical methods for assessing agreement between two methods of clinical measurements. Lancet. 1986;**1**(8476):307-310

[17] Consolazio CF, Johnson RE, Pecora LJ. Physiological variability in young men. In: Consolazio CF, Johnson RE, Pecora LJ, editors. Physiological Measurements of Metabolic Functions. New York: McGraw Hill; 1963. p. 453-480

[18] O'Brien C, Hoyt RW, Buller MJ, Castellani JW, Young AJ. Telemetry pill measurement of core temperature in humans during active heating and cooling. Medicine and Science in Sports and Exercise. 1998;**30**(3):468-472. DOI: 10.1097/00005768-199803000-00020

[19] Fujii N, Honda Y, Ogawa T, Tsuji B, Kondo N, Koga S, Nishiyasu T. Short-term exercise-heat acclimation enhances skin vasodilation but not hyperthermic hyperpnea in humans exercising in a hot environment. European Journal of Applied Physiology. 2012;**112**(1): 295-307. DOI: 10.1007/s00421-011-1980-6

Chapter 7

Application of Kalman Filtering in Dynamic Prediction for Corporate Financial Distress

Qian Zhuang

Additional information is available at the end of the chapter

http://dx.doi.org/10.5772/intechopen.71616

Abstract

This chapter aims to dynamically improve the method of predicting financial distress based on Kalman filtering. Financial distress prediction (FDP) is an important study area of corporate finance. The widely used discriminant models currently for financial distress prediction have deficiencies in dynamics. Based on the state-space method, we establish two models that are used to describe the dynamic process and discriminant rules of financial distress, respectively, that is, a process model and a discriminant model. These two models collectively are called dynamic prediction models for financial distress. The operation of the dynamic prediction is achieved by Kalman filtering algorithm, and further, a general n-step-ahead prediction algorithm based on Kalman filtering is derived for prospective prediction. We also conduct an empirical study for China's manufacturing industry, and the results have proved the accuracy and advance of predicting financial distress in such case.

Keywords: financial distress prediction, pattern recognition, state space model, stochastic process, optimal estimation

1. Introduction

Research on financial distress prediction (FDP) is an important area of corporate finance. Early prediction methods are univariate analysis (UA), multiple discriminant analysis (MDA), logistic model, probit model, and so on [1–5]. With the development of computer technology, some new methods based on artificial intelligence technology with distributed computing capabilities that can deal with problems of nonlinear systems are widely introduced into the field of financial distress prediction. These methods include neural network (NN), genetic algorithm (GA), rough set theory (RST), case-based reasoning (CBR) and support vector machine (SVM), and so on [6–14]. Each model established for financial distress prediction, whether based on statistical methods or artificial intelligence methods, has advantages and disadvantages under different conditions. Let us take the most widely used multiple discriminant analysis (MDA) and back-propagation neural network (BPNN) for example. MDA has the advantage of simplicity and good interpretation, but the deficiency in its application is limited by strict assumptions that sometimes cannot be satisfied. Besides, MDA is a static discriminant model [2, 3, 6, 15, 20]. For the application of BPNN, it does not need any probability distribution assumption. BPNN is considered as an effective tool of pattern recognition for nonlinear systems. Therefore, many researchers have tried to apply triple BPNN in financial distress prediction, using the nonlinear pattern recognition capability of BPNN for classification of different financial state [7, 8, 15].

The prediction often achieved through a cross-sectional analysis at different time points. That is, the sample data of period $t-1$, $t-2$, ... before financial distress are studied by BPNN, respectively, and the features are extracted, based on what the judgment for the financial state of next new period is made [16, 17, 20]. This treatment is a relatively complete cross-sectional analysis. But the conclusions on discrimination among different time points are lack of logistic links. Therefore, this prediction is not completely dynamic. Furthermore, BPNN is a static neural network even when directly used in time-series prediction (Neural networks can be divided into static or dynamic neural networks based on whether they contain feedback loops or delay. BPNN is a backpropagation network without feedback and belongs to static neural networks.). There are some inherent problems such as overfitting, for example, the fitting error of training data has reduced, but the prediction error has increased at the same time. Even if the data are normalized, the effect is not satisfactory when the testing data are not sufficient [13, 18–20].

Actually, corporate financial distress is a gradual and cumulative process, which is developed from a healthy state. The mutation is often the result at which the gradual change and cumulation have reached the critical condition. It is neither reasonable nor logical if only the cross-sectional data at the time point prior to the occurrence of financial distress are used to make a determination for the corporate future state. It should take into account two aspects at least when conducting the research on financial distress prediction: firstly, the alternative data for prediction should contain all the historical information; secondly, the prediction method is dynamic designed for financial distress characterized by cumulative variation [21–23]. However, the current discriminant models have some deficiencies in dynamic prediction. Also, there is a problem of massive data processing. This chapter attempts to make a dynamic improvement on prediction methods for financial distress based on Kalman filtering algorithm in order to solve the above problems.

The main contribution of the paper is that it constructs a state-space model of corporate financial distress from the perspective of the cumulative effect of historical information on current state and improves Kalman filtering algorithm for dynamic prediction. A whole process of dynamic prediction for corporate financial distress is developed from a long period of time, and time-series data of high-frequency are collected for optimal estimation of financial state, which is seen as a stochastic process. The advantage of the model is proved by an empirical research, and the result shows that it can give relatively accurate warning before the occurrence of financial distress.

The rest of this chapter is organized as follows. Dynamic prediction models consisting of a process model and a discriminant model based on Kalman filtering algorithm are described in

Section 2. Then, a whole process of dynamic prediction for corporate financial distress is elaborated in Section 3. Section 4 presents empirical analysis for China's manufacturing industry. Section 5 draws conclusions and discusses future research.

2. Dynamic prediction models based on Kalman filtering algorithm

Based on the state-space method, we establish two models, being used to describe the dynamic process and discriminant rules of financial distress, respectively, that is, a process model and a discriminant model. These two models collectively are called dynamic prediction models for financial distress. We see the evolution of financial distress for a company as a stochastic process and establish a process model, which is used to describe the dynamic process of development of the financial state. We define the financial state as a set of vectors, which summarizes all the information necessary about the past behavior of the company except for the external effects of the inputs, so that it can uniquely describe the behavior of the company in the near future [24]. The financial state of a company often cannot be observed directly, but only some signal indicators associated with the financial state can be observed. Therefore, we establish a discriminant model, which is used to describe the correlation between the financial state and the signal indicators. The discriminant model can be a recursive form of any statistical model or artificial intelligence model, theoretically. At first, we take the linear models, which are simple and intuitive as an example and establish dynamic prediction models for financial distress, as

$$X_t = A_{t|t-1}X_{t-1} + W_{t-1} \tag{1}$$

$$Z_t = H_tX_t + V_t \tag{2}$$

where X_t is the financial state of a company in period t; Z_t is the signal indicators of the company in period t; $W_{t\text{-}1}$ is the process noise of the financial state in period $t-1$; V_t is the observation noise of the signal indicators in period t; $A_{t|t-1}$ is used to describe the dynamic process of the financial state transferring from period $t-1$ to t; H_t is used to describe the mathematical relations between the financial state and the signal indicators in period t. Eq. (1) is a process model; and Eq. (2) is a discriminant model.

Assume that the process noise and the observation noise are Gaussian white noises, which are mutually independent and normally distributed, i.e.

$$\begin{cases} E[W_t] = 0, \ E\left[W_tW_j^T\right] = Q_t\delta_{tj} \\ E[V_t] = 0, \ E\left[V_tV_j^T\right] = R_t\delta_{tj} \\ E\left[W_tV_j^T\right] = 0 \end{cases} \tag{3}$$

where, Q_t is a $p \times p$-dimensional symmetric nonnegative definite covariance matrix of process noise W_t; R_t is a $m \times m$-dimensional symmetric positive definite covariance matrix of observation noise V_t; δ_{tj} is Kronecker - δ function.

The above equations can be solved by Kalman filtering algorithm. The Kalman filter is named after Rudolph E. Kalman, who in 1960 published his famous paper describing a recursive solution to the discrete-data linear filtering problem. The Kalman filter is essentially a set of mathematical equations that implement a predictor–corrector type estimator that is optimal in the sense that it minimizes the estimated error covariance, when some presumed conditions are met [25, 26]. Kalman filter is widely used for its relative simplicity and robust nature. Rarely do the conditions necessary for optimality actually exist, and yet, the filter apparently works well for many applications in spite of this situation. Application of Kalman filter in dynamic prediction for corporate financial state consists of five steps [27, 28]:

The first step is to compute the one-step prediction of the financial state $\widehat{X}_{t|t-1}$ under the conditions of known $\widehat{X}_{t-1|t-1}$, which is the optimal estimation of the financial state at time $t-1$

$$\widehat{X}_{t|t-1} = A_{t|t-1}\widehat{X}_{t-1|t-1} \tag{4}$$

The second step is to compute the error covariance matrix $P_{t|t-1}$ for one-step prediction

$$P_{t|t-1} = A_{t|t-1}P_{t-1|t-1}A^T_{t|t-1} + Q_{t-1} \tag{5}$$

The third step is to compute the Kalman gain K_t, which is a blending factor that is used to adjust the discrepancy between the predicted observation $H_t\widehat{X}_{t|t-1}$ and the actual observation Z_t, in order to obtain the optimal estimation $\widehat{X}_{t|t}$ closer to the actual financial state

$$K_t = P_{t|t-1}H_t^T\left[H_tP_{t|t-1}H_t^T + R_t\right]^{-1} \tag{6}$$

The fourth step is to correct the one-step predicted financial state $\widehat{X}_{t|t-1}$ according to the principle of minimum error covariance and thus obtain the optimal estimation $\widehat{X}_{t|t}$ of the financial state

$$\widehat{X}_{t|t} = \widehat{X}_{t|t-1} + K_t\left[Z_t - H_t\widehat{X}_{t|t-1}\right] \tag{7}$$

The fifth step is to compute the error covariance matrix $P_{t|t}$ of the updated financial state estimation $\widehat{X}_{t|t}$

$$P_{t|t} = [I - K_tH_t]P_{t|t-1} \tag{8}$$

These are the basic equations of Kalman filtering for a stochastic linear discrete financial system. The actual filtering process is an ongoing "predicting-correcting" process of a recursive nature. **Figure 1** below offers a complete picture of the operation of the Kalman filter in dynamic prediction for corporate financial state.

The Kalman filter does not require storing large amount of data in solving the problem. Once new data are observed, new filtering value can be calculated at any time. Therefore, this method facilitates real-time processing and is easy to implement on the computer.

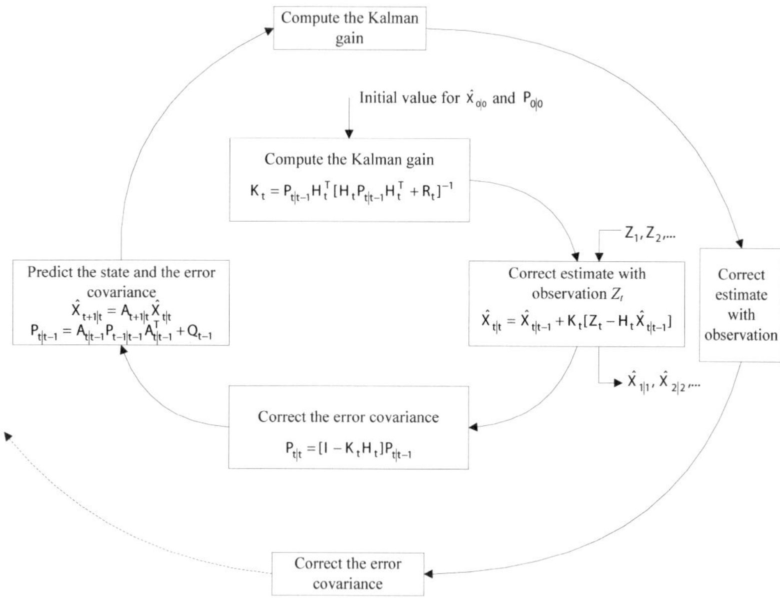

Figure 1. A complete picture of the operation of the Kalman filter in dynamic prediction for corporate financial state.

3. A whole process of dynamic prediction for corporate financial distress

As previously described, corporate financial distress is a gradual and cumulative process, which is developed from a healthy state, and so the prediction should be long-term and continuous and the continuously updated time-series data should be collected for the dynamic prediction, which could be the fresh input into the Kalman filter in order to obtain the optimal estimation closer to the actual state. The whole process of dynamic prediction for corporate financial distress is described as follows.

From **Figure 2**, we can see that if we want to predict the corporate financial state at time $t + 2$, we just need to know the optimal estimation of the corporate financial state at time t and the signal indicators observed at time $t + 1$. The rest may be deduced by analogy; if we want to predict the corporate financial state at time $t + \mathrm{n}$, we just need to know the optimal estimation of the corporate financial state at time $t + \mathrm{n}-2$ and the signal indicators observed at time $t + \mathrm{n}-1$. This continuous prediction does not require saving the observed data in the past. Every time the new signal indicators are observed, they are put into the Kalman filter as fresh. It helps solve the problems of storing, calling, and processing the massive data and thus greatly improving the speed of operation on the computer.

Further, if we want to predict the corporate financial state n-step ahead, we can obtain the n-step-ahead prediction algorithm derived from the basic Kalman filtering algorithm according to the dynamic prediction process described above.

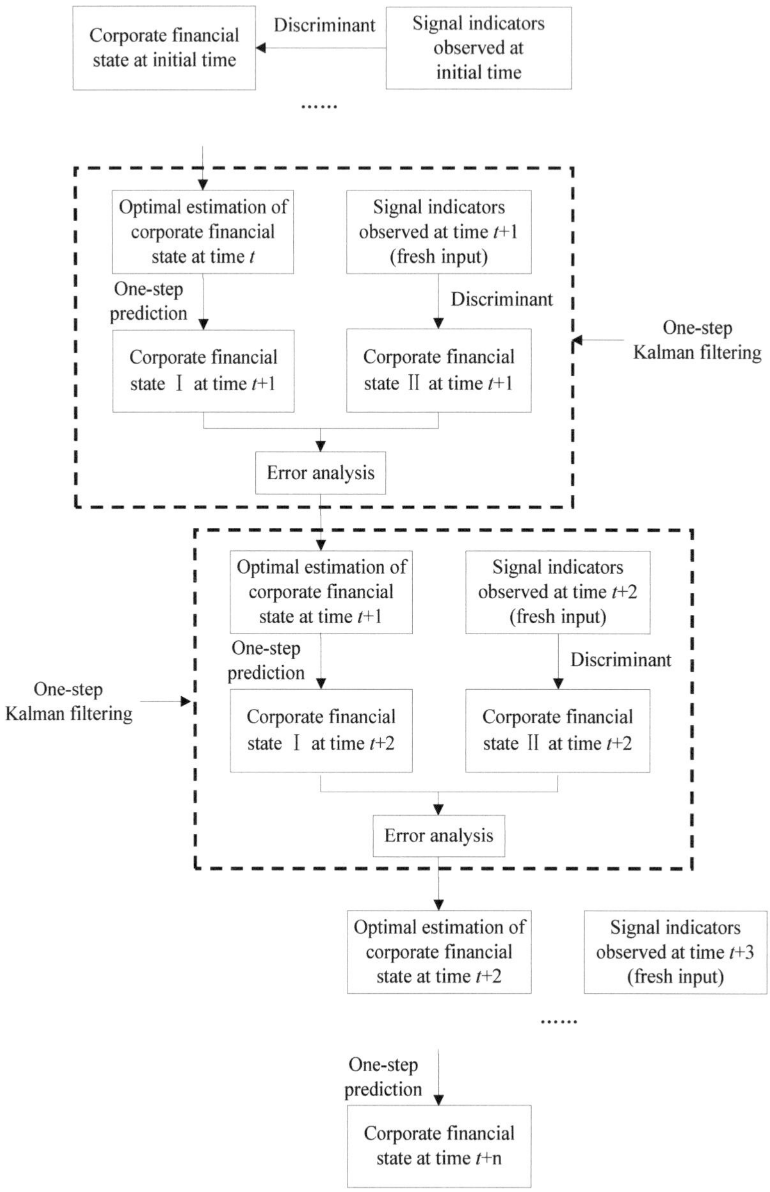

Figure 2. A whole process of dynamic prediction for corporate financial distress.

The general n-step-ahead prediction algorithm is derived as

$$\widehat{X}_{t+n|t} = A_{t+n|t+n-1} A_{t+n-1|t+n-2} \cdots A_{t+1|t} \widehat{X}_{t|t} = \prod_{i=1}^{n} A_{t+i|t+i-1} \widehat{X}_{t|t} \tag{9}$$

The n-step-ahead prediction error variance matrix is

$$P_{t+n|t} = \prod_{i=1}^{n} A_{t+i|t+i-1} P_{t|t} \prod_{i=1}^{n} A_{t+i|t+i-1}^{T} + \prod_{j=2}^{n} A_{t+j|t+j-1} Q_t \prod_{j=1}^{n} A_{t+j|t+j-1}^{T} + \cdots$$
$$+ A_{t+n|t+n-1} Q_{t+n-2} A_{t+n|t+n-1}^{T} + Q_{t+n-1} \tag{10}$$

Assume that the system parameters A_t and Q_t have nothing to do with the time, then the above equations simplify to

$$\widehat{X}_{t+n|t} = A^n \widehat{X}_{t|t} \tag{11}$$

$$P_{t+n|t} = A^n P_{t|t} \left(A^T\right)^n + A^{n-1} Q \left(A^T\right)^{n-1} + \cdots + AQA^T + Q = A^n P_{t|t} \left(A^T\right)^n$$
$$+ \sum_{j=2}^{n+1} A^{n+1-j} Q \left(A^T\right)^{n+1-j} \tag{12}$$

Based on the Eqs. (9)–(12), we could use data at shorter time interval to predict n-step ahead, but the prerequisite of sufficiently long-term data to find out the trend of development of the financial state should be satisfied.

In the dynamic prediction models for financial distress established in Section 2, we suppose that the financial state X cannot be observed. But in reality, whether the company is trapped in financial distress at time t-1 and before can be known at time t, that is part of X can be observed. We put this part of observed information into a likelihood equation in order to improve the accuracy of dynamic prediction. The probability of the company being trapped in financial distress is

$$P(X_t > S_c) = \int_S^{\infty} p(x_t) dx_t \tag{13}$$

where, S_c is the critical value; x_{t+1} equals $e_{t+1} = x_{t+1} - \widehat{x}_{t+1|t}$, and the latter is normally distributed with mean of 0 and variance of $F_{t+1} = H_{t+1} P_{t+1|t} H_{t+1}^T + R_{t+1}$. Then

$$p(x_t) = \frac{1}{\sqrt{2\pi P_t}} e^{-\frac{1}{2}\left(\frac{x_t - \hat{x}_t}{P_t}\right)^2} \tag{14}$$

If M is the last year that X can be observed, then the additional estimation equation is

$$l = -\frac{NM}{2} \lg(2\pi) - \frac{1}{2} \sum_{t=1}^{M} \lg|F_t| - \frac{1}{2} \sum_{t=1}^{M} e_t^T F e_t + \sum_{t=1}^{M} \{\lg[P(X_t > S_c)]\delta(t) + \lg[P(X_t < S_c)][1 - \delta(t)]\} \tag{15}$$

This additional estimation equation is used every year, no matter if 1 year is divided into n periods. If standing at managers' position and suppose we know the corporate financial state at time $t-1$, when we stay at time t, and then the additional estimation equation can be embedded in the general n-step-ahead prediction algorithm every time to help improve the accuracy of dynamic prediction.

4. Empirical analysis

4.1. Data description and experiment design

Manufacturing industry is a major industry in China. "Made in China" has an important impact on the global economy. Therefore, prediction of corporate financial distress for China's manufacturing industry is of great significance. Generally, the manufacturing companies have complete production processes, equilibrious production cycle, as well as a more stable trend of development of the financial state. The characteristics of these companies can be well described using the existing financial indicators, and the dynamic prediction method described above can be put into practice for these manufacturing companies.

In this research, the data for our experiment are collected from the Shanghai Stock Exchange and Shenzhen Stock Exchange databases in China. ST (special treatment) companies because of financial problems are selected as distress samples; meanwhile, companies of similar asset size that have never been special treated are selected as healthy samples. The ST time is treated as period *T*. For a 6-month interval, the data 8 years or 16 periods before ST are selected as time-series sets for the distress samples. The time span of the paired samples is the same as the distress samples.

According to the above principles, the data of 152 listed companies are collected, and the time span is year 2002 to year 2009, year 2003 to year 2010, year 2004 to year 2011, respectively. A total of 60 ST companies and 60 paired companies of the first half of year 2010 and 2011 are treated as training set, which is used to derive the model. A total of 16 ST companies and 16 paired companies of the first half of year 2012 are treated as testing set, which is used to test the effect of the model.

From the holistic perspective, we select 29 financial indicators covering four aspects of profitability, solvency, management efficiency, and market reaction as alternative signal indicators. The effect of the corporate financial problems may be amplified or reduced in information transmission mechanism of the market, and the problems may be exposed to the open market in advance or with a delay. If the problems are exposed in advance, the indicators can be used as a pilot signal of financial distress prediction; if delayed exposure, it can also be served as comprehensive evaluation of financial distress or the signal for the trend of development in the future. These are indicators of market reaction. The 29 signal indicators are listed in **Table 1**.

A three-dimensional database is established consisting of 16 periods' time-series data of the above 152 sample enterprises, the financial state of which is represented by 27 signal indicators each (As operating profit margin growth (Z_5) and interest coverage ratio (Z_{12}) have much missing data, we ignore these two subsets of the data, leaving the rest 27 subsets.). The dynamic prediction method described above is based on the trend of the time-series data. The centralized tendency of signal indicators of profitability, solvency, management efficiency, and market reaction is shown in **Figures 3–6** (Some indicators of management efficiency and market reaction show cyclical fluctuations, so we amend these indicators by smoothing. **Figures 5** and **6** have been amended.).

From **Figures 3–6**, we can see most indicators show a certain trend, which is the foundation of dynamic prediction.

http://dx.doi.org/10.5772/intechopen.71616

Type	Code	Signal indicators
Profitability	Z_1	Operating profit margin
	Z_2	Net profit margin
	Z_3	Return on assets
	Z_4	Return on equity
	Z_5	Operating profit margin growth
	Z_6	Operating revenue growth
	Z_7	Total assets growth
Solvency	Z_8	Current ratio
	Z_9	Quick ratio
	Z_{10}	Cash debt ratio
	Z_{11}	Debt coverage ratio
	Z_{12}	Interest coverage ratio
	Z_{13}	Liabilities to assets ratio
	Z_{14}	Liabilities to equity ratio
Management efficiency	Z_{15}	Total assets turnover
	Z_{16}	Fixed asset turnover
	Z_{17}	Current assets turnover
	Z_{18}	Inventory turnover
	Z_{19}	Accounts receivable turnover
	Z_{20}	Cash ratio of main business
	Z_{21}	Cash return on assets
Market reaction	Z_{22}	Earnings per share
	Z_{23}	Net assets per share
	Z_{24}	Operating revenue per share
	Z_{25}	Capital reserve per share
	Z_{26}	Retained earnings per share
	Z_{27}	Price to book ratio
	Z_{28}	Equity to invested capital ratio
	Z_{29}	Net cash flow per share

Table 1. Comprehensive signal indicators of financial distress prediction.

Then, we use nonparametric test of Mann-Whitney U to find out when the difference between distress samples and healthy samples occurs. The results show that the gap between the two is maximized 2 years before ST time and the significant difference occurs 4 years before ST time, that is, the earliest time to accurately predict the occurrence of financial distress should be 4 years before ST time.

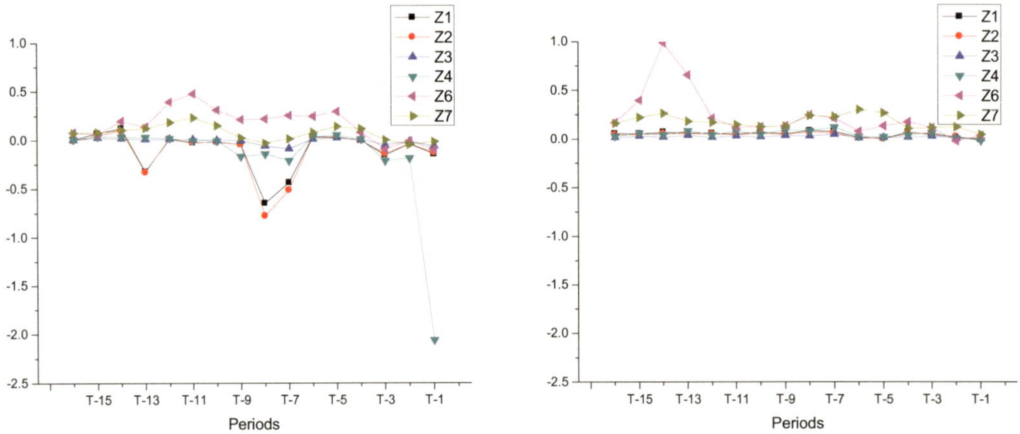

Figure 3. The centralized tendency of signal indicators of profitability for distress samples and healthy samples.

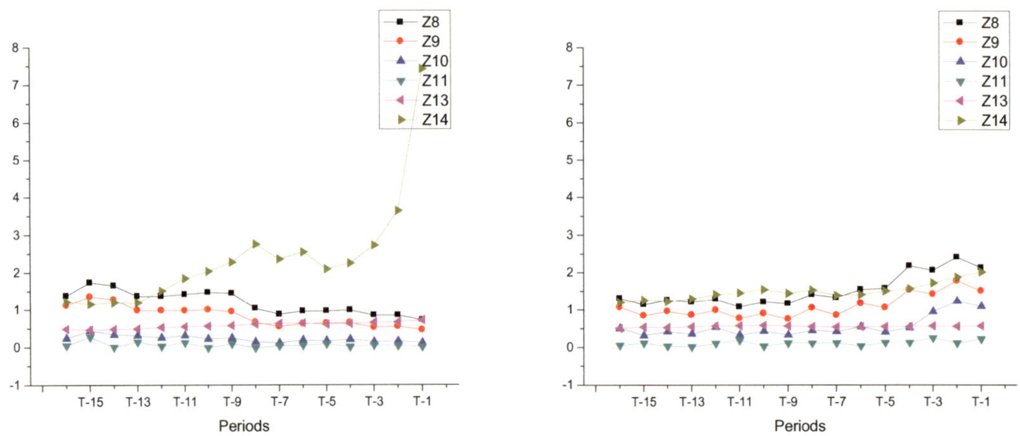

Figure 4. The centralized tendency of signal indicators of solvency for distress samples and healthy samples.

4.2. Experiment results and analysis

We use principal component analysis to eliminate the effect of multicollinearity on the original variables. We extract first 10 principal components, and the accumulative contribution rate is above 92% each for 152 companies. These principal components are linear combinations of the original signal indicators, which can be served as part of discriminant models for each company.

The parameters of process model are estimated from the data of training set and also using the data of training set, the judgment for the threshold of financial distress is set as an interval, which has lower and upper confidence limit.

The results show that the lower confidence limit is -0.796 and the upper confidence limit is 0.205. When the predictive value of a company's financial state is lower than -0.796, the

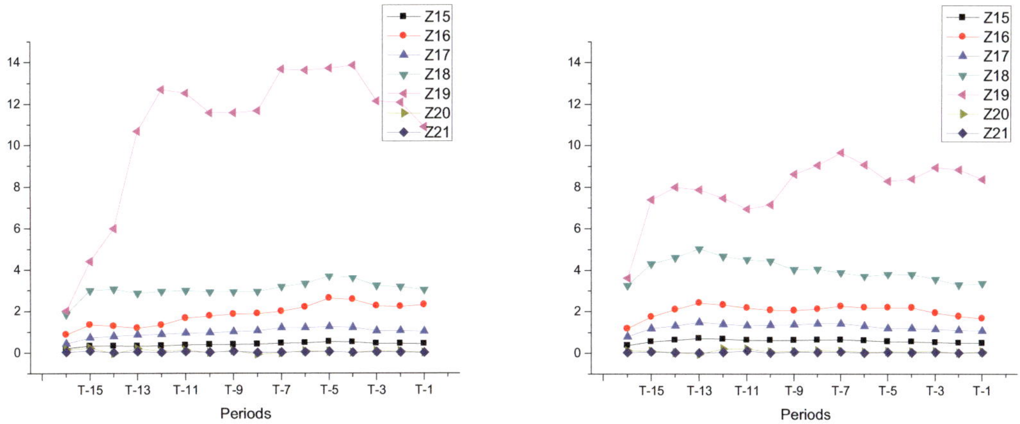

Figure 5. The centralized tendency of signal indicators of management efficiency for distress samples and healthy samples.

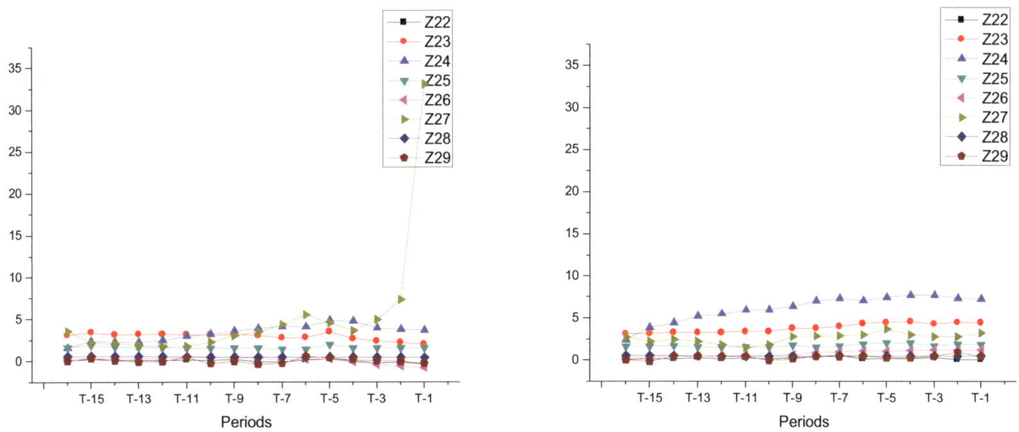

Figure 6. The centralized tendency of signal indicators of market reaction for distress samples and healthy samples.

company may fall into severe financial distress; when the predictive value is higher than 0.205, the company is well operated; and when the predictive value is between −0.796 and 0.205, it is possible that the company is getting into financial distress.

Then, we test the effect of the dynamic prediction models using the data of testing set. Subject to space restrictions, we just list dynamic prediction figures for six companies, among which first three are ST companies, while the other three are non-ST companies. Names and stock codes of the companies are Sichuan Chemical Company Limited (000155), MCC Meili Paper Industry Co., Ltd. (000815), Guangzhou Guangri Stock Co., Ltd. (600894), Xinxiang Chemical Fiber Co., Ltd. (000949), Nantong Jiangshan Agrochemical & Chemicals Co., Ltd. (600389), Nanzhi Co., Ltd., and Fujian (600163), in turn. Dynamic prediction figures for these six companies are shown in **Figure 7**.

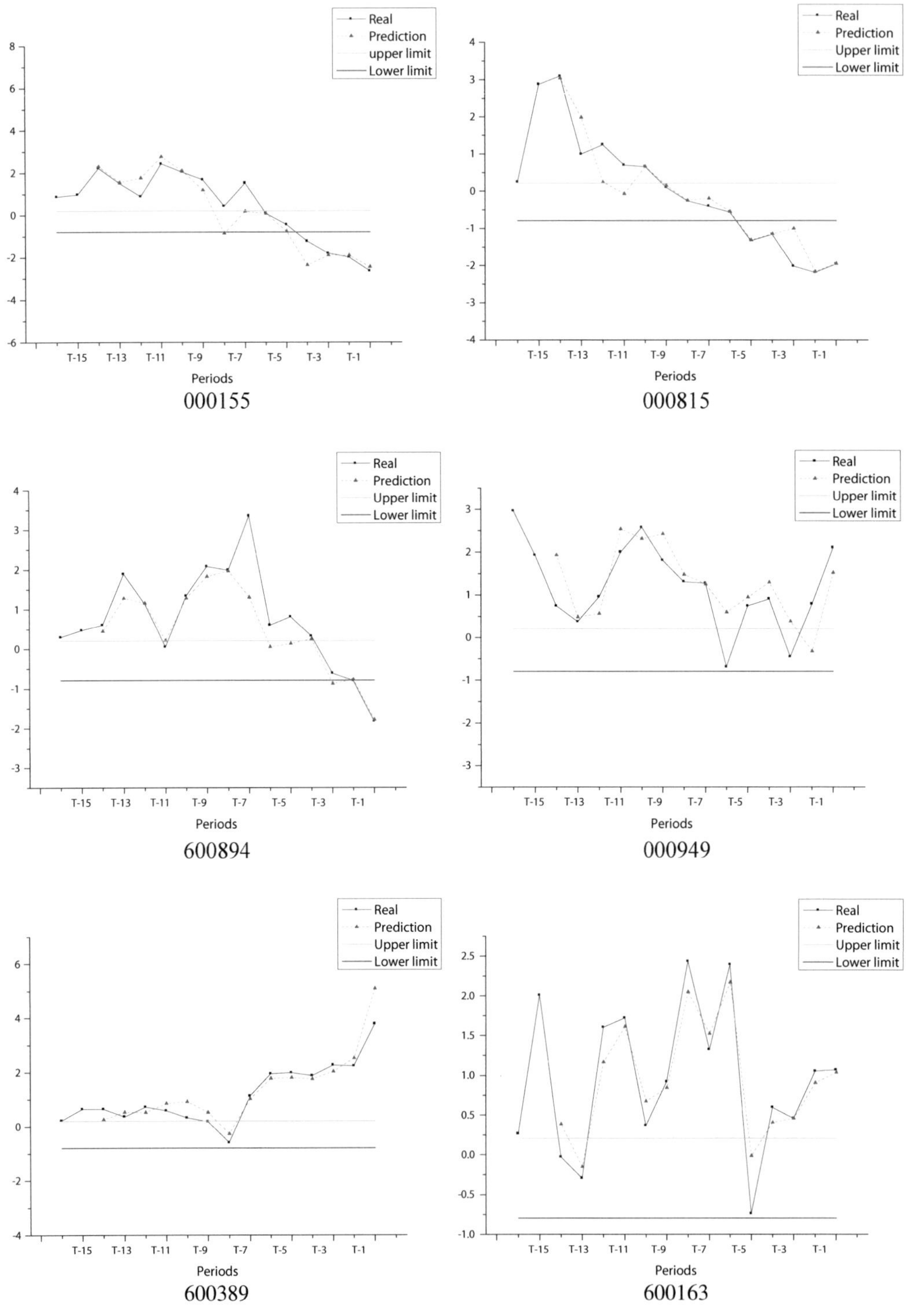

Figure 7. Dynamic prediction figures for part of testing samples.

The testing results show that almost all the curves of predictive value fits the ones of real value for 32 testing samples.

Of 16 distress testing samples, 15 companies give mild alarm in period $T-9$ (4 years ahead), and 13 companies give severe alarm in period $T-7$ and period $T-5$ (3 years ahead and 2 years ahead). All the 16 companies give severe alarm in period $T-3$ (1 year ahead). This shows that the information the dynamic model absorbed and produced almost covers the characteristics of financial distress. An accurate warning can be made 4 years before financial distress, and the accuracy is 93.8% (We also take triple BPNN to make a comparison. The results show that the accuracy of prediction 1 year ahead to 4 years ahead is 100, 93.8, 62.5, and 43.8%, respectively. The accuracy sharply declines 3 years ahead. It shows that the triple BPNN has better effect for short-term prediction rather than long-term prediction.)

For healthy testing samples, none is lower than the severe alarm limit. But sometimes, the predictive values appear slightly below the mild alarm limit, showing that there have been cases of temporary deviation from healthy state for healthy testing samples. The dynamic model conducts a track and thereafter modifies. This shows that the model can objectively track and effectively predict the overall financial state of a company from a long run.

5. Conclusions and future work

In this chapter, we focus on the dynamic nature of corporate financial distress and establish dynamic prediction models consisting of a process model and a discriminant model, which are used to describe the dynamic process and discriminant rules of financial distress, respectively. The operation of the dynamic prediction is achieved by Kalman filtering algorithm, and a general n-step-ahead prediction algorithm based on Kalman filter is derived for prospective prediction. To validate the prediction performance of this method, we conduct an empirical study for China's manufacturing industry. The empirical results have proved the accuracy and advance of predicting financial distress using this dynamic model. The accuracy of prediction 4 years before financial distress is 93.8%. In addition, this method also solves the problem of massive data processing as it does not require storing large amounts of historical data and thus can achieve real-time processing of data.

In this research, we suppose the dynamic process of financial distress is linear. The Kalman filtering algorithm will be applied to a nonlinear dynamic model in the future research, and it will offer a wider range of applications.

Acknowledgements

This research is supported by National Natural Science Foundation of China (Grant no. 71602188) and National Social Science Foundation of China (Grant no. 15ZDB167).

Author details

Qian Zhuang

Address all correspondence to: zhuangcpu@163.com

School of International Pharmaceutical Business, China Pharmaceutical University, Nanjing, Jiangsu, China

References

[1] Beaver WH. Financial ratios as predictors of failure. Journal of Accounting Research (Supplement). 1966;**4**:71-111. DOI: 10.2307/2490171

[2] Altman EI. Financial ratios, discriminant analysis and the prediction of corporate bankruptcy. The Journal of Finance. 1968;**23**:589-609. DOI: 10.1111/j.1540-6261.1968.tb00843.x

[3] Altman EI, Haldeman RG, Narayanan P. Zeta analysis: A new model to identify bankruptcy risk of corporations. Journal of Banking & Finance. 1977;**1**:29-54

[4] Martin D. Early warning of bank failure: A logit regression approach. Journal of Banking & Finance. 1977;**1**:249-276. DOI: 10.1016/0378-4266(77)90022-X

[5] Ohlson JA. Financial ratios and the probabilistic prediction of bankruptcy. Journal of Accounting Research. 1980;**18**:109-131. DOI: 10.2307/2490395

[6] DS W, Liang L, Yang ZJ. Analyzing the financial distress of Chinese public companies using probabilistic neural networks and multivariate discriminate analysis. Socio-Economic Planning Sciences. 2008;**42**:206-220. DOI: 10.1016/j.seps.2006.11.002

[7] Shen H, Cui J, Zhou ZB, Min H. BP-neural network model for financial risk warning in medicine listed company. In: Proceedings of 2011 Fourth International Joint Conference on Computational Sciences and Optimization (CSO); May 2011; Yunnan, China: IEEE; 2011. pp. 767-770. DOI: 10.1109/CSO.2011.97

[8] Zhou X, Wang JY, Xie W, Hong Y. Research on the optimal methods of financial distress prediction based on BP neural networks. In: Proceedings of the 2012 Second International Conference on Electric Information and Control Engineering; 06–08 Apr 2012; Jiangxi, China: IEEE; 2012. pp. 735-738. DOI: 10.1109/ICEICE.2012.1070

[9] Sun J, He KY, Li H. SFFS-PC-NN optimized by genetic algorithm for dynamic prediction of financial distress with longitudinal data streams. Knowledge-Based Systems. 2011;**24**: 1013-1023. DOI: 10.1016/j.knosys.2011.04.013

[10] Cao Y, Wan GY, Wang FQ. Predicting financial distress of Chinese listed companies using rough set theory and support vector machine. Asia-Pacific Journal of Operational Research. 2011;**28**:95-109. DOI: 10.1142/S0217595911003077

[11] Park CS, Han I. A case-based reasoning with the feature weights derived by analytic hierarchy process for bankruptcy prediction. Expert Systems with Applications. 2002;**23**:255-264. DOI: 10.1016/S0957-4174(02)00045-3

[12] Shin KS, Lee TS, Kim HJ. An application of support vector machines in bankruptcy prediction model. Expert Systems with Applications. 2005;**28**:127-135. DOI: 10.1016/j.eswa.2004.08.009

[13] Lee MC, To C. Comparison of support vector machine and back propagation neural network in evaluating the enterprise financial distress. International Journal of Artificial Intelligence & Applications. 2010;**1**:31-43. DOI: 10.5121/ijaia.2010.1303

[14] Chaudhuria A, De K. Fuzzy support vector machine for bankruptcy prediction. Applied Soft Computing. 2011;**11**:2472-2486. DOI: 10.1016/j.asoc.2010.10.003

[15] Altman EI, Marco G, Varetto F. Corporate distress diagnosis: Comparisons using linear discriminant analysis and neural networks. Journal of Banking & Finance. 1994;**18**:505-529. DOI: 10.1016/0378-4266(94)90007-8

[16] Giovanis E. A study of panel logit model and adaptive neuro-fuzzy inference system in the prediction of financial distress periods. World Academy of Science, Engineering and Technology. 2010;**64**:646-652

[17] Ravisankar P, Ravi V. Financial distress prediction in banks using group method of data handling neural network, counter propagation neural network and fuzzy ARTMAP. Knowledge-Based Systems. 2010;**23**:823-831. DOI: 10.1016/j.knosys.2010.05.007

[18] Bahrammirzaee A. A comparative survey of artificial intelligence applications in finance: Artificial neural networks, expert system and hybrid intelligent systems. Neural Computing and Applications. 2010;**19**:1165-1195. DOI: 10.1007/s00521-010-0362-z

[19] Tsenga FM, YC H. Comparing four bankruptcy prediction models: Logit, quadratic interval logit, neural and fuzzy neural networks. Expert Systems with Applications. 2010;**37**:1846-1853. DOI: 10.1016/j.eswa.2009.07.081

[20] Rafieia FM, Manzarib SM, Bostanianb S. Financial health prediction models using artificial neural networks, genetic algorithm and multivariate discriminant analysis: Iranian evidence. Expert Systems with Applications. 2011;**38**:10210-10217

[21] Sun J, Li H. Dynamic financial distress prediction using instance selection for the disposal of concept drift. Expert Systems with Applications. 2011;**38**:2566-2576. DOI: 10.1016/j.eswa.2010.08.046

[22] Konstantaras K, Siriopoulos C. Estimating financial distress with a dynamic model: Evidence from family owned enterprises in a small open economy. Journal of Multinational Financial Management. 2011;**21**:239-255. DOI: 10.1016/j.mulfin.2011.04.001

[23] Giarda E. Persistency of financial distress amongst Italian households: Evidence from dynamic models for binary panel data. Journal of Banking & Finance. 2013;**37**:3425-3434. DOI: 10.1016/j.jbankfin.2013.05.005

[24] Durbin J, Koopman SJ. Time Series Analysis by State Space Methods. 2nd ed. Oxford, UK: Oxford University Press; 2012. DOI: 10.1093/acprof:oso/9780199641178.001.0001

[25] Kalman RE. A new approach to linear filtering and prediction problems. Journal of Basic Engineering. 1960;**82**:35-45. DOI: 10.1115/1.3662552

[26] Brown RG, PYC H. Introduction to Random Signals and Applied Kalman Filtering. 3rd ed. USA: John Wiley & Sons, Inc.; 1997

[27] Arnold T, Bertus M, Godbey JM. A simplified approach to understanding the Kalman filter technique. The Engineering Economist. 2008;**53**:140-155. DOI: 10.2139/ssrn.715301

[28] Shi Y, Fang HZ. Kalman filter-based identification for systems with randomly missing measurements in a network environment. International Journal of Control. 2010;**83**: 538-551. DOI: 10.1080/00207170903273987

Chapter 8

Predicting Collisions in Mobile Robot Navigation by Kalman Filter

Angel Sánchez, Homero Ríos,
Gustavo Quintana and Antonio Marín

Additional information is available at the end of the chapter

http://dx.doi.org/10.5772/intechopen.71653

Abstract

The growing trend of the use of robots in many areas of daily life makes it necessary to search for approaches to improve efficiency in tasks performed by robots. For that reason, we show, in this chapter, the application of the Kalman filter applied to the navigation of mobile robots, specifically the Time-to-contact (TTC) problem. We present a summary of approaches that have been taken to address the TTC problem. We use a monocular vision-based approach to detect potential obstacles and follow them over time through their apparent size change. Our approach collects information about obstacle data and models the behavior while the robot is approaching the obstacle, in order to predict collisions. We highlight some characteristics of the Kalman filter applied to our problem. Finally, we show of our results applied to sequences composed of 210 frames in different real scenarios. The results show a fast convergence of the model to the data and good fit even with noisy measures.

Keywords: Kalman filter, Time-to-contact, avoiding collisions, robot navigation, forecasting

1. Introduction

Nowadays, many robotic entities have been developed in order to make many human activities more efficient. In fact, it is common to find robots in hospitals, factories, and even in homes, which help to automate many tasks autonomously or semi-autonomously. However, many problems can arise when mobile autonomous robots must travel in uncertain and unknown environments. For this reason, techniques, programs, and sensors have been developed over time to address such tasks as localization, communication, path planning, and collision avoidance. Faced with these problems, robots must be able to sense their environment and understand it to make the best decisions. From these tasks, we will analyze the task of avoiding obstacles by mobile robots, using the well-known term "Time-to-contact" or TTC.

TTC is a biologically inspired method for obstacle detection and reactive control of motion. TTC was first studied and defined in [1] as *"the distance to an obstacle divided by the relative velocity between them."* In other words, TTC is the elapsed time before an observer (the center of projection) makes contact with the surface being viewed if the current relative motion between the observer (e.g., a robot's camera) and the surface were to continue without changes, i.e., under constant relative velocity. TTC is usually expressed in terms of the speed and the distance of the considered obstacle. The classical equation to compute TTC is given by Eq. (1)

$$TTC = -\frac{Z}{\frac{dZ}{dt}} \tag{1}$$

where Z is the distance between the observer and the obstacle, and $\frac{dZ}{dt}$ is the velocity of the robot with respect to the obstacle. **Figure 1** shows the camera model and perception of obstacles from a mobile robot using monocular vision, where t represents the time, Z is the distance, f referring to the focal length, r is defined as the distance between the center of projection and the obstacle, and S representing the height of an obstacle.

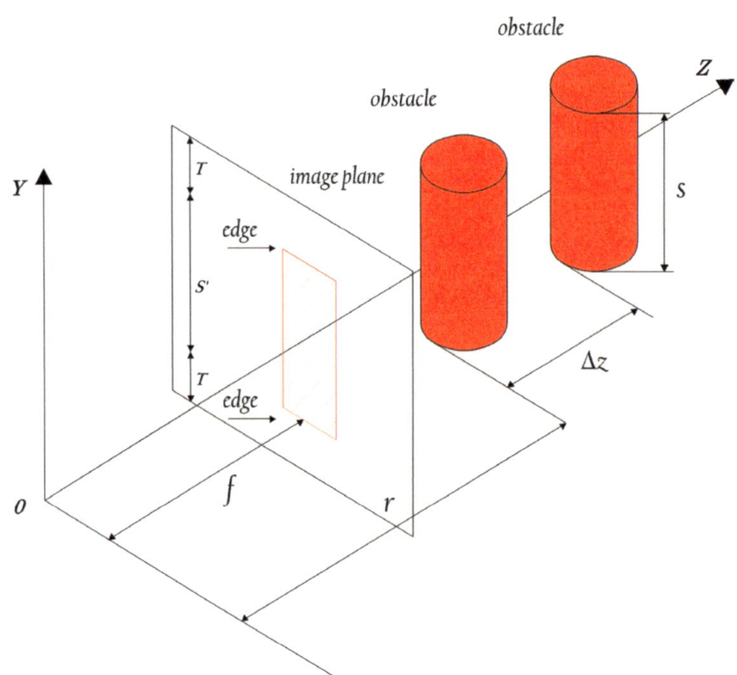

Figure 1. Isometric view of the model of perception.

The main idea of this chapter is to address this problem, modeling the growth of objects as the camera approaches the obstacle, in order to avoid detection obstacles in each frame once a precise model is built. First, we highlight some important works in the literature to give an overview of the different approaches that have been used to address this problem, discussing advantages and disadvantages of them. In the next section, we will give a brief description of the whole process that we use and how we address the problem using the Kalman Filter. Subsequently, we describe some experiments and results that we have made, to conclude with some conclusions and future work.

2. Background

The problem of estimating the Time-to-contact continues to be approached with different techniques. This is due to the fact that there are some factors that can prevent the robot from recognizing its environment reliably, such as the change of light intensity in the environment, erroneous segmentation of obstacles, or errors due to sinuous floors (especially with mobile robots on wheel).

Perhaps, the most practical and accurate approach is to use specialized sensors embedded in robots, such as sonars [2, 3]. Depth sensing used in the literature for robotic navigation and egomotion has been performed by binocular vision (stereo vision) [4, 5]. Other studies [6] have estimated TTC using paracatadioptric sensors with good results in real time. Although ultrasonic range sensors have large field of view, cross-talk problems might appear if more than one sensor is used simultaneously. As a result, "*the frequency of obstacle detection is limited by the number of sensors in use and the time required for an echo to return from an obstacle*" [7].

In order to minimize the energy consumption and cost, we decided to use passive sensors such as one camera only, i.e., monocular vision. Several studies have employed monocular vision to estimate TTC. For example, motion has been computed from images in space and spectral domains, and specifically TTC has been estimated using temporal change in power spectra between successive images [8]; however, switching from time domain to frequency domain involves extra processing. Additionally, TTC can be estimated for different goals (e.g., docking and landing) [9, 10] from the focus of expansion (FOE), however, these approaches are based on estimating optical flow, and hence they "*are iterative, need to work at multiple scales, tend to be computationally expensive and require a significant effort to implement properly*" [11].

Other method to estimate TTC is by "Direct method," which works directly with the derivatives of image brightness and does not require feature detection, feature tracking, or estimate of the optical flow [12]. Despite this method has achieved good results to approach surfaces, there are cases where the accuracy is compromised (e.g., when the robot approaches untextured walls and thus changing of the brightness is zero).

Finally, TTC has been computed using changes in the obstacle's size. For instance, studies in [13, 14] have used the fact that animals and insects obtain information from the apparent size S of objects and the temporal changes in the size. This information is usually called the

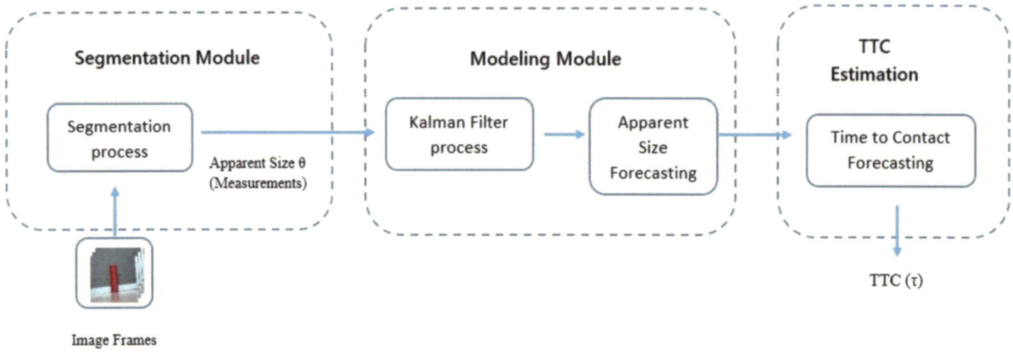

Figure 2. Methodology to performance the TTC forecasting.

"tau-margin" defined as Eq. (2). Tau-margin is derived from Eq. (1) by using a characteristic size of the obstacle on the image [15] and the approximation that the obstacle is planar and parallel to the image plane.

$$\tau = -\frac{S}{\frac{dS}{dt}} \tag{2}$$

However, a last focus on which we are interested is based on modeling the robot's movement. The study [16] has calculated TTC in vehicular motion using several scenarios. The idea is promising; however, due to the use of interest points, a mechanism needs to be implemented for grouping these points into different regions representing different obstacles.

In this brief summary, we can see that many approaches have been proposed, however, more approaches are still emerging because there are factors that condition the accuracy of the TTC estimate. But, we decided to use monocular vision to minimize costs of specialized sensors (in terms of energy and money).

The approach we have worked on is to model the apparent sizes of the segmented obstacle in each number of frames, in order to predict the Time-to-contact, that is, in how many frames the robot could collide with that obstacle. Since the TTC is estimated by analyzing how the apparent size of the object is changing with respect to time, we can find errors if the segmentation is not correct, which would lead to an incorrect acceleration or deceleration of the robot and to a possible collision.

When constructing models of a phenomenon, we base the predictions on previous data, which leads the robot to "understand" the behavior of its environment, if it has a constant speed. Having explained the importance of modeling, we decided to incorporate the Kalman Filter to address this problem and have reliable predictions. **Figure 2** shows the methodology used to estimate and forecast TTC.

3. Estimating Time-to-contact

In the following sections, we will give an outline of the processes used in each module, shown in **Figure 2**.

3.1. Segmentation process

In a controlled experimental environment, we detect obstacles by color, and analyze the height of the segmented region by enclosing it in a rectangle. The colors are pre-calibrated. **Figure 3** shows examples of two experimental cases and their segmentation by color. In the first scenario, the robot approaches an opaque red cylinder. In the second case, the robot approaches a bright green sphere.

3.2. Calculating apparent size

Figure 4 represents a mobile robot model, where S is representing the size of the detected object, r represents the distance between the mobile robot and the detected obstacle, and θ the proportional the aperture of the angle to the object's size projected on the image called "apparent size."

Figure 5 illustrates how the apparent size of the objects is expanded while it is approaching the camera. From a perspective view, T regions decrease in the image and the apparent size S′ (S projected on the image) grows proportionally as the robot approaches the obstacle. The

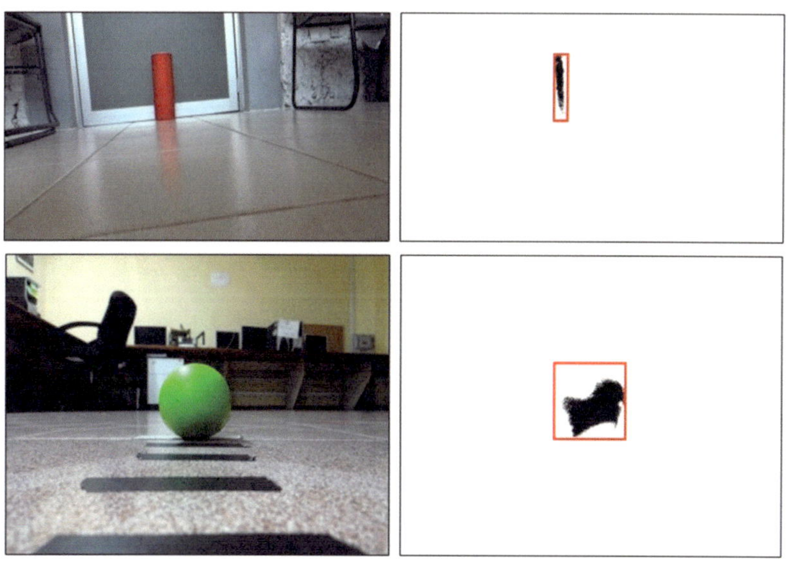

Figure 3. Example of segmentation by color process with experimental scenarios.

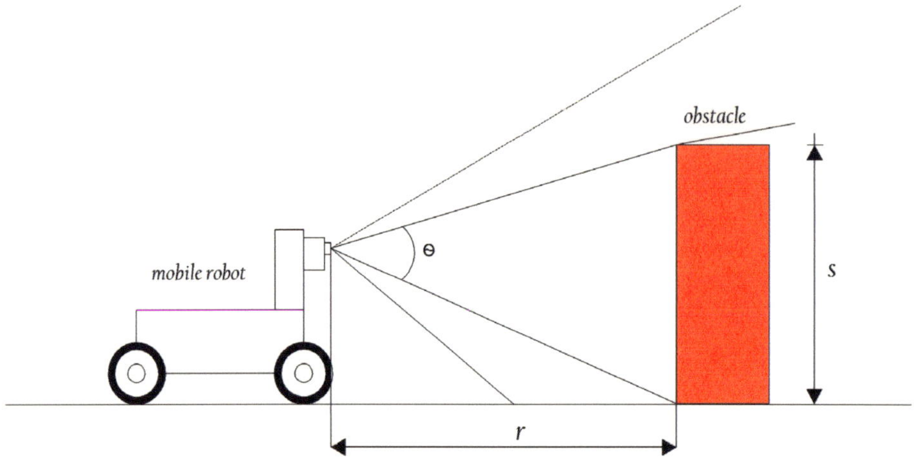

Figure 4. View of the mobile robot approaching an obstacle.

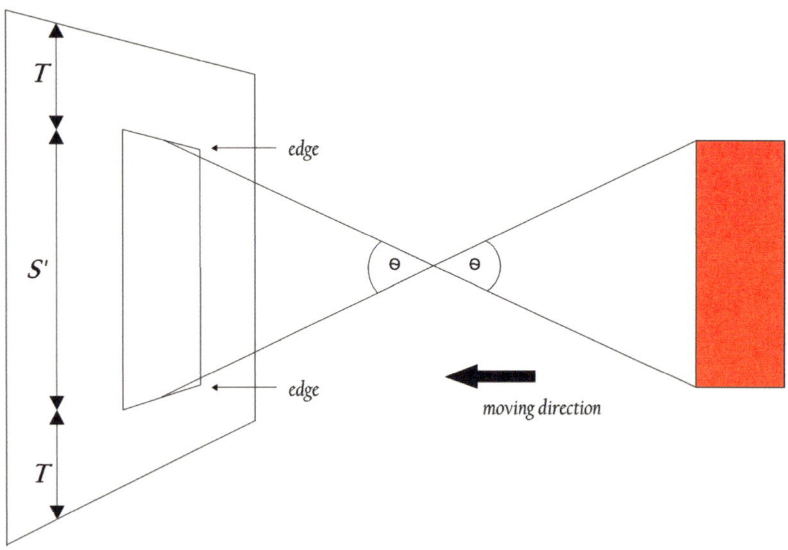

Figure 5. Perspective view of the approaching process.

best case is when the obstacle approaching the camera is alienated with it. In this case, to find the value of θ, the triangle ACD (see **Figure 6**) is divided into two right triangles (ABD and BCD) and from opposite angles by a vertex, we can estimate θ_1 by Eq. (3). Since $\theta_1 = \theta_2$, rearranging Eq. (3) we obtain θ as is shown in Eq. (4)

$$tan(\theta_1) = \frac{\frac{S'}{2}}{f} \tag{3}$$

$$\theta = 2\left[arc\left(\frac{\frac{S'}{2}}{f}\right)\right] \quad (4)$$

3.3. Modeling process

So, suppose we have a mobile robot moving towards an obstacle and we need to be generating a model of the size of the obstacle, that is, as it approaches the obstacle, the camera detects that the obstacle is growing. This is a scenario for the Kalman Filter, because the method is a part of the temporal and tracking models, which when applied to our problem means the tracking of the apparent size of any obstacle over time. The main characteristic of the temporal models is that they relate the state of the system to time $t - 1$ and t as shown by Eq. (5)

$$w_t = \mu_p + Fw_{t-1} + \epsilon \quad (5)$$

In Eq. (4), W_t is an n-dimensional vector of the state components, F is an n-by-n matrix called the transfer matrix or transition matrix, which relates the mean of the state at time t to the state at time $t - 1$, and ε is a random variable (usually called the process noise) associated with random events or forces that directly affect the actual state of the system, and which is normally distributed and determines how closely related the states are at times t and $t - 1$ [16].

In Eq. (4), we can see that it is a recursive model, because we assume that each state depends only upon its predecessor. We assume that W_t is conditionally independent of the states $W_1, \ldots, W_{t-2}$ given its immediate predecessor W_{t-1}, and just model the conditional relationship $Pr(W_t | W_{t-1})$ [17].

Below, we give a brief description of the specific case Kalman filter. It is not our intention to explain in detail the Kalman filter, we only want to highlight some characteristics and explain the application to our problem.

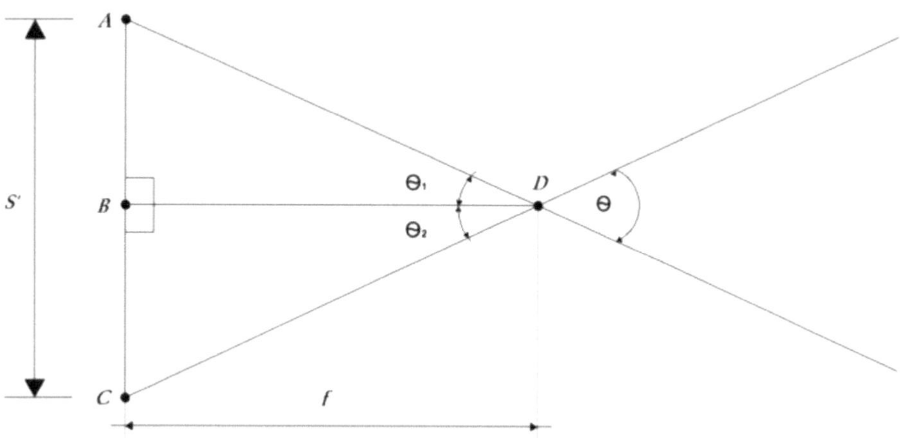

Figure 6. Geometrical view of the camera model.

3.3.1. Description of Kalman filter

The Kalman filter is a set of mathematical equations, described first time in [18], where a recursive solution to the discrete data linear filtering problem is presented. This method has been extensively researched and applied in various fields because it provides us an efficient computational (recursive) mechanism to estimate the state of a process. The filter is powerful because it involves estimations of past, present, and even future states. Kalman filter involves these elements with the use of knowledge of the system and measurement device, the statistical description of the system noises and any available information about initial conditions of the variables of interest [19].

To begin, we must remember that the basic idea of the Kalman Filter is, under a set of assumptions, it will be possible, given a history of measurements of a system, construct a model for the state of the System that maximizes the a posteriori probability of previous measurements. We can maximize *a posteriori* probability without having a long history of measurements above. Instead, we can iteratively update our state model of a system and maintain only that model for the next iteration [16]. These iterations are formed mainly by processes of prediction, measurement, and updating of the state.

Before explaining the process used, we must remember that Kalman Filter is based on three assumptions:

1. The evolution of state space is linear.
2. The errors or noise subject to the measurements are "white."
3. This noise is also Gaussian.

In other words, the first assumption means that the state of the system at time t can be modeled as a matrix multiplied by the state at time $t - 1$. That is good because linear systems are more easily manipulated and practical than nonlinear. The additional assumptions that the noise is white and Gaussian means that the noise is not correlated over time and that its amplitude can be accurately modeled using a mean and a covariance (i.e., noise is fully described by its first and second moments) [16].

3.3.2. Kalman filter applied to TTC

Below, a brief explanation of how the Kalman Filter is used to predict new apparent size measurements of the obstacle to avoid will be given. We emphasize that we do not model the TTC as such, but we model the behavior of the apparent size of the projected obstacle on the image because the TTC depends on this growth.

So, at some time t_1, we determine the apparent size θ to be θ_1. However, because of inherent measuring device inaccuracies (such as changes of light intensity or non-smooth floor mentioned above), the result of the measurements is somewhat uncertain. Then, we decide that the precision is such that the standard deviation involved is σ_1 (only one variable). Thus, we can establish the conditional probability, the value at time t_1, conditioned on the observed value of the measurement θ_1, that is, we have the probability that θ has a value, based upon the measurement we took. At this moment, we best estimate of the $\hat{\theta}_1 = \theta_1$ and the variance $\hat{\sigma}^2_{\theta_1} = \sigma^2_{\theta_1}$.

After, the robot takes another measurement based on color segmentation at time t_2, and therefore, is obtained θ_2 with a variance $\sigma_{\theta 2}$ (which is assumed to be less than the first). To combine these measures, and to obtain a new one with its own variation (Gaussian distribution) Eqs. (6) and (7) are used, where can be seen that the new value is just a weighted combination of the two measured means and the weighting is determined by the relative uncertainties of the two measurements (conditional mean). The weight in these equations can be seen as: if $\sigma_{\theta 1}$ is greater than $\sigma_{\theta 2}$ (that is, more variability), $\sigma_{\theta 2}$ would have more weight because $\sigma_{\theta 2}$ has less variability. Also, the uncertainty in the estimate of new θ has been decreased by combining the two pieces of information [19]

$$\mu = \theta_{12} = \left(\frac{\sigma_{\theta_2}^2}{\sigma_{\theta_1}^2 + \sigma_{\theta_2}^2}\right)\theta_1 + \left(\frac{\sigma_{\theta_1}^2}{\sigma_{\theta_1}^2 + \sigma_{\theta_2}^2}\right)\theta_2 \tag{6}$$

$$\sigma_{12}^2 = \frac{\sigma_{\theta_1}^2 \sigma_{\theta_2}^2}{\sigma_{\theta_1}^2 + \sigma_{\theta_2}^2} \tag{7}$$

Now that we know how to obtain a next measure, we can continue with this process N times (N measurements). This is because we can combine the first two, then the third with the combination of the first two, the fourth with the combination of the first three, and so on [16]. This is what happens when we are tracking the θ over time, we obtain one measure followed by another followed by another.

Usually, Eq. (6) is rewritten as Eq. (8), and Eq. (7) as Eq. (9) because with this new forms, we can separate the old information from de new information. The new information $(\theta_2 - \theta_1)$ is called *innovation*

$$\widehat{\theta_2} = \theta_1 + \frac{\sigma_{\theta_1}^2}{\sigma_{\theta_1}^2 + \sigma_{\theta_2}^2}(\theta_2 - \theta_1) \tag{8}$$

$$\widehat{\sigma_2^2} = \left(1 - \frac{\sigma_{\theta_1}^2}{\sigma_{\theta_1}^2 + \sigma_{\theta_2}^2}\right)\sigma_{\theta_1}^2 \tag{9}$$

Finally, Eq. (10) shows our optimal iterative update factor, which is known as the *update gain K*, and so, we obtain the recursive form described in Eqs. (11) and (12). For a more detailed explanation, we suggest to the reader have a look in [19]

$$K = \frac{\sigma_{\theta_1}^2}{\sigma_{\theta_1}^2 + \sigma_{\theta_2}^2} \tag{10}$$

$$\widehat{\theta_2} = \theta_1 + K(\theta_2 - \theta_1) \tag{11}$$

$$\widehat{\sigma_{\theta_2}^2} = (1 - K)\,\sigma_{\theta_1}^2 \tag{12}$$

Finally, once the behavior has been modeled, it will be possible to forecast TTC by Eq. (2).

Figure 7. Robot based on Raspberry pi used for experiments.

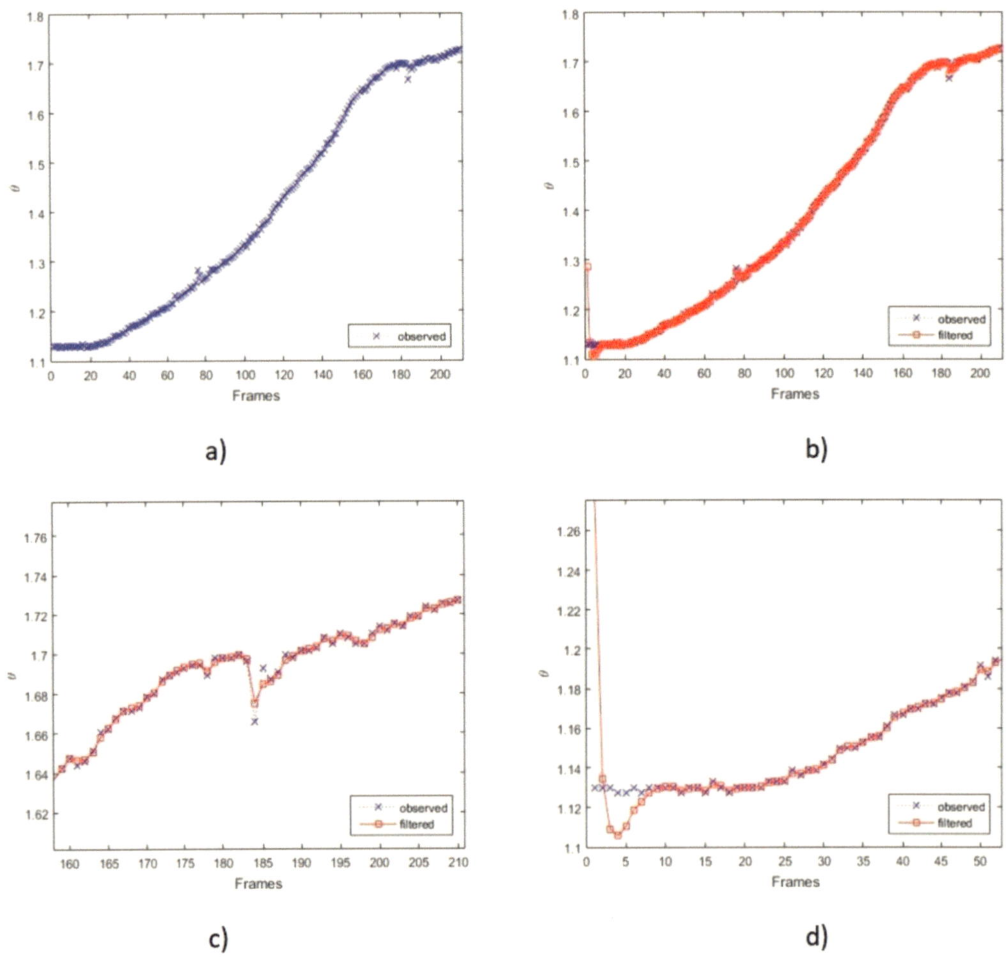

Figure 8. Results of experiments of scenario 1 with a red cylinder. (a) Measurements. (b) Filtered data. (c) Window example, where there is greater variation. (d) Initial convergence process.

4. Experiments and results

In order to test our proposal, we design experiments in which a robot approaches two obstacles mentioned in the segmentation section (see **Figure 3**) at constant velocity and we took 210 frames (equivalent to 7 s) of these real scenarios. A camera was used as a sensor to locate the objects in the robot environment.

Figure 7 shows the robot used for the experiments. In **Figures 8** and **9**, the results of the filtering process are shown for the experiments with the red cylinder and green sphere, respectively. For both figures, (a) shows blue, the measurements obtained by the sensor (specifically by the process of segmentation by color in each image). As can be seen, growth of the θ (obtained by the camera) over time is not "smooth," due to various factors in the environment mentioned above. (b) Indicates red color, the Kalman filtering on the measurements, where it can be seen that they are very close. (c) Shows, for each case, a window of the results where there is greater error in the measurements, which leads to have errors in the prediction but it getting closer to the measurements. (d) Shows an expansion at the beginning of (b), where

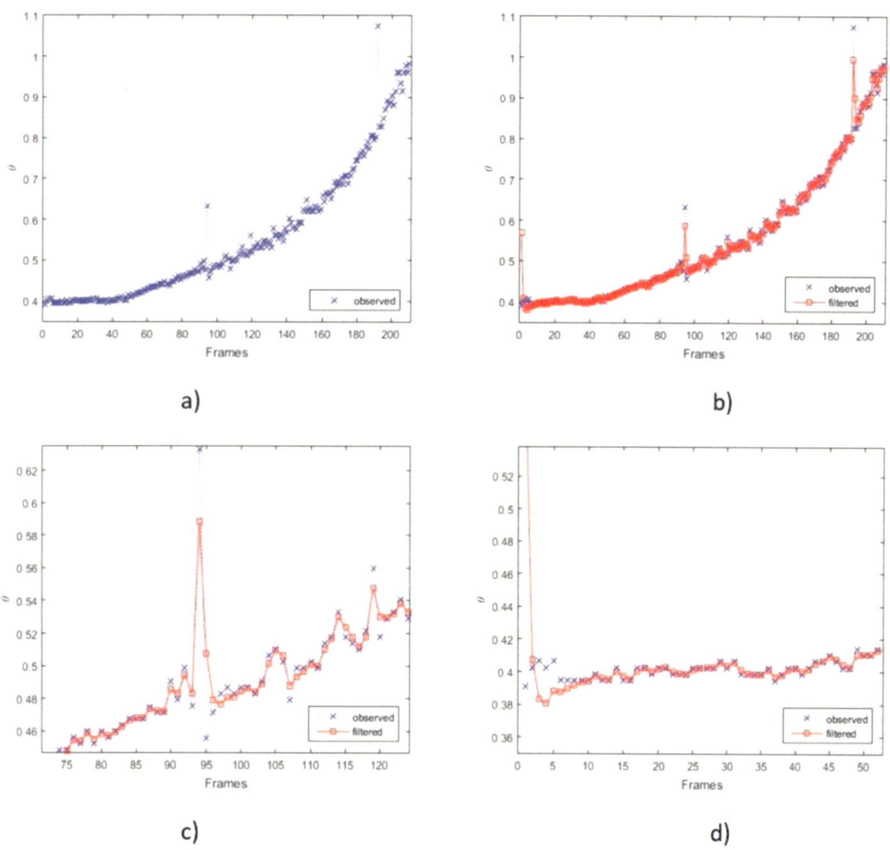

Figure 9. Results of experiments of scenario 2 with a green sphere. (a) Measurements. (b) Filtered data. (c) Window example, where there is greater variation. (d) Initial convergence process.

it can be seen that in both cases it takes little less than 10 frames (about 0.3 s) to converge to obtain estimates close to the measurements.

5. Conclusions and future work

In this chapter, an approach for estimating possible collisions was presented. A brief description of different approaches used to address the TTC problem was analyzed. Taking into account the advantages and disadvantages of these approaches, we present a way to handle the problem by modeling the behavior of the apparent size of the segmented obstacles, which the robot senses in each frame. This apparent size calculation is formally described and used to obtain measures of obstacles. We also apply Kalman filtering as a mechanism to model and predict θ, which will ultimately serve to predict the TTC and therefore avoid collisions. Some features of the Kalman Filter are also highlighted and we describe how it does the estimation for our problem. Finally, our approach is applied to two real cases where the modeling process is observed and the proximity to the measures taken, reducing noise of the measurements. This approach is gaining strength because it is easier to predict (given some measures taken previously) than to be looking for obstacles in each frame. In addition, the Kalman filter will correct errors if measures are incorporated sometimes. As future work, we will continue working on several scenarios and comparing this approach with some others, such as modeling and predicting using system identification techniques or time series.

Author details

Angel Sánchez[1*], Homero Ríos[1], Gustavo Quintana[2] and Antonio Marín[1]

*Address all correspondence to: angelsg89@hotmail.com

1 Research Center of Artificial Intelligence, University of Veracruz, Mexico

2 ELEC Department, Vrije Universiteit Brussel, Belgium

References

[1] Lee DNA. Theory of visual control of braking based on information about time-to-collision. Perception. 1976;5(4):1005020. DOI: 10.1068/p050437

[2] Byoung-Kyun S, Jun-Seok Y, Eok-Gon K, Yang-Keun J, Jong Bum W, Sung-Hyun H. A travelling control of mobile robot based on sonar sensors. In: 15th International Conference of Control, Automation and Systems (ICCAS); October 13-16, 2015; Busan, South Korea. IEEE; 2015. pp. 13-16. DOI: 10.1109/ICCAS.2015.7364821

[3] Vaščák J, Hvizdoš J. Vehicle navigation by fuzzy cognitive maps using sonar and RFID technologies. In: 14th International Symposium on Applied Machine Intelligence and Informatics (SAMI); January 21-23, 2016; Herlany, Slovakia. IEEE; 2016. DOI: 10.1109/SAMI.2016.7422985

[4] Stefan KG, Felix E, Thomas M. A real-time low-power stereo vision engine using semi-global matching. In: Mario Fritz, Bernt Schiele and Justus H. Piater, editors. In: 7th International Conference on Computer Vision Systems; October 13-15, 2009; Liege, Belgium. Springer, Berlin, Heidelberg; 2009. p. 134-143. DOI: 10.1007/978-3-642-04667-4_14

[5] Muffert M, Milbich T, Pfeiffer D, Franke U. May I enter the roundabout? A time-to-contact computation based on stereo-vision. In: Intelligent Vehicles Symposium (IV) 2012; June 3-7, 2012; Alcala de Henares, Spain. IEEE; 2012. pp. 565-570. DOI: 10.1109/IVS.2012.6232178

[6] Benamar F, El Fkihi S, Demonceaux C, Mouaddib E, Aboutajdine D. Gradient-based time to contact on paracatadioptric camera. In: IEEE International Conference on Image Processing, ICIP'2013; September 15-18, 2013; Melbourne, Australia. IEEE; 2013. pp. 5-9. DOI: 10.1109/ICIP.2013.6738002

[7] Alenya G, Negre A, Crowly JL. Time to Contact for Obstacle Avoidance. In: Proceedings of the 4th European Conference on Mobile Robots (ECMR) 09; September 23-25, 2009; Dubrovnik, Croatia. 2009. pp. 19-24

[8] Izumi S, Yamaguchi T. Time-to-contact estimation on the scaled-matching of power spectra. In: SICE, 2007 Annual Conference; September 17-20, 2007; Takamatsu, Japan. IEEE; 2008. pp. 2885-2889. DOI: 10.1109/SICE.2007.4421482

[9] McCarthy C, Barnes N, Mahony R. A robust docking strategy for a mobile robot using flow field divergence. IEEE Transactions on Robotics. 2008;**24**(4). DOI: 10.1109/TRO.2008.926871

[10] McCarthy C, Barnes N. A unified strategy for landing and docking using spherical flow divergence. IEEE Transactions on Pattern Analysis and Machine Intelligence. 2012;**34**(5):1024-1031. DOI: 10.1109/TPAMI.2012.27

[11] Horn BKP, Fang Y, Masaki I. Hierarchical framework for direct gradient-based time-to-contact estimation. In: IEEE Intelligent Vehicles Symposium, 2009; June 3-5, 2009; Xian, China. IEEE; 2009. pp. 1394-1400. DOI: 10.1109/IVS.2009.5164489

[12] Horn BKP, Fang Y, Masaki I. Time to contact relative to a planar surface. In: IEEE Intelligent Vehicles Symposium, 2007; June 13-15, 2007, Istanbul, Turkey. IEEE. 2007. pp. 68-74. DOI: 10.1109/IVS.2007.4290093

[13] Kaneta Y, Hagisaka Y, Ito K. Determination of time to contact and application to timing control of mobile robot. In: IEEE International Conference on Robotics and Biomimetics (ROBIO), 2010; December 14-18, 2010; Tianjin, China. IEEE; 2010. pp. 161-166. DOI: 10.1109/ROBIO.2010.5723320

[14] Sanchez A, Rios H, Marin A, Verdin K, Contreras G. Estimation of time-to-contact from Tau-margin and statistical analysis of behavior. In: International Conference on Systems,

Signals and Image Processing (IWSSIP), 2016; May 23-25, 2016; Bratislava, Slovakia. IEEE; 2016. pp. 1-6. DOI: 10.1109/IWSSIP.2016.7502702

[15] Muller D, Pauli J, Nunn C, Gormer S, Muller-Schneiders S. Time to contact estimation using interest points. In: 12th International Conference on Transportation Systems, 2009. ITSC '09; October 4-7, 2009; St. Louis, MO, USA. IEEE; 2009. pp. 1-6. DOI: 10.1109/ITSC.2009.5309851

[16] Bradski G, Kaehler A. Learning OpenCV. 1st ed. USA: O'Reilly Media, Inc.; 2008. p. 555

[17] Prince S. Computer Vision: Models, Learning and Inference. USA: Cambridge University Press; 2012. p. 580

[18] Kalman REA. New approach to linear filtering and prediction problems. Transactions of the ASME – Journal of Basic Engineering. 1960;**82**(1):35-45. DOI: 10.1115/1.3662552

[19] Maybeck P. Introduction. In: Stochastic Models, Estimation and Control. 1st ed. London: Academic Press; 1979. p. 1-16

Efficient Matrix-Free Ensemble Kalman Filter Implementations: Accounting for Localization

Elias David Niño Ruiz, Rolando Beltrán Arrieta and Alfonso Manuel Mancilla Herrera

Additional information is available at the end of the chapter

http://dx.doi.org/10.5772/intechopen.72465

Abstract

This chapter discusses efficient and practical matrix-free implementations of the ensemble Kalman filter (EnKF) in order to account for localization during the assimilation of observations. In the EnKF context, an ensemble of model realizations is utilized in order to estimate the moments of its underlying error distribution. Since ensemble members come at high computational costs (owing to current operational model resolutions) ensemble sizes are constrained by the hundreds while, typically, their error distributions range in the order of millions. This induces spurious correlations in estimates of prior error correlations when these are approximated via the ensemble covariance matrix. Localization methods are commonly utilized in order to counteract this effect. EnKF implementations in this context are based on a modified Cholesky decomposition. Different flavours of Cholesky-based filters are discussed in this chapter. Furthermore, the computational effort in all formulations is linear with regard to model resolutions. Experimental tests are performed making use of the Lorenz 96 model. The results reveal that, in terms of root-mean-square-errors, all formulations perform equivalently.

Keywords: ensemble Kalman filter, modified Cholesky decomposition, sampling methods

1. Introduction

The ensemble Kalman filter (EnKF) is a sequential Monte Carlo method for parameter and state estimation in highly nonlinear models. The popularity of the EnKF owes to its simple formulation and relatively easy implementation. In the EnKF, an ensemble of model realizations is utilized in order to, among other things, estimate the moments of the prior error distribution (forecast distribution). During this estimation and the assimilation of observations, many challenges are present in practice. Some of them are addressed in this chapter: the first one is related to the proper estimation of background error correlations, which has a high impact on the quality of the analysis corrections while the last one is related to the proper estimation of the posterior ensemble when the observation operator is nonlinear. In all cases, background error correlations are estimated based on the Bickel and Levina estimator [1].

2. Preliminaries

We want to estimate the state of a dynamical system $\boldsymbol{x}^* \in \mathbb{R}^{n\times 1}$, which (approximately) evolves according to some (imperfect) numerical model operators $\boldsymbol{x}_k^* = \boldsymbol{\mathcal{M}}_{t_{k-1}\to t_k}\left(\boldsymbol{x}_{k-1}^*\right)$, for instance, a model which mimics the behavior of the atmosphere and/or the ocean where n is the model dimension (typically associated with the model resolution). The estimation process is based on two sources of information: a prior estimate of $\boldsymbol{x}^*$, $\boldsymbol{x}^b = \boldsymbol{x}^* + \boldsymbol{\xi}$ with $\boldsymbol{\xi} \sim \boldsymbol{\mathcal{N}}(\mathbf{0}_n, \boldsymbol{B})$, and a noisy observation $\boldsymbol{y} = \boldsymbol{\mathcal{H}}(\boldsymbol{x}^*) + \boldsymbol{\epsilon}$ with $\boldsymbol{\epsilon} \sim \boldsymbol{\mathcal{N}}(\mathbf{0}_m, \boldsymbol{R})$, where m is the number of observed model components, $\boldsymbol{R}^{m\times m}$ is the estimated data error covariance matrix, $\boldsymbol{\mathcal{H}} : \mathbb{R}^{n\times 1} \to \mathbb{R}^{m\times 1}$ is the (nonlinear) observation operator that maps the information from the model domain to the observation space, $\mathbf{0}_n$ is the nth dimensional vector whose components are all zeros. In practice, $m \ll n$ or $m < n$. The assimilation process can be performed by using sequential data assimilation methods, and one of the most widely used methods is the ensemble Kalman filter, which is detailed in the following paragraphs.

2.1. The ensemble Kalman filter

In the ensemble Kalman filter (EnKF) [2], an ensemble of model realizations

$$\boldsymbol{X}^b = \left[\boldsymbol{x}^{b[1]}, \boldsymbol{x}^{b[2]}, \dots, \boldsymbol{x}^{b[N]}\right] \in \mathbb{R}^{n\times N}, \tag{1}$$

is utilized in order to estimate the moments of the background error distribution,

$$\boldsymbol{x}^{b[i]} \sim \boldsymbol{\mathcal{N}}\left(\boldsymbol{x}^b, \boldsymbol{B}\right), \tag{2}$$

via the empirical moments of the ensemble (1) and therefore,

$$\boldsymbol{x}^b \approx \overline{\boldsymbol{x}}^b = \frac{1}{N} \cdot \sum_{i=1}^{N} \boldsymbol{x}^{b[i]} \in \mathbb{R}^{n\times 1}, \tag{3}$$

and

$$B \approx P^b = \frac{1}{N-1} \cdot \Delta X^b \cdot \left[\Delta X^b\right]^T \in \mathbb{R}^{n \times n}, \tag{4}$$

where N is the ensemble size, $x^{b[i]} \in \mathbb{R}^{n \times 1}$ is the ith ensemble member, for $1 \le i \le N$, $x^b \in \mathbb{R}^{n \times 1}$ is well known as the background state, $B \in \mathbb{R}^{n \times n}$ stands for the background error covariance matrix, $\overline{x}^b$ is the ensemble mean, and P^b is the ensemble covariance matrix. Likewise, the matrix of member deviations $\Delta X^b \in \mathbb{R}^{n \times N}$ reads,

$$\Delta X^b = X^b - \overline{x}^b \cdot \mathbf{1}_N^T \in \mathbb{R}^{n \times N}. \tag{5}$$

A variety of EnKF implementations have been proposed in the current literature, most of them rely on basic assumptions over the moments of two probability distributions: the background (prior) and the analysis (posterior) distributions. In most of EnKF formulations, normal assumptions are done over both error distributions.

In general, depending on how the assimilation process is performed, the EnKF implementation will fall in one of two categories: stochastic filter [3] or deterministic filter [4]. In the context of stochastic filters, for instance, the assimilation of observations can be performed by using any of the next formulations,

$$X^a = A \cdot \left[\left[P^b\right]^{-1} \cdot X^b + H^T \cdot R^{-1} \cdot Y^s\right] \in R^{n \times N}, \tag{6}$$

$$X^a = X^b + A \cdot H^T \cdot R^{-1} \cdot \Delta Y \in \mathbb{R}^{n \times N}, \tag{7}$$

and

$$X^a = X^b + P^b \cdot H^T \cdot \left[R^{-1} + H \cdot P^b \cdot H^T\right]^{-1} \cdot \Delta Y \in \mathbb{R}^{n \times N}, \tag{8}$$

where the analysis covariance matrix $A \in \mathbb{R}^{n \times n}$ is given by $A = \left[\left[P^b\right]^{-1} + H^T \cdot R^{-1} \cdot H\right]^{-1}$, the matrix of innovations (on observations) is denoted by $\Delta Y = Y^s - H \cdot X^b \in \mathbb{R}^{n \times N}$, and the matrix of perturbed observations reads $Y^s = y \cdot \mathbf{1}_N^T + E \in \mathbb{R}^{m \times N}$, where the columns of $E \in \mathbb{R}^{m \times N}$ are formed by samples from a normal distribution with moments $\mathcal{N}(\mathbf{0}, R)$. In practice, one of the most utilized formulations is given by (8).

2.2. Cholesky and modified Cholesky decomposition

The forms of the Cholesky and the modified Cholesky decomposition were discussed by Golub and Van Loan in [5]. If $A \in \mathbb{R}^{n \times n}$ is symmetric positive definite, then there exist a unique lower triangular $L \in \mathbb{R}^{n \times n}$ with positive diagonal entries such that $A = L \cdot L^T$. If all the leading principal submatrices of $A \in \mathbb{R}^{n \times n}$ are non-singular, then there exist unique lower triangular matrices L and M and a unique diagonal matrix $D = \boldsymbol{diag}(d_1, d_2, \dots, d_n)$ such that $A = L \cdot D \cdot M^T$. If A is symmetric, then $L = M$ and $A = L \cdot D \cdot L^T$. Golub and Van Loan provide proofs of those decompositions, as well as various ways to compute them. We can use the Cholesky factor of a covariance matrix to quickly solve linear systems that involve a covariance matrix. Note that if we compute the Cholesky factorization and solve the triangular system $\mathbf{L} \cdot \mathbf{y} = \mathbf{b}$ and $\mathbf{L}^T \cdot \mathbf{x} = \mathbf{y}$, then

$$\mathbf{A} \cdot \mathbf{x} = \left(\mathbf{L} \cdot \mathbf{L}^T\right) \cdot \mathbf{x} = \mathbf{L} \cdot \left(\mathbf{L}^T \cdot \mathbf{x}\right) = \mathbf{L} \cdot \mathbf{y} = \mathbf{b}. \tag{9}$$

For a symmetric matrix $\mathbf{A} \in \mathbf{R}^{n \times n}$, the modified Cholesky decomposition involves finding a non-negative diagonal matrix $\mathbf{E}$ such that $\mathbf{A} + \mathbf{E}$ is positive definite. In particular, $\mathbf{E} = \mathbf{0}$ whenever $\mathbf{A}$ is positive definite, otherwise $\|\mathbf{E}\|$ should be sufficiently small. This allows applying the usual Cholesky factorization to $\mathbf{A} + \mathbf{E}$, i.e. finding matrices $\mathbf{L}, \mathbf{D} \in \mathbb{R}^{n \times n}$ such that $\mathbf{A} + \mathbf{E} = \mathbf{L} \cdot \mathbf{D} \cdot \mathbf{L}^T$.

Recall $\Delta \mathbf{X}^b = \mathbf{X}^b - \overline{\mathbf{x}}^b \otimes \mathbf{1}_N^T \in \mathbb{R}^{n \times N}$. Thus, $\Delta \mathbf{x}^{b[i]} \sim \mathcal{N}(\mathbf{0}_n, \mathbf{B})$ for $\mathbf{1} \leq \boldsymbol{i} \leq \boldsymbol{n}$. Let $\mathbf{x}^{b[i]} \in \mathbb{R}^{N \times 1}$, the vector holding the ith row across all columns of $\Delta \mathbf{X}^b$ for $\mathbf{1} \leq \boldsymbol{i} \leq \boldsymbol{n}$. Bickel and Levina in [1] discussed a modified Cholesky decomposition for the estimation of precision covariance matrix, and this allows in our context to estimate of $\mathbf{B}^{-1}$ by fitting regressions of the form:

$$\mathbf{x}^{b[i]} = \sum_{j=1}^{i-1} \mathbf{x}^{b[j]} \cdot \beta_{i,j} + \xi^{[i]} \in \mathbb{R}^{N \times 1}, \tag{10}$$

then, $\{L\}_{ij} = -\beta_{i,j}$, for $1 \leq i < j \leq n$.

3. Filters based on modified Cholesky decomposition

3.1. Ensemble Kalman filter based on modified Cholesky

In the ensemble Kalman filter based on a modified Cholesky decomposition (EnKF-MC) [6], the analysis ensemble is estimated by using the following equations,

$$\mathbf{X}^a = \mathbf{X}^b + \left[\mathbf{B}^{-1} + \mathbf{H}^T \mathbf{R}^{-1} \cdot \mathbf{H}\right]^{-1} \cdot \mathbf{H}^T \cdot \mathbf{R}^{-1} \cdot \Delta Y, \tag{11}$$

and

$$\Delta Y = \left[\mathbf{Y}^S - \mathbf{H}.\mathbf{X}^b\right] \in \mathbb{R}^{n \times N}. \tag{12}$$

In this context, error correlations are estimated via a modified Cholesky decomposition. This provides an estimate of the inverse background error covariance matrix of the form,

$$\mathbf{B}^{-1} \approx \widehat{\mathbf{B}}^{-1} = \mathbf{L}^T \cdot \mathbf{D}^{-1} \cdot \mathbf{L} \in \mathbb{R}^{n \times n}, \tag{13}$$

or equivalently

$$B \approx \widehat{B} = L^{-1} \cdot D \cdot L^{-T} \in \mathbb{R}^{n \times n}, \tag{14}$$

where $L \in \mathbb{R}^{n \times n}$ is a lower-triangular matrix whose diagonal elements are all ones, and $D \in \mathbb{R}^{n \times n}$ is a diagonal matrix. Even more, when only local effects are considered during the estimation of $\widehat{B}^{-1}$, sparse estimators of the precision background can be obtained. Furthermore, the matrix L can be sparse with only a few non-zero elements per row. Typically, the number of non-zero elements depends on the radius of influence during the estimation of background error correlations. For instance, in the one-dimensional case, the radius of influence denotes the maximum number of non-zero elements, per row, in L. The EnKF-MC is then obtained by plugging in the estimator (3) in (6). Given the structure of the Cholesky factors, the EnKF-MC can be seen as a matrix-free implementation of the EnKF.

Recall that the precision analysis covariance matrix reads,

$$A^{-1} = B^{-1} + H^{T} \cdot R^{-1} \cdot H \cdot \mathbb{R}^{n \times n}, \tag{15}$$

moreover, since $H^{T} \cdot R^{-1} \cdot H \in \mathbb{R}^{n \times n}$ can be written as a sum of m rank-one matrices, the factors (3) can be updated in order to obtain an estimate of the inverse analysis covariance matrix. The next section discussed such approximation in detail.

3.1.1. Posterior ensemble Kalman filter stochastic (PEnKF-S)

In the stochastic posterior ensemble Kalman filter (PEnKF-S) [7, 8], we want to estimate the moments of the analysis distribution,

$$x \sim \mathcal{N}(x^{a}, A), \tag{16}$$

based on the background ensemble $\left(X^{b}\right)$, where x^{a} is the analysis state and $A \in \mathbb{R}^{n \times n}$ is the analysis covariance matrix. Consider the estimate of the inverse background error covariance matrix Eq. (3), the precision analysis covariance matrix Eq. (4) can be approximated as follows:

$$A^{-1} \approx \widehat{A}^{-1} = \widehat{B}^{-1} + X \cdot X^{T}, \tag{17}$$

where $X = H^{T} \cdot R^{-\frac{1}{2}} \in \mathbb{R}^{n \times m}$. The matrix (17) can be written as follows:

$$\widehat{A}^{-1} = L^{T} \cdot D^{-1} \cdot L + \sum_{i=1}^{m} x_i \cdot x_i^{T},$$

where x_i denotes the ith column of matrix X, for $1 \le i \le m$. Consider the sequence of factor updates,

$$\left[L^i\right]^T \cdot D^i \cdot L^i = \left[L^{(i-1)}\right]^T \cdot D^{i-1} \cdot L^{i-1} + x_i \cdot x_i^T$$
$$\left[L^i\right]^T \cdot D^i \cdot L^i = \left[L^{(i-1)}\right]^T \cdot \left[D^{(i-1)} + p_i \cdot p_i^T\right] \cdot L^{(i-1)}$$
$$\left[L^i\right]^T \cdot D^i \cdot L^i = \left[\tilde{L}^{(i-1)} \cdot L^{(i-1)}\right]^T \cdot \tilde{D}^{(i-1)} \cdot \left[\tilde{L}^{(i-1)} \cdot L^{(i-1)}\right],$$

where $L^{i-1} \cdot p_i = x_i \in \mathbb{R}^{n \times 1}$, for $1 \leq i \leq m$, $\widehat{B}^{-1} = \left[L^{(0)}\right]^T \cdot D^{(0)} \cdot L^{(0)}$ and

$$D^{(i)} + p_i \cdot p_i^T = \left[\tilde{L}^{(i)}\right]^T \cdot \tilde{D}^{(i)} \cdot \tilde{L}^{(i)} \in \mathbb{R}^{n \times n}, \tag{18}$$

We can use of Dolittle's method in order to compute the factors $\tilde{D}^{(i)}$ and $\tilde{L}^{(i)}$ in (9), and it is enough to note that,

$$\underbrace{\begin{bmatrix} 1 & \tilde{l}_{21} & \tilde{l}_{31} & \cdots & \tilde{l}_{n1} \\ 0 & 1 & \tilde{l}_{32} & \cdots & \tilde{l}_{n2} \\ 0 & 0 & 1 & \cdots & \tilde{l}_{n3} \\ \vdots & \vdots & \vdots & \ddots & \vdots \\ 0 & 0 & 0 & \cdots & 1 \end{bmatrix}}_{\left[\tilde{L}^{(i)}\right]^T} \underbrace{\begin{bmatrix} \tilde{d}_1 & 0 & 0 & 0 & 0 \\ 0 & \tilde{d}_2 & 0 & 0 & 0 \\ 0 & 0 & \tilde{d}_3 & 0 & 0 \\ \vdots & \vdots & \vdots & \ddots & 0 \\ 0 & 0 & 0 & 0 & \tilde{d}_n \end{bmatrix}}_{\left[\tilde{D}^{(i)}\right]} \underbrace{\begin{bmatrix} 1 & 0 & 0 & \ldots & 0 \\ \tilde{l}_{21} & 1 & 0 & \ldots & 0 \\ \tilde{l}_{31} & \tilde{l}_{21} & 1 & \ldots & 0 \\ \vdots & \vdots & \vdots & \ddots & 0 \\ \tilde{l}_{n1} & \tilde{l}_{n2} & \tilde{l}_{n3} & \ldots & 1 \end{bmatrix}}_{\left[\tilde{L}^{(i)}\right]}$$
$$= \underbrace{\begin{bmatrix} d_1 + p_1^2 & p_1 \cdot p_2 & p_1 \cdot p_3 & \cdots & p_1 \cdot p_n \\ p_2 \cdot p_1 & d_2 + p_2^2 & p_2 \cdot p_3 & \cdots & p_2 \cdot p_n \\ p_3 \cdot p_1 & p_3 \cdot p_2 & d_3 + p_3^2 & \cdots & p_3 \cdot p_n \\ \vdots & \vdots & \vdots & \ddots & \vdots \\ p_n \cdot p_1 & p_n \cdot p_2 & p_n \cdot p_3 & \cdots & d_n + p_n^2 \end{bmatrix}}_{\left[D^{(i)} + p_i \cdot p_i^T\right]}$$

After some math simplifications, the next equations are obtained,

$$\tilde{d}_k = p_k^2 + d_k - \sum_{q=k+1}^{n} \tilde{d}_q \cdot \tilde{l}_{qi}^2, \tag{19}$$

and

$$\tilde{l}_{kj} = \frac{1}{\tilde{d}_k} \cdot \left[p_k \cdot p_j - \sum_{q=k+1}^{n} \tilde{d}_q \cdot \tilde{l}_{qi} \cdot \tilde{l}_{qj}\right], \tag{20}$$

for $1 \le k < n$ and $1 \le j \le k-1$. The set of Eqs. (19) and (20) can be used in order to derive an algorithm for rank-one update of Cholesky factors, and the updating process is shown in Algorithm 1.

Algorithm 1. Rank-one update for the factors $\boldsymbol{L}^{(i-1)}$ and $\boldsymbol{D}^{(i-1)}$

---

function UPD_CHOLESKY_FACTORS $\left(\boldsymbol{L}^{(i-1)}, \boldsymbol{D}^{(i-1)}, \boldsymbol{x}_i\right)$

  Compute $\boldsymbol{p}_i$ from $\left[\boldsymbol{L}^i\right]^T \cdot \boldsymbol{p}_i = \boldsymbol{x}_i$

  for $k = n \sim 1$ **do**

    Compute $\tilde{d}_k$ via Eq. (10).

    Set $l_{kk} \leftarrow 1$.

    for $j = 1 \rightarrow k-1$ **do**

      Compute $\tilde{l}_{kk}$ according to (11).

    end for

  end for

  Set $\boldsymbol{L}^{(i)} \leftarrow \boldsymbol{L}^{(i-1)} \cdot \boldsymbol{L}^{(i-1)}$ and $\boldsymbol{D}^{(i)} \leftarrow \tilde{\boldsymbol{D}}^{(i)}$.

  return $\boldsymbol{L}^{(i)}, \boldsymbol{D}^{(i)}$

end function

Algorithm 1 can be used in order to update the factors of $\widehat{\boldsymbol{B}}^{-1}$ for all column vectors in $\boldsymbol{X}$, and this process is detailed in Algorithm 2.

Algorithm 2. Computing the factors $\boldsymbol{L}^{(m)}$ and $\boldsymbol{D}^{(m)}$ of $\left[\boldsymbol{L}^{(m)}\right]^T \cdot \boldsymbol{D}^{(m)} \cdot \boldsymbol{L}^{(m)}$

---

function COMPUTE_ANALYSIS_FACTOR $\left(\boldsymbol{L}^{(0)}, \boldsymbol{D}^{(0)}, \boldsymbol{H}, \boldsymbol{R}\right)$

  Set $\boldsymbol{X} \leftarrow \boldsymbol{H}^T \cdot \boldsymbol{R}^{-\frac{1}{2}}$.

  for $i = 1 \rightarrow m$ do

    Set $\left[\boldsymbol{L}^{(i)}, \boldsymbol{D}^{(i)}\right] \leftarrow$ UPD_CHOLESKY_FACTORS $\left(\boldsymbol{L}^{(i-1)}, \boldsymbol{D}^{(i-1)}, \boldsymbol{x}_i\right)$

  end for

  return $\boldsymbol{L}^{(m)}, \boldsymbol{D}^{(m)}$

end function

Once the updating process has been performed, the resulting factors form an estimate of the inverse analysis covariance matrix,

$$\widehat{\boldsymbol{A}}^{-1} = \left[\boldsymbol{L}^{(m)}\right]^{T} \cdot \boldsymbol{D}^{(m)} \cdot \boldsymbol{L}^{(m)} \in \mathbb{R}^{n \times n}, \tag{21}$$

From this covariance matrix, the posterior mode of the distribution can be approximated as follows:

$$\overline{\boldsymbol{x}}^{a} = \overline{\boldsymbol{x}}^{b} + \boldsymbol{z} \in \mathbb{R}^{n \times 1}, \tag{22}$$

where

$$\left[\boldsymbol{L}^{(m)}\right]^{T} \cdot \boldsymbol{D}^{(m)} \cdot \boldsymbol{L}^{(m)} \cdot \boldsymbol{z} = \boldsymbol{q}, \tag{23}$$

with $\boldsymbol{q} = \boldsymbol{H}^{T} \cdot \boldsymbol{R}^{-1} \cdot \left[\boldsymbol{y} - \boldsymbol{H} \cdot \overline{\boldsymbol{x}}^{b}\right] \in \mathbb{R}^{n \times 1}$. Note that the linear system (23) involves lower and upper triangular matrices, and therefore, $\overline{\boldsymbol{x}}^{a}$ can be estimated without the need of matrix inversion. Once the posterior mode is computed, the analysis ensemble is built about it. Note that $\widehat{\boldsymbol{A}}$ reads $\widehat{\boldsymbol{A}} = \left[\boldsymbol{L}^{(m)}\right]^{-1} \cdot \left[\boldsymbol{D}^{(m)}\right]^{-1} \cdot \left[\boldsymbol{L}^{(m)}\right]^{-T}$, and therefore, a square root of $\widehat{\boldsymbol{A}}$ can be approximated as follows:

$$\widehat{\boldsymbol{A}}^{\frac{1}{2}} = \left[\boldsymbol{L}^{(m)}\right]^{-1} \cdot \left[\boldsymbol{D}^{(m)}\right]^{-\frac{1}{2}} \in \mathbb{R}^{n \times n}, \tag{24}$$

which can be utilized in order to build the analysis ensemble,

$$\boldsymbol{X}^{a} = \overline{\boldsymbol{x}}^{a} \cdot \boldsymbol{1}_{N}^{T} + \Delta \boldsymbol{X}^{a}, \tag{25}$$

where $\Delta \boldsymbol{X}^{a} \in \mathbb{R}^{n \times N}$ is given by the solution of the linear system,

$$\boldsymbol{L}^{(m)} \cdot \left[\boldsymbol{D}^{(m)}\right]^{-\frac{1}{2}} \cdot \Delta \boldsymbol{X}^{a} = \boldsymbol{W} \in \mathbb{R}^{n \times N}, \tag{26}$$

In addition, the columns of $\boldsymbol{W} \in \mathbb{R}^{n \times N}$ are formed by samples from a multivariate standard normal distribution. Again, since $\boldsymbol{L}^{(m)}$ is lower triangular, the solution of (26) can be obtained readily.

3.1.2. Posterior ensemble Kalman filter deterministic (PEnKF-D)

The deterministic posterior ensemble Kalman filter (PEnKF-D) is a square root formulation of the PEnKF-S. The main difference between them lies on the use of synthetic data. In both methods, the matrices $\boldsymbol{A}^{-1}$ and $\boldsymbol{B}^{-1}$ are computed similarly (i.e., by using Dolittle's method and the algorithms described in the previous section).

In the PEnKF-D, the computation of $\widehat{\boldsymbol{B}}^{-1}$ is performed (13) based on the empirical moments of the ensemble. The analysis update is performed by using the perturbations

$$\widehat{B}^{-\frac{1}{2}} \cdot \left(X^b - \overline{x}^b \cdot \mathbf{1}_N^T\right) \sim \mathcal{N}(\mathbf{0}, I), \tag{27}$$

which are consistent with the dynamics of the numerical model, for instance, samples from (27) are driven by the physics and the dynamics of the numerical model. Finally, the analysis equation of the PEnKF-D

$$X^a = \overline{x}^a \cdot \mathbf{1}_N^T + \widehat{A}^{\frac{1}{2}} \cdot \widehat{B}^{-\frac{1}{2}} \cdot \Delta X^b \in \mathbb{R}^{n \times N}, \tag{28}$$

where

$$\overline{x}^a = \overline{x}^b + A \cdot H^T \cdot R^{-1} \cdot \left[y - H \cdot \overline{x}^b\right], \tag{29}$$

$$A^{\frac{1}{2}} \approx \widehat{A}^{\frac{1}{2}} = \left[\tilde{L}^T \cdot \tilde{D}^{\frac{1}{2}}\right]^{-1}, \tag{30}$$

and

$$B^{-\frac{1}{2}} \approx \widehat{B}^{-\frac{1}{2}} = L \cdot D^{\frac{1}{2}}, \tag{31}$$

Since the moments of the posterior distribution are unchanged, we expect both methods PEnKF-S and PEnKF-D to perform equivalently in terms of errors during the estimation process.

3.2. Markov Chain Monte Carlo-based filters

Definitely, the EnKF represents a breakthrough in the data assimilation context, perhaps its biggest appeal is that we can obtain a closed form expression for the analysis members. Nevertheless, as can be noted, during the derivation of the analysis equations, some constrains are imposed, for instance, the observation operator is assumed linear, and therefore, the likelihood $\mathcal{P}(y|x)$ obeys a Gaussian distribution. Of course, in practice, this assumption can be easily broken, and therefore, bias can be introduced on the posterior prediction; this is notoriously significant if one considers the fields in which data assimilation lives are submerged. From a statistical point of view, more specifically, under the Bayesian framework, this situation can be summarized as follows:

- The prediction is obtained from the posterior distribution: $\mathcal{P}(x|y)$.
- The information from numerical model is known like the prior of background: $\mathcal{P}^b(x)$.
- The information from the observations is incorporated through the likelihood: $\mathcal{P}(y|x)$.
- The posterior distribution is calculated: $\mathcal{P}(x|y) \propto \mathcal{P}^b(x) \cdot \mathcal{P}(y|x)$.
- If $\mathcal{P}^b(x)$ and $\mathcal{P}(y|x)$ are Gaussian, $\mathcal{P}(y|x)$ is Gaussian too, and the analysis is equal to the mean of $\mathcal{P}(y|x)$.

A significant number of approaches have been proposed in the current literature in order to deal with these constraints; for instance, the particle filters (PFs) and the maximum likelihood ensemble filter (MLEF) are ensemble-based methods, which deal with nonlinear observation operators. Unfortunately, in the context of PF, its use is questionable under realistic weather forecast scenarios since the number of particles (ensemble members) increases exponentially regarding the number of components in the model state. Anyways, an extended analysis of these methods exceeds the scope of this document, but its exploration is highly recommended.

Taking samples directly from the posterior distribution is a strategy that can help to remove the bias induced by wrong assumptions on the posterior distribution (i.e., normal assumptions). We do not put the sights on finding the mode of the posterior distribution; instead of this, we want a set of state vectors that allow us to create a satisfactory representation of the posterior error distribution, then, based on these samples, it is possible to estimate moments of the posterior distribution from which the analysis ensemble can be obtained [9].

Hereafter, we will construct a modification of the sequential scheme of data assimilation; first, we will describe how to compute the analysis members by using variations in Markov Chain Monte Carlo (MCMC)-based methods, and then, we will include the modified Cholesky decomposition in order to estimate a precision background covariance.

An overview of the proposed method is as follows:

1. The forecast step is unchanged; the forecast ensemble is obtained by applying the model to the ensemble of states of the previous assimilation cycle.
2. The analysis step is modified, so that the analysis is not obtained anymore by, for instance (28), but we perform k iterations of an algorithm from the Markov Chain Monte Carlo (MCMC) family in order to obtain samples from the posterior distribution.

In order to be more specific in the explanation, let us define Metropolis-Hastings (MH) as the selected algorithm from the MCMC family. Now let us focus on MCMC in the most intuitive way possible. Let us define $J(x, y)$ equal to the posterior pdf obtained from the expression $\mathcal{P}(x|y) \propto \mathcal{P}^b(x) \cdot \mathcal{P}(y|x)$. MH explores the state space in order to include model states in a so-called Markov Chain. The selection of candidates is based on the condition $\left(x_{(c)}, y\right) > J\left(x_{(t-1)}, y\right)$, that is, if the value of $J(\cdot)$ for a candidate $x_{(c)}$ is greater than that for a previous vector $x_{(t-1)}$. Candidates are generated from a pre-defined proposal distribution $Q(\cdot)$; generally, a Gaussian multivariate distribution with mean $x_{(t-1)}$ is chosen, and a covariance matrix is defined in order to handle the step sizes. Concisely, MCMC methods proceed as follows: at first iteration, the chain is started with x^b vector. At end, the sample is obtained by extracting the last t vectors on the chain, and the other vectors are dismissed or "burned". In (32), we can see the expression to calculate the acceptance probability that relates the target pdf, $J(\cdot)$, with the proposal $Q(\cdot)$

$$\alpha = \frac{J\left(x_{(c)}|y\right) Q\left(x_{(t-1)}|x_{(c)}\right)}{J\left(x_{(t-1)}|y\right) Q\left(x_{(c)}|x_{(t-1)}\right)}, \tag{32}$$

The terms that involve $\boldsymbol{Q}(\cdot)$ can be canceled if a symmetric distribution is chosen as the proposal one, for instance, in the Gaussian distribution case. The procedure for the calculation of the analysis applying MH is described below:

1. Initialize the Markov Chain, $\boldsymbol{C}$, assigning the background value $\boldsymbol{x}^b$ to $\boldsymbol{C}_{(0)}$.
2. Generate a candidate vector state $\boldsymbol{x}_{(c)}$, from$\boldsymbol{Q}$.
3. Obtain$\mathcal{U}$ from a uniform (0, 1) distribution.
4. If $\mathcal{U} \leq \alpha$ Then: $\boldsymbol{C}_{(t+1)} = \boldsymbol{x}_{(c)}$ Else: $\boldsymbol{C}_{(t+1)} = \boldsymbol{C}_{(t)}$
5. Repeat Steps 2 through 4, for $t = 1$ until $t = k - 1$
6. Remove the first p vectors of the chain, the burned ones.

The analysis is computed over the sample:

$$\boldsymbol{x}^a = \frac{1}{(k-p)} \cdot \sum_{i=k-p}^{k} C_{(i)},$$

MCMC methods are straightforward to implement; when an enough number of iterations is performed, the behavior of the target function is captured on the chain [10]. This is true for even, complex density functions such as those with multiple modes. Briefly, let us focus on the fact that, generally, simulation-based methods such as MCMC explore a discretized grid, and as the mesh is refined, a huge number of iterations are needed before a high probability zone of the posterior distribution is reached [11]. Concretely, Hu et al. [12] proposed a family of modified MCMC dimension-independent algorithms under the name of preconditioned Crank-Nicolson (pCN) MCMC. These methods are robust regarding the curse of dimensionality in the statistical context. Initially, the Crank-Nicolson discretization is applied to a stochastic partial differential equation (SPDE) in order to obtain a new expression for the proposal distribution:

$$\boldsymbol{x}_{(c)} = \boldsymbol{x}_{(t-1)} - \frac{1}{2}\delta\mathcal{K}\mathcal{L}\left(\boldsymbol{x}_{(t-1)} + \boldsymbol{x}_{(c)}\right) + \sqrt{2\mathcal{K}\delta}\boldsymbol{\epsilon}_0, \tag{33}$$

where $\mathcal{L}$ is the precision matrix of $\boldsymbol{B}$, $\mathcal{K}$ is a preconditioning matrix, $\boldsymbol{\epsilon}_0$ is white noise, if $\mathcal{K} = \boldsymbol{B}$ and $\delta \in [0, 2]$ in (33), we get the pCN proposal described in (34):

$$\boldsymbol{x}_{(c)} = \boldsymbol{x}_{(t-1)} - \frac{1}{2}\delta\mathcal{K}\mathcal{L}\left(\boldsymbol{x}_{(t-1)} + \boldsymbol{x}_{(c)}\right) + \sqrt{2\mathcal{K}\delta}, \tag{34}$$

where $\omega \tilde{} \mathcal{N}(\boldsymbol{0}, \boldsymbol{B})$ and $\beta = 8\delta/(2+\delta)^2$. Of course, we use $\boldsymbol{P}^b$ like a computationally efficient estimation of $\boldsymbol{B}$. The acceptance probability is defined in (35):

$$\alpha = min\{1, \exp(\boldsymbol{J}(\boldsymbol{x}_{t-1}, \boldsymbol{y}) - \boldsymbol{J}(\boldsymbol{x}_c, \boldsymbol{y}))\}, \tag{35}$$

The procedure for the calculation of the analysis applying pCN Metropolis-Hastings is described as follows:

1. Initialize the Markov Chain, $\boldsymbol{C}$, assigning the background value $\boldsymbol{x}^b$ to $\boldsymbol{C}_{(0)}$.
2. Generate a candidate vector state using (3): $\boldsymbol{x}_{(c)} = \left(\mathbf{1} - \beta^2\right)^{\frac{1}{2}} \boldsymbol{x}_{(t-1)} + \beta\boldsymbol{\omega}$.
3. Obtain $\mathcal{U}$ from a uniform (0, 1) distribution.
4. If $\mathcal{U} \leq \alpha$, then $\boldsymbol{C}_{(t+1)} = \boldsymbol{x}_{(c)}$. Else: $\boldsymbol{C}_{(t+1)} = \boldsymbol{C}_{(t)}$.
5. Repeat Steps 2 through 4, for $t = 1$ until $t = k - 1$.
6. Remove the first p vectors of the chain, the burned ones.
7. The analysis is calculated over the sample: $\boldsymbol{x}^a = \frac{1}{(k-p)} \cdot \sum_{i=k-p}^{k} C_{(i)}$.

Finally, in the data assimilation context, by using the modified Cholesky decomposition described in the earlier sections, the pCN-MH filter reads:

1. The forecast step remains unchanged; the forecast ensemble is obtained by applying the model to the ensemble of states of the previous iteration, but the estimation of background covariance matrix is calculated by modified Cholesky.
2. The update step is modified, so that the analysis is obtained by run k iterations of pCN-MH, to obtain a sample of the posterior error distribution.

Now we are ready to numerically test the methods discussed in this chapter.

4. Experimental results

4.1. Stochastic filter PEnKF-S

We assess the accuracy of the PEnKF-S and compare it against that of the LETFK implementation proposed by Hunt [13]. The numerical model is the Lorenz 96 model [11], which mimics the behavior of the atmosphere:

$$\frac{dx_k}{dt} = -x_{k-1} \cdot \left(x_{k-2} - x_{k+1}\right) - x_k + F, \textit{for } 1 \leq k \leq n, \tag{36}$$

where n is the number of components in the model and F is an external force; with the value of the parameter, $F = 8$, the model presents a great entropy.

The experimental settings are described below:

1. An initial random solution $\boldsymbol{x}_{-3}^{+}$ is propagated for a while in time by using Lorenz 96 model and a fourth-order Runge Kutta method in order to obtain a vector state $\boldsymbol{x}_{-2}^{+}$ whose physics are consistent with the dynamics of such numerical model. This vector state serves as our reference solution.

http://dx.doi.org/10.5772/intechopen.72465

2. The reference solution is perturbed by using samples from a normal distribution with parameters $\mathcal{N}(0_n, \sigma_B \cdot I)$. Three different values for σ_B are considered during the numerical experiments $\sigma_B \in \{0.05, 0.10, 0.15\}$. This perturbed state is propagated in time in order to make it consistent with the physics and dynamics of the numerical model. From here, an initial background state $\boldsymbol{x}_{-1}^{b}$ is obtained.

3. A similar procedure is performed in order to build a perturbed ensemble about $\boldsymbol{x}_{-1}^{b}$. The ensemble members are propagated in time from where an ensemble of model realizations $\left\{\boldsymbol{x}_0^{b[i]}\right\}_{i=1}^{N}$ of model is obtained.

4. The assimilation windows consist of 15 equidistant observations. The frequency of observations is 0.5 time units, which represents 3.5 days in the atmosphere.

5. The dimension of the vector state is $n = 40$. The external force of the numerical model is set to $\boldsymbol{F} = \mathbf{80}$.

6. The number of observed components is 50% of the dimension of the vector state.

7. Three ensemble sizes are tried during the experiments $\in \{20, 40, 60\}$.

8. As a measure of quality, the L_2 norm of the analysis state and the reference solution are computed across assimilation steps.

9. A total of 100 runs are performed for each pair (N, σ_B). For each run, a different initial random vector is utilized in order to build the initial perturbed reference solution $\boldsymbol{x}_{-3}^{+}$ (before the model is applied x_{-2}^{*}). This yields to different initial ensembles as well as synthetic data for the different runs of each configuration (pair).

The average of the error norms of each pair (N, σ_B) for the LETKF and the PEnKF implementations is shown in **Table 1**. As can be seen, on average across 100 runs, the

σ_B	N	LETKF	PEnKF-S
0.05	20	22,6166	21,2591
	40	20,5671	18,2548
	60	20,0567	17,8824
0.10	20	23,1742	21,0725
	40	20,9513	18,3542
	60	18,5048	17,8240
0.15	20	24,8201	20,9059
	40	21,1314	18,1731
	60	20,8487	17,7590

Table 1. Average of L-2 norm of errors for 100 runs of each configuration (σ_B, N) for the compared filter implementations.

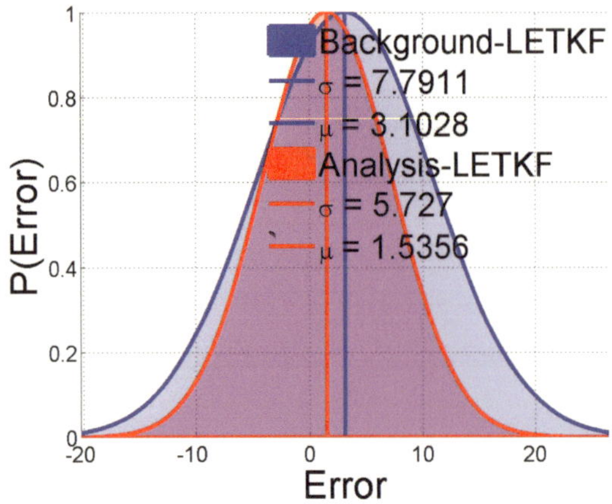

Figure 1. Local ensemble transform Kalmar filter (LETKF).

performance of the proposed EnKF implementation outperforms that of the LETKF in terms of L_2 norm of the error. Even more, the PEnKF-S seems to be invariant to the initial background error σ_B since, in all cases, when the ensemble size is increased a better estimation of the reference state $\boldsymbol{x}^*$ at different observation times is obtained. This can also obey to the estimation of background error correlations via the modified Cholesky decomposition since it is drastically improved whenever the ensemble size is increased as pointed out by Bickel and Levina in [1]. In such case, the error decreases by $\mathcal{O}(\log(n)/N)$. This is crucial in the PEnKF-S formulation since estimates of the precision analysis covariance matrix are obtained by rank-one updates on the inverse background error covariance matrix. On the other hand, in the LETKF context, increasing the ensemble size can improve the accuracy of the method but that is not better than the one shown by the PEnKF.

Some plots of the L_2 norm of error for the PEnKF and the LETKF across different configurations and runs are shown in **Figure 1**. Note that the error of the PEnKF decreases aggressively since the earlier iterations. In the LETKF context, the accuracy is similar to that of the PEnKF only at the end of the assimilation window.

4.2. Deterministic filter PEnKF-D

Holding the settings from the previous section, experiments are performed by using the PEnKF-D. The results have similar behavior to that of the LETFK and the PEnKF-S implementation proposed by Niño et al. in [8]. This can be appreciated in **Figures 1–3**. As can be seen, the error distributions reveal similar behavior across all compared filters.

4.3. MCMC filter based on modified Cholesky decomposition

We will describe experiments of our proposed filter based on a modified Cholesky decomposition and the preconditioned Crank-Nicolson Metropolis-Hastings. Again, the numerical

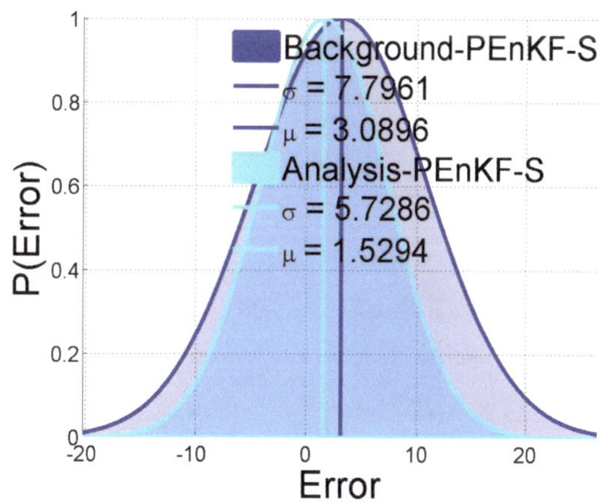

Figure 2. Posterior ensemble Kalmar filter stochastic (PEnKF-S).

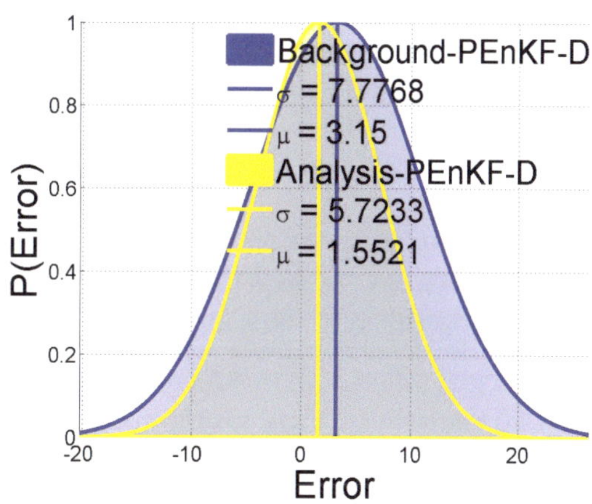

Figure 3. Posterior ensemble Kalmar filter deterministic (PEnKF-D).

model is the Lorenz 96 model. In this section, we are mainly interested on describing the behavior of the method, especially when the observational operator is nonlinear.

The experimental settings are as follows:

1. The observational operator is quadratic: $\{\mathcal{H}(x)\}_i = x_i^2$, where x_i denotes the ith component of the vector state x.

2. The vector of states has $n = 40$ components.

3. The ensemble size is $N = 20$.

4. The length of the Markov Chain is 1000.

5. The proposal is pCN with $\beta = 0.09$.

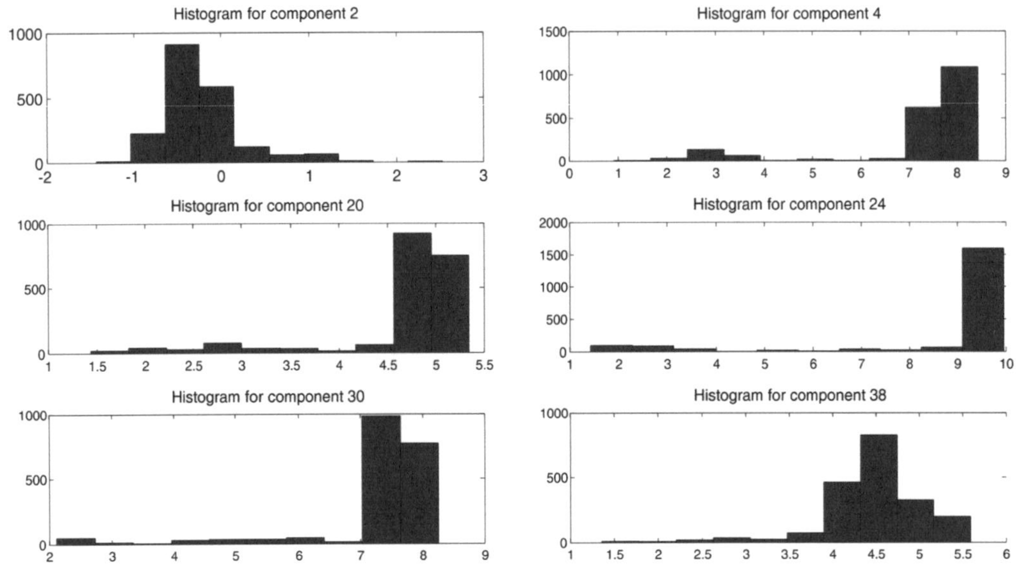

Figure 4. Histogram of six components of the state vector after 1000 iterations of the algorithm, the dashed lines indicate the position of the true value for the respective component.

Figure 4 describes the distribution of six of the 40 components of the state vector after 1000 iterations of the algorithm, attempting to visualize if the actual value is within or near the region of the highest probability density described by the sample contained in the Markov Chain.

Note that, not in all cases, the actual values of the model components were located at the peaks of the histogram, but most of them were within or near to zones of high probability described by the sample. This result is important if we take into account that probably the posterior distribution is not normal as is the case of quadratic observation operators.

5. Conclusions

In this chapter, efficient EnKF implementations were discussed. All of them are based on a modified Cholesky decomposition wherein a precision background covariance is obtained in terms of Cholesky factors. In the first filter, the PEnKF-S, synthetic data are utilized in order to compute the posterior members, as done in stochastic formulations of the filter. Even more, a sequence of rank-one updates can be applied over the factors of the prior precision matrix in order to estimate those of the posterior precision. In the second filter, the PEnKF-D, synthetic data are avoided by using perturbations obtained from the physics and the dynamics of the numerical model. Finally, a MCMC-based filter is obtained in order to reduce the impact of bias when Gaussian assumptions are broken during the assimilation of observations, for instance, when the observation operator is nonlinear. Numerical experiments with the Lorenz 96 model reveal that the proposed filters are comparable to filters from the specialized literature.

Acknowledgements

This work was supported by the Applied Math and Computational Science Lab (AML-CS) at Universidad del Norte in Barranquilla, Colombia.

Author details

Elias David Niño Ruiz*, Rolando Beltrán Arrieta and Alfonso Manuel Mancilla Herrera

*Address all correspondence to: enino@uninorte.edu.co

Applied Math and Computer Science Laboratory, Department of Computer Science, Universidad Del Norte, Barranquilla, Colombia

References

[1] Bickel PJ, Levina E. Covariance regularization by thresholding. The Annals of Statistics. 2008;**36**(6):2577-2604. DOI: 10.1214/08-AOS600

[2] Evensen G. The ensemble Kalman filter: Theoretical formulation and practical implementation. Ocean Dynamics. 2003;**53**(4):343-367

[3] Burgers G, Jan van Leeuwen P, Evensen G. Analysis scheme in the ensemble Kalman filter. Monthly Weather Review 1998;**126**(6):1719-1724

[4] Sakov P, Oke PR. A deterministic formulation of the ensemble Kalman filter: An alternative to ensemble square root filters. Tellus A. 2008;**60**(2):361-371

[5] Coleman TF, Van Loan C. Handbook for matrix computations. Society for Industrial and Applied Mathematics, 1988

[6] Nino-Ruiz ED, Sandu A, Deng X. A parallel implementation of the ensemble Kalman filter based on modified Cholesky decomposition. Journal of Computational Science. 2017

[7] Nino-Ruiz ED, Mancilla A, Calabria JC. A posterior ensemble Kalman filter based on a modified Cholesky decomposition. Procedia Computer Science. 2017;**108**:2049-2058

[8] Nino-Ruiz ED. A matrix-free posterior ensemble Kalman filter implementation based on a modified Cholesky decomposition. Atmosphere. 2017;**8**(7):125

[9] Attia A, Sandu A. A hybrid Monte Carlo sampling filter for non-gaussian data assimilation. AIMS Geosciences. 2015;**1**:41-78

[10] David H, Shane Reese C, David Moulton J, Vrugt JA, Colin F. Posterior exploration for computationally intensive. In: Steve B, Andrew G, Jones GL, Meng X-L, editors. Handbook of Markov Chain Monte Carlo. Chapman & Hall; 2011. p. 401-418

[11] Cotter SL, Roberts GO, Stuart AM, White D, et al. MCMC methods for functions: Modifying old algorithms to make them faster. Statistical Science. 2013;**28**(3):424-446

[12] Hu Z, Yao Z, Li J. On an adaptive preconditioned Crank–Nicolson MCMC algorithm for infinite dimensional Bayesian inference. Journal of Computational Physics. 2017;**332**:492-503

[13] Fatkullin I, Vanden-Eijnden E. A computational strategy for multiscale systems with applications to Lorenz 96 model. Journal of Computational Physics. 2004;**200**(2):605

Chapter 10

Kalman Filters for Reference Current Generation in Shunt Active Power Filter (APF)

Ahmad Shukri Bin Abu Hasim,
Syed Mohd Fairuz Bin Syed Mohd Dardin and
Zulkifilie Bin Ibrahim

Additional information is available at the end of the chapter

http://dx.doi.org/10.5772/intechopen.72467

Abstract

Shunt active power filter (APF) method have been used by many researchers as a solution in reducing the harmonics creating by the non-liner loads. Therefore, this research is targeted to design and implement a three-phase shunt APF employing Kalman filter estimator. Conventionally, low-pass filter (LPF) is used to filter out the unwanted DC component of the non-linear load to produce the sinusoidal waveform called the reference current. However, when applying LPF it contributes with the phase shift and high transient at the supply current. Therefore, to reduce these problems, the digital Kalman filter estimator is used to replace the LPF for generating the reference current. Details on the investigation between conventional and proposed methods under simulation based on Matlab Simulink platform and experimental that are made for two types of load, namely, three-phase rectifier with RC-load and three-phase induction motor, are presented. The performance criteria of the shunt APF are determined by the supply current waveform, total harmonic distortion (THD), harmonic spectrum and power quality measurements, which were also obtained by simulation and experimental. In conclusion, by employing Kalman filter estimator for generating the reference current, it reduces the time delay and high transient current at the power supply and, thus, improved the overall THD from 0.1 to 0.42% compared to the LPF.

Keywords: three-phase system, harmonic reduction, active power filter (APF), reference current generation, Kalman filter

1. Introduction

Electrical power is essential to people's modern lifestyle. In recent five decades, due to the development of the industry contributed to the increase of the types and capacity of the grid connected load drastically. For that reason, all the electrical consumers at all levels of usage have facing an issue of power quality problems. Both industrial/commercial sector and domestic environment commonly use sensitive equipment and non-linear loads (NLL). Inadvertently, these result in a non-sinusoidal current being drawn from the supply, which contains the harmful harmonic component and fed back to the supply system on the same point of common coupling (PCC). Passive filter is one of the common methods that have been used to overcome this problem. The passive filter is connected in parallel between the supply and NLL for improvement of power factor and harmonic suppression and thus exhibits lower impedance at tuned harmonic frequency. However, this approach does not solve the problem effectively due to its inability to compensate random frequency variation in the current, tuning problem and parallel resonant. Among the techniques, the *d-q* algorithm has been widely used to eliminate the harmonics due to its simplicity of control design relative to the rest. Commonly, the *d-q* algorithm is using LPF to generate the reference current. However, time delay introduced when applying the LPF will contribute to the phase shift in harmonics and higher transient current. Therefore, a new proposed technique of the current reference generator embedded with Kalman filter for shunt APF system is proposed where it reduces the time delay, thus producing improvement of the overall total harmonic distortion (THD) in the system.

1.1. State of the art

The active power filter (APF) technology is now mature in providing compensation for harmonics, reactive power and neutral current in AC networks. It has evolved for the past quarter century of development with varying configurations, control strategies and solid-state devices. Commonly, the APFs are used to eliminate the voltage harmonics, regulate terminal voltage, suppress voltage flicker and improve voltage balance in three-phase systems. This wide range of objectives can be achieved either individually or in combination depending upon the requirements, control strategy and configuration, which have to be selected appropriately. This section describes the history of development and the present status of the APF technology.

With the proliferation of power electronics in energy conversions, power quality is fast becoming an issue of an increasingly important aspect of electrical consumers at all levels of usage. A large number of publications have been covering the power quality survey, measurements, analysis, cause and effects of harmonics and reactive power in the electric networks [1–9]. APFs can be categorized into three types, namely, two-wire (single-phase), three-wire and four-wire three-phase configurations, to meet the requirements of the three types of NLL on supply systems. Domestic lights and ovens, TVs, computer power supplies, air conditioners, laser printers and Xerox machines behave as NLL and cause power quality problems for single-phase loads. For this type of load, the APFs are investigated in varying configurations and control strategies [10–19]. Starting in 1971, many configurations of APF have been developed for improving the power system quality. It can be categorized into four basic types, namely, series, parallel (shunt), hybrid APFs and unified power quality conditioner (universal AF). The series APF operates mainly as a voltage regulator and a harmonic isolator between NLL and utility system [20–23]. In other words, it allows only fundamental component of the current to flow in the system, suppressing other higher-frequency components. It can also be used to regulate the negative sequence voltage at the load. The series active filter is ideal for eliminating and/or maintaining the output voltage while balancing three-phase voltages [20, 21, 23–26]. On the other hand, shunt APF has been widely used to mitigate the harmonics. It cancels the load-current harmonics and provides reactive compensation to the supply, through the act of injecting equal but opposite harmonic compensating current to the supply [27–38]. Shunt APF has the advantage of carrying only the compensation current plus a small amount of active fundamental current to compensate for system losses. It is also possible to connect several filters in parallel to cater for higher currents, making this type of circuit suitable for a wide range of power ratings [26, 39, 40]. The most common configuration of shunt APF is the inverter type where the role of the filter inductor is to suppress the high frequency at tuned current generated at tuned frequency, while the converter provides complementary filtering on others that includes any random variations through switching techniques [28, 34, 41, 42]. The shunt APF controller can be used in direct or indirect connection. Hybrid APF can be characterized by a combination of passive filter and APF in series or parallel. The combination between series APF with parallel passive filter is the most popular arrangement because the solid-state devices used in active series part help in reducing the size and cost, to about 60–80% of load size [43, 44]. Furthermore, the passive parallel LC filter is used to eliminate lower-order harmonics at reasonable cost [26, 37, 44–49]. Another arrangement is the combination of active filter in series with a parallel passive filter, which is used especially for medium- and high-voltage applications [26]. Further arrangements also include a combination of parallel active and passive filters where the APF part is designed to eliminate the lower order of harmonics, while the passive filter works to eliminate the bulk load-current harmonic [26]. The combination of series active and parallel APF will produce unified power quality conditioner (also known as universal AF). The DC-link element of either inductor or capacitor is shared between two current sources or voltage-source bridges operating as active series and active parallel compensator [50, 51]. This universal AF is considered as an ideal AF, which eliminates voltage and current harmonics, thus capable of providing clean power to critical and harmonic-prone loads, such as computer, medical equipment and others. The main drawbacks are large costs and complex control due to dependency on the number of solid-state devices involved [26, 50, 51].

Many control approaches have been developed to extract and estimate the harmonics in the system. Instantaneous reactive power theory (*p-q* theory), modified *p-q* theory [52–54], *p-q-r* theory [55, 56], vectorial theory [57] and *d-q* theory [58–60] are the techniques that fall into the extraction technique. Due to its simplicity of control design relative to the rest, for that reason this *d-q* algorithm has been widely used to eliminate the harmonics [61]. On the other hand, estimation approach is used to estimate harmonics of frequency component present in the signal and measurement or estimation of the amplitude and phases of those frequencies [62]. This approach can be divided into two classes, non-parametric and parametric methods. The non-parametric methods are based on transformation of the given time-series data sequence. During the estimation process, these methods are not capable of incorporating with any available information about the system. Frequency domain approach using Fourier transform is most commonly used for spectrum analysis in this harmonic estimation [62]. In addition, parametric methods use an appropriate model to represent the signal and then estimate the parameters of the model from the available data points. Estimated parameters are applied to the selected model to determine harmonic contents in the signal. This parametric methods offer higher resolution and better accuracy than the non-parametric methods [62]. Kalman filter (KF) estimator is one of the methods that fall into the parametric method category which have been widely studied and used for different applications [62–69].

1.2. Main contribution

There are three main contributions regarding with this research:

a. Developed a new design of shunt APF employing Kalman filter estimator.

- The new design of shunt APF for generating the reference currents using Kalman filter estimator was proposed to reduce the delay time and high transient current when applying the conventional technique. In addition, the developed system was tested for two different types of loads such as three-phase rectifier with RC-load and three-phase induction motor.

b. Investigation of the details of performance based on simulation and experimental for conventional and the proposed technique.

- The investigation criteria are on the harmonic spectrum, THD and power quality for different three types of load because these criteria affect directly the performance of system that used active power filter.

c. Comparative studies between the conventional and the proposed technique upon experimental implementation.

d. An analysis is carried out in terms of harmonic spectrum, THD and power quality as well to validate the advantages offered by employing the new techniques relative to the common implementation of an active power filter.

1.3. Proposal of the research

Power quality problems have becoming a critical issue when dealing with power electronic converter and NLL due to the effects of the harmonic contamination in power system. Many techniques have been proposed to overcome these problems such as passive filter which contribute to improve the power factor and harmonic suppression and exhibit lower impedance at a tuned harmonic frequency. However this approach provides incomplete solutions particularly when compensating random frequency variations in the current, tuning and parallel resonant problems. Therefore, various active power filter (APF) configurations with their respective control strategies have been proposed and have been recognized as a viable solution to the problem created by harmonics. Among the technique, the *d-q* algorithm has been widely used to eliminate the harmonics due to its simplicity of control design relative to the rest. Commonly, the *d-q* algorithm is using LPF to generate the reference current. However, time delay introduced when applying the LPF will contribute to the phase shift in harmonics and higher transient current. A new proposed technique of the current reference generator embedded with Kalman filter (KF) for shunt APF system is proposed as shown in **Figure 1**. The KF in the system was used as a LPF to produce a reference current in three-phase system

http://dx.doi.org/10.5772/intechopen.72467

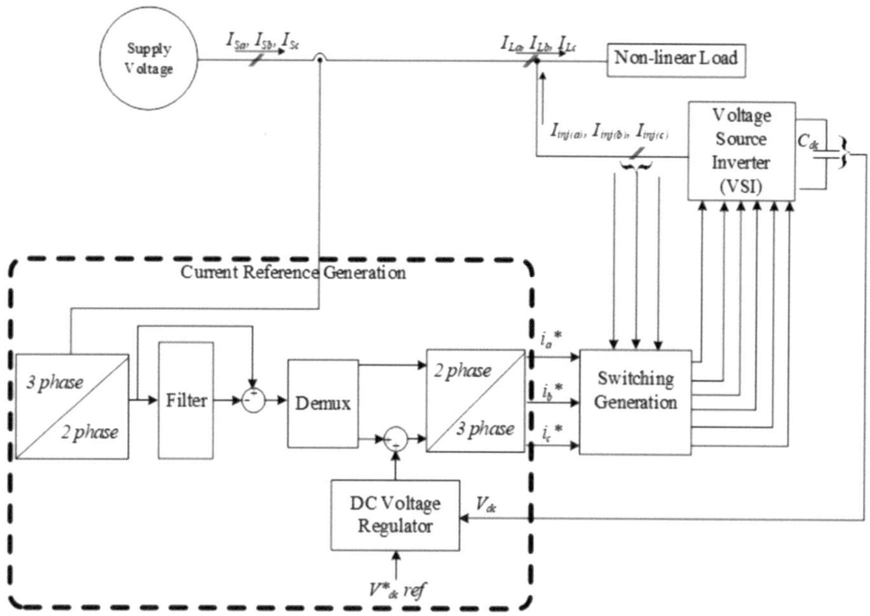

Figure 1. Overall system of shunt APF.

as shown in **Figure 2**. The KF used a form of feedback control in which the filter estimates the process at any time and then obtains feedback in the form of noisy measurements. These noisy measurements can be further exploited to improve the next estimates in which KF is able to perform because it has both time update and measurement update equations. The time update also known as predictor equation is responsible for projecting forward (in time) current state and error covariance estimate to obtain the estimation in the next time step, while the measurement update equation also called corrector equation is responsible for the feedback such as for incorporating a new measurement into the estimator to improve the

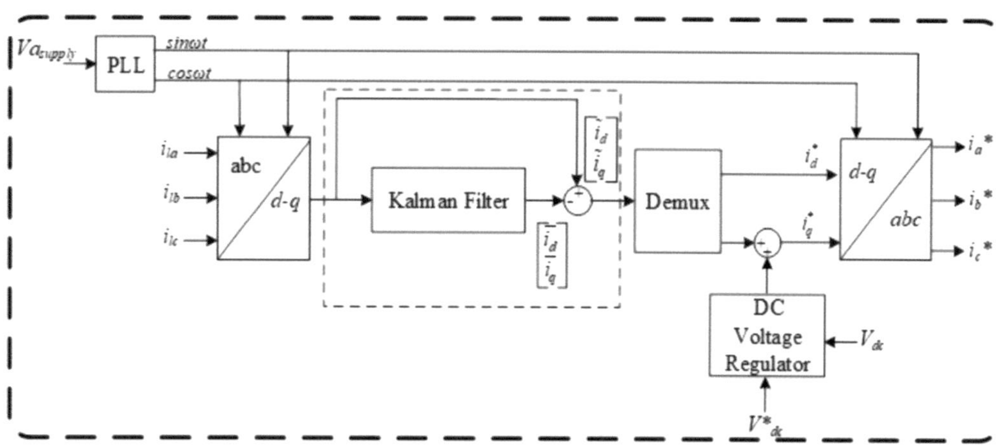

Figure 2. New technique of three-phase reference current generator employing Kalman filter estimator.

estimation. Therefore the estimation resembles the combination of predictor-corrector algorithm, which is used in the system. By applying KF, it improves the overall performance and also reduces the time delay and transient current which occurs in the conventional technique. Furthermore, this technique uses every measurement that the system has to further improve on the results by giving a better estimate at each time epoch. The significant improvement can be observed at the total harmonic distortion (THD) reduction at 2.38% compared to when the shunt APF is not implemented at all which performs at 168.39%. The TDH of the source current after the compensation is at 2.18% which is way below the IEEE 519 Standard which imposed a limit at less than 5% of the overall harmonics. In fact, for comparison, the use of KF also performed better than the established low-pass filter, which performs at 2.8% of the THD.

2. Mathematical formulation

There are three elements that involved in generating the required current reference that is used to compensate the undesirable load current components as shown in **Figure 2**. These elements are stationary reference frame, Kalman filter (KF) and DC voltage regulator. The mathematical formulation for each element is further explained in the next subtopics.

2.1. Stationary reference frame

Stationary reference frame also known as *d-q* algorithm was developed based on Park transformation. This method transforms three-phase into *d-q* coordinates (rotating reference frame with fundamental frequency) using Park transformations. In this case, the load currents are measured and transformed into *d-q* coordinates. The equations to transform *a-b-c* coordinate into *α-β-0* coordinate is presented in Eq. (1):

$$\begin{bmatrix} i_o \\ i_\alpha \\ i_\beta \end{bmatrix} = \sqrt{\frac{2}{3}} \begin{bmatrix} 1/\sqrt{2} & 1/\sqrt{2} & 1/\sqrt{2} \\ 1 & -1/2 & -1/2 \\ 0 & \sqrt{3}/2 & -\sqrt{3}/2 \end{bmatrix} \begin{bmatrix} i_{La} \\ i_{Lb} \\ i_{Lc} \end{bmatrix} \tag{1}$$

By employing Park transformation, the *α-β-0* coordinate is transformed into *d-q* coordinate as shown in Eq. (1):

$$\begin{bmatrix} i_d \\ i_q \end{bmatrix} = \begin{bmatrix} \cos\theta & \sin\theta \\ -\sin\theta & \cos\theta \end{bmatrix} \begin{bmatrix} i_\alpha \\ i_\beta \end{bmatrix} \tag{2}$$

where $\theta = \tan^{-1}\left(\frac{i_\beta}{i_\alpha}\right)$.

The phase angle, θ, in *d-q* frame is the same with fundamental frequency which makes the DC fundamental current component $(i_{\bar{d}}, i_q)$ and harmonic AC component $(i_{\tilde{d}}, i_{\tilde{q}})$ arise due to harmonics at the load [61]. Low-pass filter (LPF) is normally used to determine the DC component. Nevertheless, for such system, phase shift in harmonics and high transient response is unavoidable before attaining its steady state. This is where KF estimator is used to replace the LPF and further improve the overall performance of the system. In order to stabilize the voltage on the DC side of the VSI, the measurement voltage, V_{dc}, measure must follow the reference voltage, V_{dc} *ref*. Therefore, DC voltage regulator loop is designed by integrating a suitable PI controller.

2.2. DC voltage regulator

The DC voltage regulator is controlled with a traditional PI controller. The DC voltage, V_{dc}, is measured and then compared with a constant reference value V_{dc}*. The error is processed by a PI controller with two gains: K_p and K_i. Both gains are calculated and tuned accordingly to the dynamic response in which the values of both gain are set to 4 for K_p and 91 for K_i.

2.3. Kalman filter

The use of Kalman filter (KF) provides an efficient computational means to estimate the state of a process which is able to minimize the means of the squared error. This is achieved by keeping tracks of the estimated state of the system as well as the variance of the estimates via two distinct phases: predict and update. The basic KF can be defined as.

$$x_k = A\,x_{k-1} + B\,u_k + w_k \tag{3}$$

and

$$z_k = H\,x_k + v_k \tag{4}$$

where A is the state transition matrix, B is the control matrix that is applied to u_k, which is the control vector of the system, and H is defined as observation matrix with x_k the state of the system and y_k the measurement or sometimes called observation vector. w_k and v_k are the process noise vector and observation noise vector, respectively, and it is assumed to be mutually independent and normally distributed. Relative to the system, since the fundamental positive sequence components of the non-linear load current appears as DC quantities of the synchronous reference frame rotating at 50 Hz, it can then be separated from the load currents using KF as depicted in **Figure 3**.

In this case, the state transition matrix is the differential equation that relates the state at the previous time step $k - 1$ to the current step k. Therefore, the state vector x_k can be further defined as.

$$x_k = \begin{bmatrix} i_{d(k)} \\ i_{q(k)} \end{bmatrix} \tag{5}$$

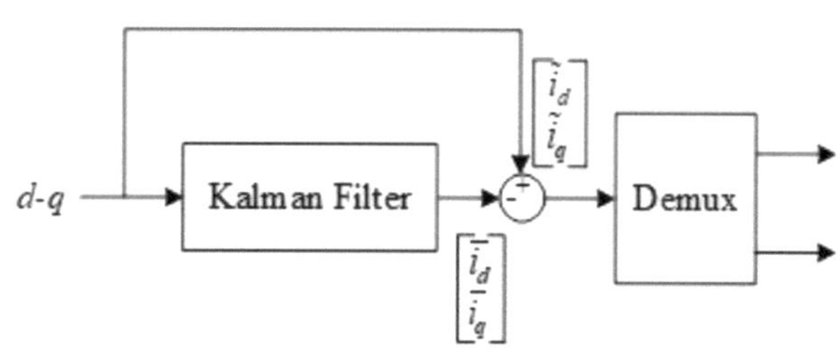

Figure 3. Kalman filter.

Furthermore, the optional control input which defined the control matrix B can be neglected. Since the system only measured two parameters, respectively, therefore the measurement matrix can be simply represented by two-by-two identity matrix. Therefore, the implementation of the KF can be rewritten as.

$$\begin{aligned} x_k &= A\,x_{k-1} + w_k \\ P_k^- &= A\,P_{k-1}\,A^T + Q \end{aligned} \tag{6}$$

for the predictor:

$$\begin{aligned} z_k &= H\,\hat{x}_k^- + v_k \\ S_k &= H\,P_k^-\,H^T + R_k \\ \hat{x}_k &= \hat{x}_k^- + K_k(z_k - H\,\hat{x}_k^-) \end{aligned} \tag{7}$$

The measurement update equation $\hat{x}_k$ is the estimate reference current of i_d and i_q, $\hat{x}_k^-$ is the predicted state, z_k is the measurement of actual current, P_k is the estimate error covariance, R_k is the observation covariance matrix and K_k is the Kalman gain. In this representation, matrix P is the variance matrix of the error $x_k - \hat{x}_k$ where the goal is to minimize this value. Here the Kalman gain calculation will be based on the conventional calculation defined in Eq. (8):

$$\begin{aligned} K_k &= P_k^-\,H^T\,S^{-1} \\ P_k &= (I - K_k\,H)\,P_k^- \end{aligned} \tag{8}$$

The process noise covariance matrix Q and observation noise covariance matrix R are tuned manually in order to achieve the optimal performance of the filter. **Figure 4** shows the cycle of KF.

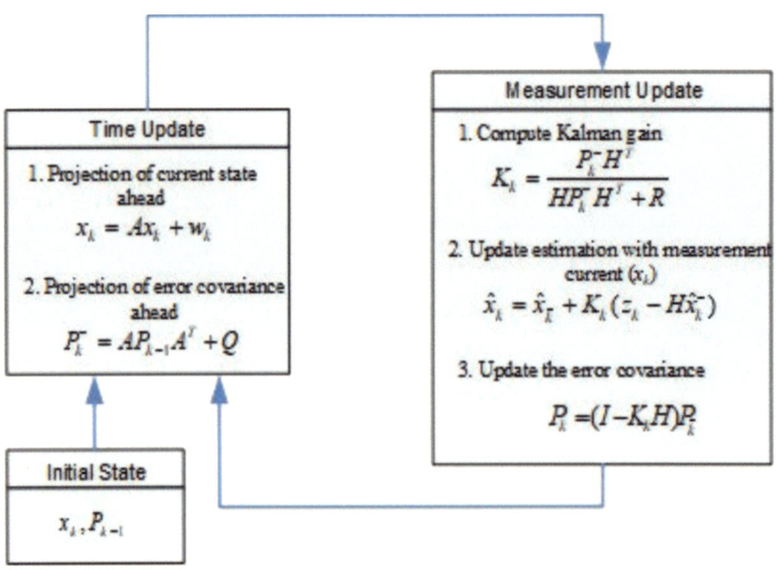

Figure 4. Cycle of discrete Kalman filter.

3. Simulation and experimental result

The proposed simulation and experimental results designed for the three-phase reference current generation employing KF estimator for three-phase shunt APF are presented. The work is simulated and implemented using Matlab Simulink and dSPACE.

3.1. Non-linear load

The results for the APF before and after compensation are simulated using Matlab Simulink, while Fluke Power Quality Analyzer captures the results for the experimental. **Figure 5(a)** and **(b)** shows the supply current waveform before the compensation for simulation and experimental result; thus, the harmonic spectrum of both simulation and experimental is shown in **Figure 6**, respectively.

From the harmonic spectrum results, the total harmonic distortion (THD) can be determined by using the formula defined as.

$$\%THD = \frac{\sqrt{\sum_{n=2}^{\infty} I_h^2}}{I_f} \tag{9}$$

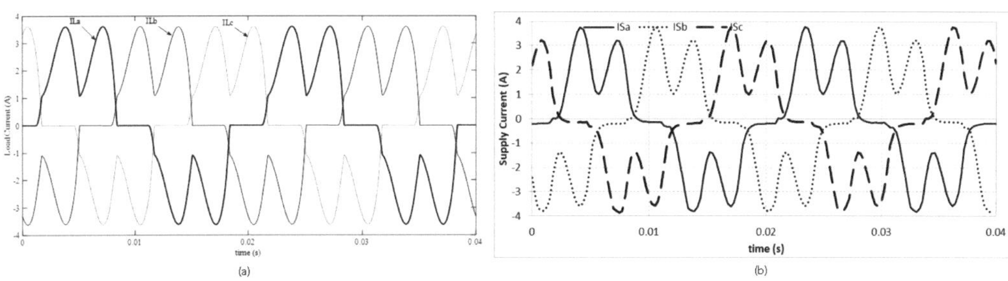

Figure 5. Simulation and experimental result without shunt APF (a) simulation and (b) Experiment.

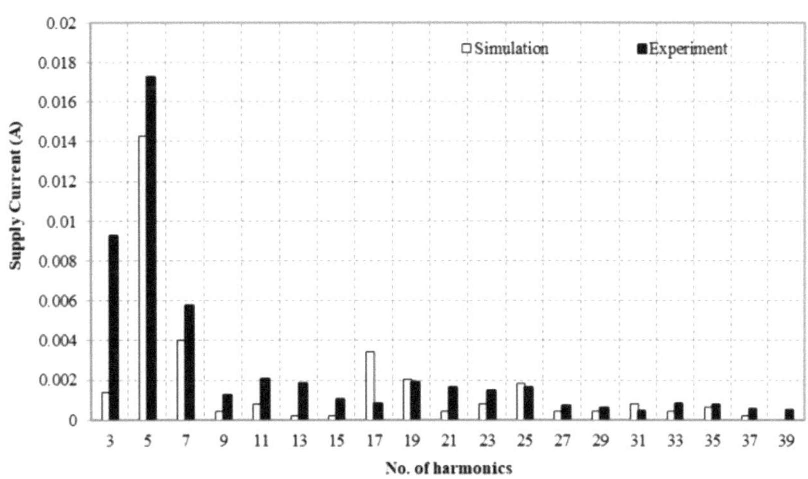

Figure 6. Harmonic spectrum before the compensation.

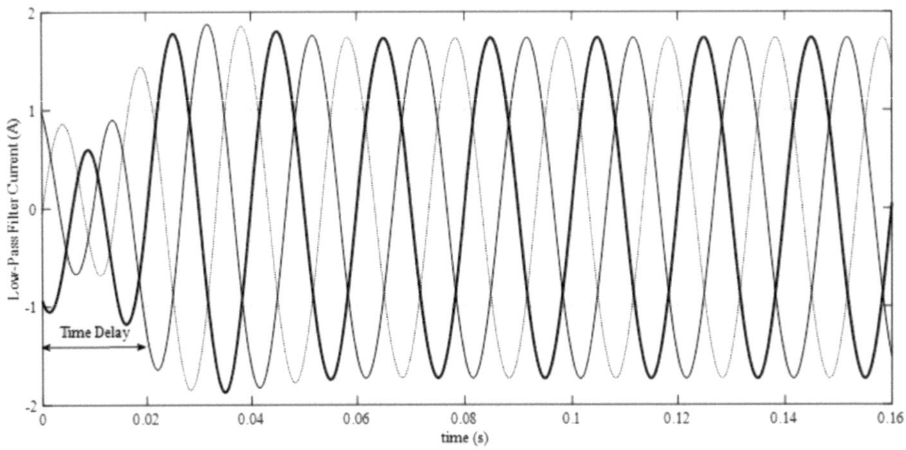

Figure 7. Simulation Butterworth low-pass filter.

where I_h is the harmonic component, I_f is the fundamental component, and n is the harmonic number: 2, 3, 4, etc.

Therefore, the THD of the line current obtained by the simulation is 56.14%, while the experimental obtains about 47.26%. There are slightly different between the simulation and experimental results because the simulation is simulated at an ideal condition.

3.2. Kalman filter estimator result versus low-pass filter

Commonly, a Butterworth low-pass filter (LPF) was applied to filter out the unwanted DC component for *d-q* algorithm to ensure that the correct reference currents are generated in the system. Failure to obtain the correct reference current resolves reduction of the overall

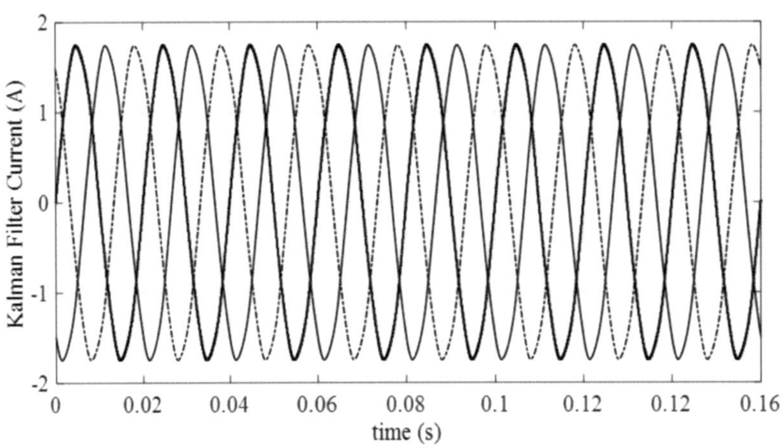

Figure 8. Simulation of Kalman filter.

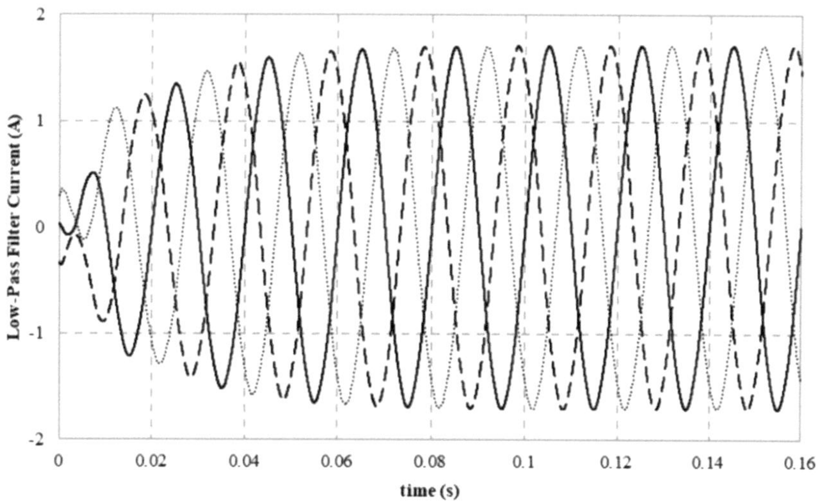

Figure 9. Experimental Butterworth low-pass filter.

performance of the active power filter (APF). But, time delay which contributes to the phase shift in harmonics and high transient current is the common effect when applying the LPF. **Figure 7** shows the shunt APF when applying Butterworth LPF. It is clearly shown that from the figure, the time delay is recorded at 0.02 s with 43.36% of the THD. On the other hand, there is no time delay when applying the shunt APF using KF estimator which is shown in **Figure 8**. Therefore, the THD produced by the KF estimator is 98% improvement compared to LPF. On the other hand, the experimental results for low-pass and KF are shown in **Figures 9** and **10**, respectively.

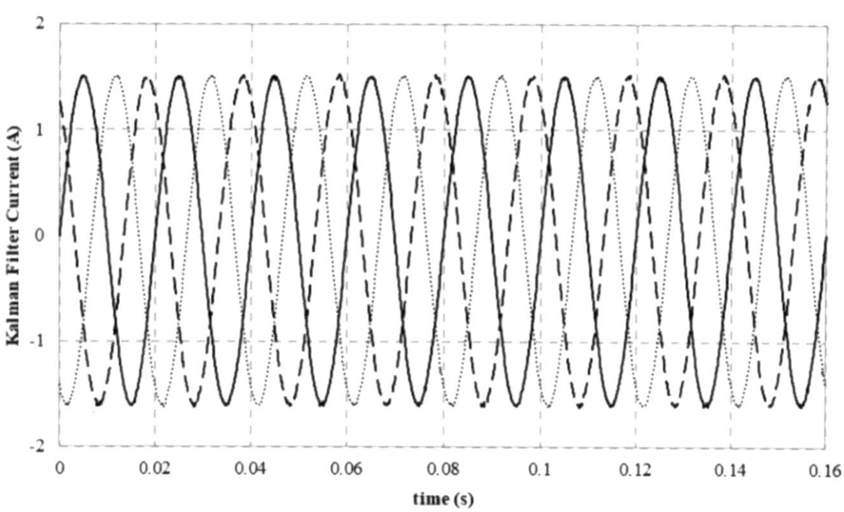

Figure 10. Experiment of Kalman filter.

3.3. Three-phase shunt active power filter

In this shunt active power filter feeding a non-linear load, the results between Butterworth LPF and KF estimator are compared between simulation and experimental, which are shown in **Figures 11** and **12**, respectively.

From the results obtained, it can be concluded that almost the same waveform was produced for both simulation and experimental approaches. Furthermore, the harmonic spectrum form the experimental is shown in **Figure 13**.

Both methods have demonstrated a harmonic reduction with almost identical fundamental current between simulation and experimental.

The THD results obtained show that the new technique shunt APF abides the regulation of IEEE 519–1992 standard. **Tables 1** and **2** show the THD after simulation and experimental results, respectively.

From the observation, the shunt APF using KF estimator technique produces about 0.1% better THD than LPF either in simulation or experimental.

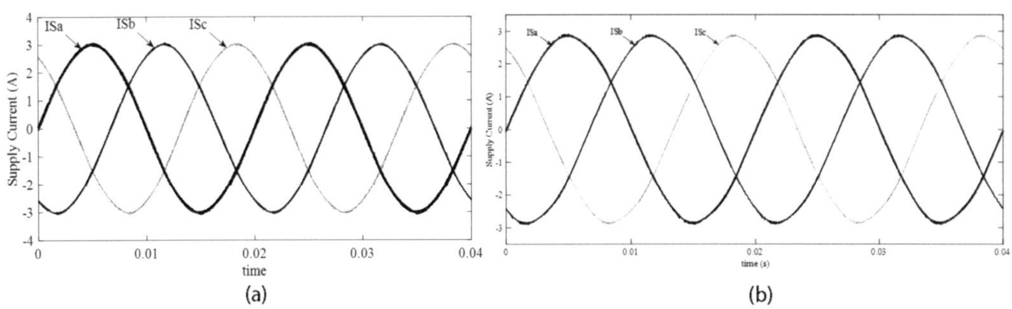

Figure 11. Simulation result for shunt APF (a) low-pass filter and (b) Kalman filter.

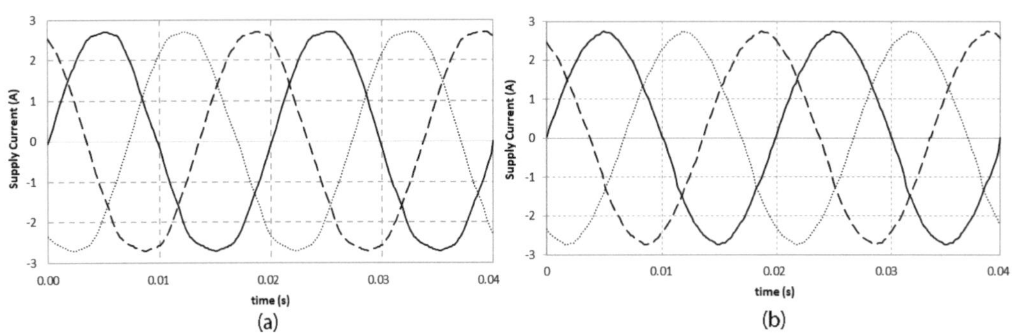

Figure 12. Experimental results for shunt APF: (a) low-pass filter and (b) Kalman filter.

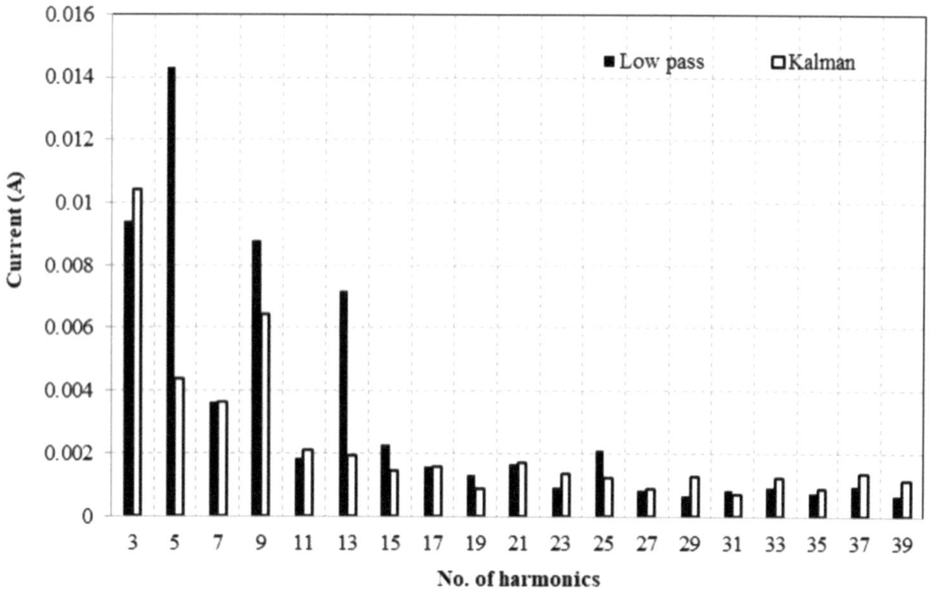

Figure 13. Harmonics spectrum.

Types of reference current generation	THD before (%)	THD after (%)
Low-pass filter	55.88	2.09
Kalman filter estimator		1.99

Table 1. Simulation results.

Types of reference current generation	THD before (%)	THD after (%)
Low-pass filter	47.27	2.30
Kalman filter estimator		2.18

Table 2. Experimental results.

3.4. Operation with three-phase induction motor speed drive

A 1.5 kW, 380 V variable speed induction motor (IM) drive is connected in parallel to the APF and the three-phase supply voltages. The motor is operated as a non-linear load and starts to accelerate from standstill at time, $t = 0.06$ s, until it reached the required reference speed which is set at 1400 rpm. **Figures 14** and **15** show the supply current when the IM starts to accelerate without and with shunt APF, respectively. Fluke Power Quality Analyzer was used to measure the THD at steady-state condition (t = 0.28 s). THD obtained before applying shunt APF is 168.39%, whereas when applying the shunt APF using both KF and LFP, the THD reduced to 2.38 and 2.80%. Furthermore, the harmonic spectrum with or without shunt APF for both KF and LPF is shown in **Figures 16–18**, respectively. It can conclude that from the results, the shunt APF employing KF-based estimator produced lower THD than LPF for an induction motor drive application.

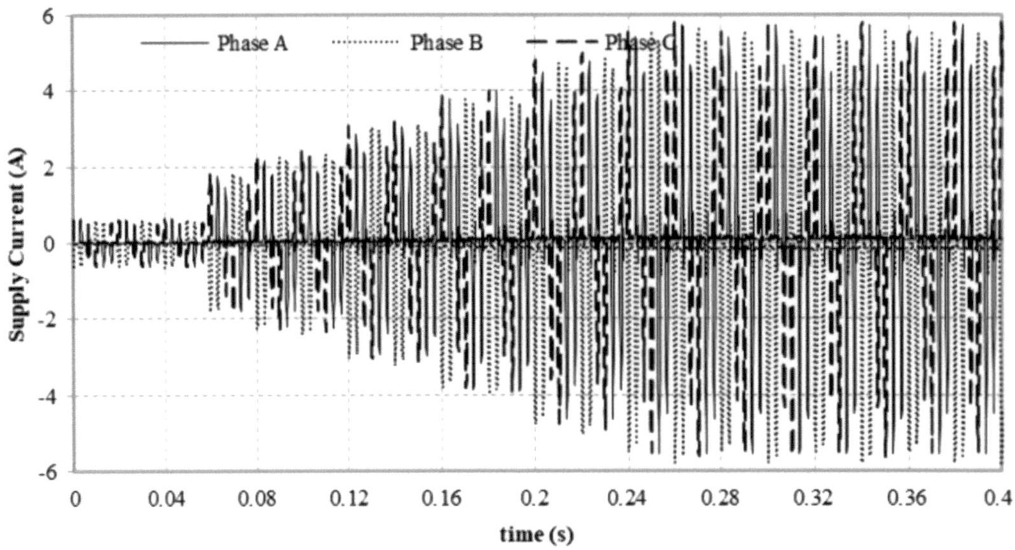

Figure 14. Supply current waveform without shunt APF.

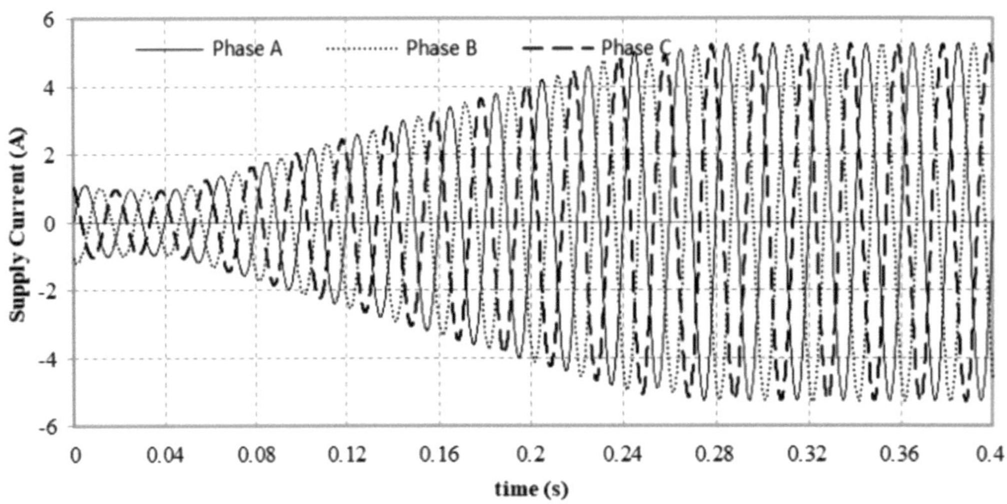

Figure 15. Supply current when applying shunt APF with Kalman filter estimator.

http://dx.doi.org/10.5772/intechopen.72467

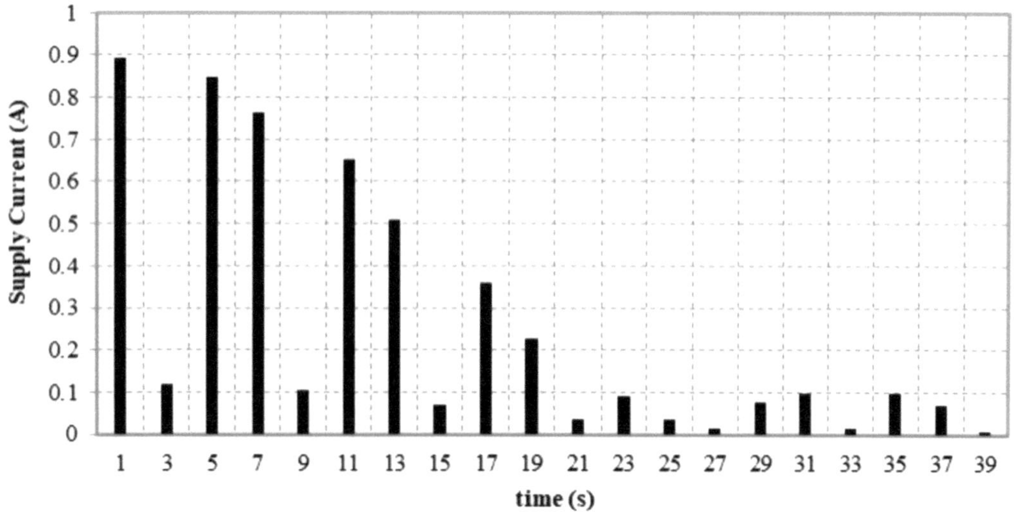

Figure 16. Harmonic spectrum without shunt APF.

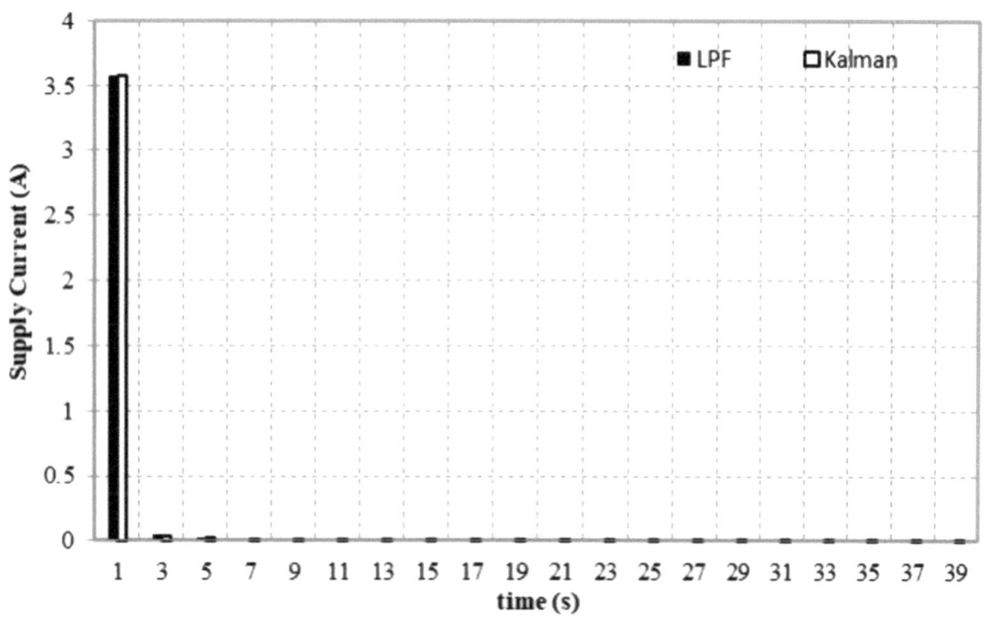

Figure 17. Harmonics spectrum after applying shunt APF.

The overall total harmonic distortion with or without shunt APF is shown tabulated in **Table 3**, which shows that the KF estimator produces lower THD than LPF for three-phase induction motor.

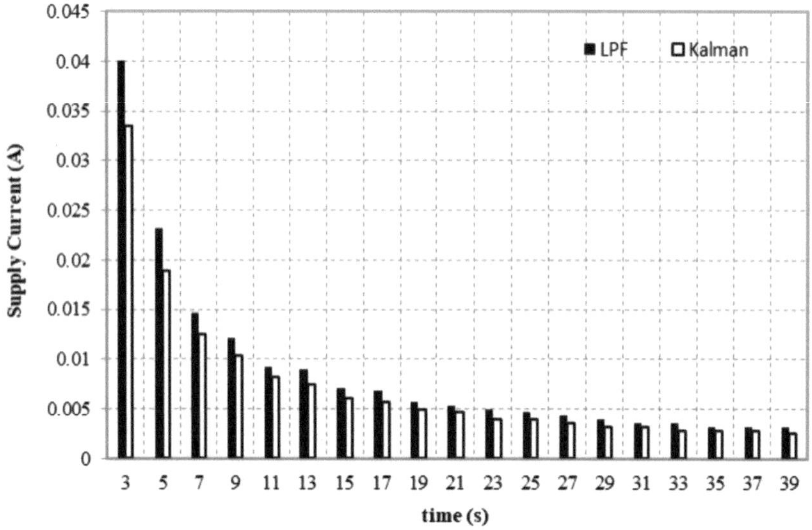

Figure 18. Harmonics spectrum without fundamental.

Reference current generation	THD before (%)	THD after (%)
Low-pass filter	168.39	2.80
Kalman filter estimator		2.38

Table 3. THD of supply current before and after applying shunt APF.

4. Conclusion

The new techniques of reference current generator by employing KF estimator for shunt APF technique have been presented. The results of the proposed technique in generating the three-phase reference current towards reducing the THD are established using simulation and experimental. For the three-phase rectifier connected with RC load, the performance of the proposed technique is comparable with those based on the LPF reference current generation. The THD of the source current from the experimental result after the compensation is 2.18% which is less than 5% of the harmonic limit imposed by the IEEE 519 standard. Furthermore, nearly 0.1% THD improvement was gained by the proposed techniques compared to LPF. Thus, the comparison of different reference current grid generations for shunt APF is also presented. The performance of KF estimator reference current generation was also studied for induction motor variable speed drive. In induction motor, almost 0.42% improvement of THD was gathered when applying KF estimator compared to LPF.

5. Future works

Although the proposed technique improved the overall performance of shunt APF, there is a room of improvement and suggestion for further research work such as:

1. The hysteresis band plays a significant influence to the THD reduction. With the lower hysteresis band, more accurate PWM generated, thereby improving the THD. In this thesis, the hysteresis band is set to ±0.08 while in simulation at ±0.001. Therefore, in order to have faster and more accurate result, the combination of dSpace and FPGA can be implemented to reduce the computational time from the dSpace.

2. Further combination between shunt APF and passive filter (hybrid APF) can be used to improve the performance of the APF. Where the passive filter used to filter the higher order harmonic while shunt APF filter the lower order harmonic.

3. Apply the Kalman filter to the instantaneous real and reactive algorithm. Conventionally, the technique used *p-q* algorithm that combined with high-pass filter. However, Kalman filter can be used to replace it, and the performance of the system can be further investigated.

Author details

Ahmad Shukri Bin Abu Hasim[1]*, Syed Mohd Fairuz Bin Syed Mohd Dardin[1] and Zulkifilie Bin Ibrahim[2]

*Address all correspondence to: shukri@upnm.edu.my

1 Department of Electrical and Electronics Engineering, Faculty of Engineering, National Defence University of Malaysia, Kuala Lumpur, Malaysia

2 Faculty of Electrical Engineering, Universiti Teknikal Malaysia Melaka, Melaka, Malaysia

References

[1] Duffey CK, Stratford RP. Update of harmonic standard IEEE-519-IEEE recommended practices and requirements for harmonic control in electric power systems. In: Conference Record of the IEEE Industry Applications Society Annual Meeting. 1989;**2**:1618-1624

[2] Shuter TC, Vollkommer HT, Kirkpatrick TL. Survey of harmonic levels on the American electric power distribution system. IEEE Transactions on Power Delivery. 1989;**4**:2204-2213

[3] Subjak JS Jr, McQuilkin JS. Harmonics-causes, effects, measurements, and analysis: An update. IEEE Transactions on Industry Applications. 1990;**26**:1034-1042

[4] Beides HM, Heydt GT. Dynamic state estimation of power system harmonics using Kalman filter methodology. IEEE Transactions on Power Delivery. 1991;**6**:1663-1670

[5] Gruzs TM. Uncertainties in compliance with harmonic current distortion limits in electric power systems. IEEE Transactions on Industry Applications. 1991;**27**:680-685

[6] Emanuel AE, Orr JA, Cyganski D, Gulachenski EM. A survey of harmonic voltages and currents at the customer's bus. IEEE Transactions on Power Delivery. 1993;**8**:411-421

[7] Henderson RD, Rose PJ. Harmonics: The effects on power quality and transformers. IEEE Transactions on Industry Applications. 1994;**30**:528-532

[8] Mansoor A, Grady WM, Staats PT, Thallam RS, Doyle MT, Samotyj MJ. Predicting the net harmonic currents produced by large numbers of distributed single-phase computer loads. IEEE Transactions on Power Delivery. 1995;**10**:2001-2006

[9] Clark SL, Famouri P, Cooley WL. Elimination of supply harmonics. IEEE Industry Applications Magazine. 1997;**3**:62-67

[10] Kazerani M, Ziogas PD, Joos G. A novel active current waveshaping technique for solid-state input power factor conditioners. IEEE Transactions on Industrial Electronics. 1991;**38**:72-78

[11] Enjeti P, Shireen W, Pitel I. Analysis and design of an active power filter to cancel harmonic currents in low voltage electric power distribution systems. In: Proceedings of the 1992 International Conference on Industrial Electronics, Control, Instrumentation, and Automation. 1992;**1**:368-373

[12] Duke RM, Round SD. The steady-state performance of a controlled current active filter. IEEE Transactions on Power Electronics. 1993;**8**:140-146

[13] Round SD, Mohan N. Comparison of frequency and time domain neural network controllers for an active power filter. In: Proceedings of the IECON '93, International Conference on Industrial Electronics, Control, and Instrumentation; 1993. 1993;**2**:1099-1104

[14] Jou HL, Wu JC, Chu HY. New single-phase active power filter. IEE Proceedings—Electric Power Applications. 1994;**141**:129-134

[15] Nastran J, Cajhen R, Seliger M, Jereb P. Active power filter for nonlinear AC loads. IEEE Transactions on Power Electronics. 1994;**9**:92-96

[16] Jae-Ho C, Ga-Woo P, Dewan SB. Standby power supply with active power filter ability using digital controller. In: Applied Power Electronics Conference and Exposition, 1995. APEC '95. Conference Proceedings 1995, Tenth Annual. 1995;**2**:783-789

[17] Torrey DA, Al-Zamel AMAM. Single-phase active power filters for multiple nonlinear loads. IEEE Transactions on Power Electronics. 1995;**10**:263-272

[18] Kim YJ, Kim JS, Kim YS. Single-phase active power filter based on rotating reference frame method. In: Proceedings of the Eighth International Conference on Electrical Machines and Systems, 2005. ICEMS 2005. 2005. pp. 1428-1431

[19] Khan MM, Feng JF, Chen C, Zhiming W. Single-phase dynamically decoupled active power filter for system-integrated application. IEEE Proceedings—Electric Power Applications. 2006;**153**:625-631

[20] Rudnick H, Dixon J, Moran L. Delivering clean and pure power. IEEE Power and Energy Magazine. 2003;**1**:32-40

[21] Rahim NA, Mekhilef S, Zahrul I. A single-phase active power filter for harmonic compensation. In: IEEE International Conference on Industrial Technology, 2005. ICIT 2005. 2005. pp. 1075-1079

[22] Wang X, Liu J, Yuan C, Wang Z. A comparative study on voltage-source control and current-source control of series active power filter. In: Applied Power Electronics Conference and Exposition, 2006. APEC '06. Twenty-First Annual IEEE. 2006. 6 pp

[23] Chang GW, Chen SK, Chin YC, Chen WC. An a-b-c reference frame-based compensation strategy for series active power filter control. In: 2006 1ST IEEE Conference on Industrial Electronics and Applications. 2006. pp. 1-4

[24] Hurng-Liahng J, Jinn-Chang W, Yao-Jen C, Ya-Tsung F. A novel active power filter for harmonic suppression. IEEE Transactions on Power Delivery. 2005;**20**:1507-1513

[25] Moran LT, Mahomar JJ, Dixon JR. Careful connection. IEEE Industry Applications Magazine. 2004;**10**:43-50

[26] El-Habrouk M, Darwish MK, Mehta P. Active power filters: A review. IEE Proceedings—Electric Power Applications. 2000;**147**:403-413

[27] Golwala H, Chudamani R. Comparative study of switching signal generation techniques for three phase four wire shunt active power filter. In: Electric Machines & Drives Conference (IEMDC), 2011 IEEE International. 2011. pp. 1409-1414

[28] Zeng FP, Tan GH, Wang JZ, Ji YC. Novel single-phase five-level voltage-source inverter for the shunt active power filter. IET Power Electronics. 2010;**3**:480-489

[29] Rahmani S, Mendalek N, Al-Haddad K. Experimental design of a nonlinear control technique for three-phase shunt active power filter. IEEE Transactions on Industrial Electronics. 2010;**57**:3364-3375

[30] Golwala H, Chudamani R. Simulation of three-phase four-wire shunt active power filter using novel switching technique. In: 2010 Joint International Conference on Power Electronics, Drives and Energy Systems (PEDES) & 2010 Power India. 2010. pp. 1-7

[31] Abdalla II, Rao KSR, Perumal N. Harmonics mitigation and power factor correction with a modern three-phase four-leg shunt active power filter. In: 2010 IEEE International Conference on Power and Energy (PECon). 2010. pp. 156-161

[32] Vodyakho O, Mi CC. Three-level inverter-based shunt active power filter in three-phase three-wire and four-wire systems. IEEE Transactions on Power Electronics. 2009;**24**:1350-1363

[33] Vodyakho O, Kim T, Kwak S, Edrington CS. Comparison of the space vector current controls for shunt active power filters. IET Power Electronics. 2009;**2**:653-664

[34] Vodyakho O, Kim T. Shunt active filter based on three-level inverter for three-phase four-wire systems. IET Power Electronics. 2009;**2**:216-226

[35] Uyyuru KR, Mishra MK, Ghosh A. An optimization-based algorithm for shunt active filter under distorted supply voltages. IEEE Transactions on Power Electronics. 2009;**24**:1223-1232

[36] Singh B, Solanki J. An implementation of an adaptive control algorithm for a three-phase shunt active filter. IEEE Transactions on Industrial Electronics. 2009;**56**:2811-2820

[37] Rahmani S, Hamadi A, Mendalek N, Al-Haddad K. A new control technique for three-phase shunt hybrid power filter. IEEE Transactions on Industrial Electronics. 2009;**56**:2904-2915

[38] Peng X, Venayagamoorthy GK, Corzine KA. Seven-level shunt active power filter for high-power drive systems. IEEE Transactions on Power Electronics. 2009;**24**:6-13

[39] Akagi H. The state-of-the-art of active filters for power conditioning. In: 2005 European Conference on Power Electronics and Applications. 2005. pp. 15

[40] Akagi H. Active harmonic filters. Proceedings of the IEEE. 2005;**93**:2128-2141

[41] Pini SH, Barbi I. A single-phase high-power-factor rectifier, based on a two-quadrant shunt active filter. IEEE Transactions on Power Electronics. 2011;**26**:3131-3143

[42] Karuppanan P, Mahapatra K. PLL with PI, PID and fuzzy logic controllers based shunt active power line conditioners. In: 2010 Joint International Conference on Power Electronics, Drives and Energy Systems (PEDES) & 2010 Power India. 2010. pp. 1-6

[43] Singh B, Verma V, Chandra A, Al-Haddad K. Hybrid filters for power quality improvement. IEE Proceedings—Generation, Transmission and Distribution. 2005;**152**:365-378

[44] Zhao W, Luo A, Shen ZJ, Wu C. injection-type hybrid active power filter in high-power grid with background harmonic voltage. IET Power Electronics. 2011;**4**:63-71

[45] A. Varschavsky, J. Dixon, M. Rotella, Moran L. Cascaded nine-level inverter for hybrid-series active power filter, using industrial controller, IEEE Transactions on Industrial Electronics; 2010;**57**:2761-2767

[46] Salmeron P, Litran SP. Improvement of the electric power quality using series active and shunt passive filters. IEEE Transactions on Power Delivery. 2010;**25**:1058-1067

[47] Luo A, Shuai Z, Zhu W, Shen ZJ, Tu C. Design and application of a hybrid active power filter with injection circuit. IET Power Electronics. 2010;**3**:54-64

[48] Litran SP, Salmeron P, Vazquez JR, Herrera RS, Perez A. Control strategy for hybrid power filter to compensate unbalanced and non-linear, three-phase loads. In: 13th European Conference on Power Electronics and Applications, 2009. EPE '09. 2009. pp. 1-10

[49] An L, Ci T, Zhi Kang S, Wei Z, Fei R, Ke Z. A novel three-phase hybrid active power filter with a series resonance circuit tuned at the fundamental frequency. IEEE Transactions on Industrial Electronics. 2009;**56**:2431-2440

[50] Pal Y, Swarup A, Singh B. A review of compensating type custom power devices for power quality improvement. In: Joint International Conference on Power System Technology and IEEE Power India Conference, 2008. POWERCON 2008. 2008. pp. 1-8

[51] Kazemi A, Sarlak M, Barkhordary M. An adaptive noise canceling method for single-phase unified power quality conditioner. In: 2006 1ST IEEE Conference on Industrial Electronics and Applications. 2006. pp. 1-6

[52] Herrera RS, Salmeron P, Vazquez JR, Litran SP, Perez A. GENERALIZED instantaneous reactive power theory in poly-phase power systems. In: 13th European Conference on Power Electronics and Applications, 2009. EPE '09. 2009. pp. 1-10

[53] Kelesidis K, Adamidis G, Tsengenes G. Investigation of a control scheme based on modified p-q theory for single phase single stage grid connected PV system. In: 2011 International Conference on Clean Electrical Power (ICCEP). 2011. pp. 535-540

[54] Mulla MA, Rajagopalan C, Chowdhury A. Hardware implementation of series hybrid active power filter using a novel control strategy based on generalised instantaneous power theory. IET Power Electronics. 2013;**6**:592-600

[55] Esfandiari A, Parniani M, Mokhtari H. A new control strategy of shunt active filters for power quality improvement of highly and randomly varying loads. In: 2004 IEEE International Symposium on Industrial Electronics. 2004;**2**:1297-1302

[56] Sawant RR, Chandorkar MC. Methods for multi-functional converter control in three-phase four-wire systems. IET Power Electronics. 2009;**2**:52-66

[57] Salmeron P, Herrera RS, Vazquez JR. Mapping matrices against vectorial frame in the instantaneous reactive power compensation. IET Electric Power Applications. 2007;**1**:727-736

[58] Bhattacharya A, Chakraborty C, Bhattacharya S. Parallel-connected shunt hybrid active power filters operating at different switching frequencies for improved performance. IEEE Transactions on Industrial Electronics. 2012;**59**:4007-4019

[59] Suresh Y, Panda AK, Suresh M. Real-time implementation of adaptive fuzzy hysteresis-band current control technique for shunt active power filter. IET Power Electronics. 2012;**5**:1188-1195

[60] Quoc-Nam T, Hong-Hee L. An advanced current control strategy for three-phase shunt active power filters. IEEE Transactions on Industrial Electronics. 2013;**60**:5400-5410

[61] Asiminoaei L, Blaabjerg F, Hansen S. Detection is key—harmonic detection methods for active power filter applications. IEEE Industry Applications Magazine. 2007;**13**:22-33

[62] Jain SK, Singh SN. Harmonics estimation in emerging power system: Key issues and challenges. Electric Power Systems Research. 2011;**81**:1754-1766

[63] Moreno VM, Lopez AP, Garcias RID. Reference current estimation under distorted line voltage for control of shunt active power filters. IEEE Transactions on Power Electronics. 2004;**19**:988-994

[64] Petit JF, Robles G, Amaris H. Predictive algorithm for harmonic mitigation in non-linear loads based on active filters. In: Power Tech, 2005 IEEE Russia. 2005. pp. 1-6

[65] Rosendo JA, Bachiller A, Gomez A. Application of self-tuned Kalman filters to control of active power filters. In: Power Tech, 2007 IEEE Lausanne. 2007. pp. 1262-1265

[66] Cardoso R, de Camargo RF, Pinheiro H, Grundling HA. Kalman filter based synchronisation methods. IET *Generation, Transmission & Distribution*. 2008;**2**:542-555

[67] Panigrahi R, Panda PC, Subudhi BD. Comparison of performances of hysteresis and dead beat controllers in active power filtering. In: 2012 IEEE Third International Conference on Sustainable Energy Technologies (ICSET). 2012. pp. 287-292

[68] Regulski P, Terzija V. Estimation of frequency and fundamental power components using an unscented Kalman filter. IEEE Transactions on Instrumentation and Measurement. 2012;**61**:952-962

[69] Kanieski JM, Cardoso R, Pinheiro H, Grundling HA. Kalman filter-based control system for power quality conditioning devices. IEEE Transactions on Industrial Electronics. 2013;**60**:5214-5227

Applications of Kalman Filters for Coherent Optical Communication Systems

Lalitha Pakala and Bernhard Schmauss

Additional information is available at the end of the chapter

http://dx.doi.org/10.5772/intechopen.71617

Abstract

In this chapter, we review various applications of Kalman filtering for coherent optical communication systems. First, we briefly discuss the principles of Kalman filter and its variations including extended Kalman filter (EKF) and adaptive Kalman filter (AKF). Later on, we illustrate the applicability of Kalman filters for joint tracking of several optical transmission impairments, simultaneously, by formulating the state space model (SSM) and detailing the principles. A detailed methodology is presented for the joint tracking of linear and nonlinear phase noise along with amplitude noise using EKF. Also, approaches to enhance the performance obtained by EKF by combining with other existing digital signal processing (DSP) techniques are presented. Frequency and phase offset estimation using a two stage linear Kalman filter (LKF)/EKF is also discussed. A cascaded structure of LKF and EKF by splitting the SSM to jointly mitigate the effects of polarization, phase and amplitude noise is also presented. The numerical analysis concludes that the Kalman filter based approaches outperform the conventional methods with better tracking capability and faster convergence besides offering more feasibility for real-time implementations.

Keywords: optical communication systems, coherent optical transmission, digital signal processing, nonlinear mitigation, phase noise, amplitude noise, QAM

1. Introduction

In order to meet the yearning demands of bandwidth and capacity due to ever increasing data traffic, the contemporary research in the field of optical transmission, is focused on developing 400 Gbps and above, Ethernet transmission [1–5]. The achievable information rates using optical fiber as communication channel have been rapidly increased over the past few decades. Some of the technology breakthroughs behind this rapid increase in the transmission capacity, can be listed as the invention and development of the erbium doped fiber amplifiers (EDFA), wavelength division multiplexing (WDM) systems, coherent detection, digital signal processing (DSP) techniques and forward error correction (FEC) schemes ensuring reliable transmission. The advent of coherent detection along with subsequent DSP made it possible to deploy spectrally efficient higher order modulation formats and multiplexing techniques [6, 7]. Moreover, it has also made feasible to digitally equalize the optical fiber transmission impairments [8], which are the main hurdle to increase the bandwidth-distance product. The transmission capacity can be increased several times by employing complex modulation formats like m-ary quadrature amplitude modulation (with m = 4, 16, 64 and so on), and multiplexing techniques like polarization division multiplexing (PDM) and WDM. However, they are more vulnerable to the optical transmission impairments as well as to the carrier phase and frequency offset (FO). Hence, effective DSP algorithms for combatting with the channel impairments were under active research over the past decade [8–23]. Consequently, coherent optical receivers are well developed and employ digital filters that allow for effective equalization of fiber linear impairments like chromatic dispersion (CD) and polarization mode dispersion (PMD) in the electric domain [9]. Typically, CD can be compensated by either frequency or time domain filters using finite impulse response (FIR) or infinite impulse response (IIR) design. Optical receivers exploiting polarization diversity should also compensate for the random fluctuations of the polarization state caused by the stochastic change of fiber birefringence. PMD compensation is widely performed using constant modulus algorithm (CMA) [15] or multi modulus algorithm (MMA) [16]. Attributed to these well-developed linear equalization techniques, fiber nonlinearity still remains a bottleneck for increasing the capacity and transmission reach [24].

Although, multiple information bits being encoded in a single symbol significantly increase the spectral efficiency, the signal becomes more sensitive to the amplified spontaneous emission (ASE) noise that is added in the optical amplifiers along the transmission link. Therefore, a reliable transmission over long distance demands the signal to be launched into the optical fiber at higher power, to ensure a sufficiently high optical signal to noise ratio (OSNR) at the receiver. However, the maximum transmittable launch power per fiber span is constrained by the Kerr nonlinear effects, including self-phase modulation (SPM) and cross phase modulation (XPM) in case of WDM systems, which results in signal degradation [25]. This degrading impact of Kerr nonlinearity is much more severe in multi-channel systems with increasing number of channels [26]. Moreover, the nonlinear phase noise (NLPN) resulting from the signal and ASE noise interactions at high launch powers, deteriorates the signal quality further. On the other hand, signal transmission at low launch powers is limited by the ASE noise. Therefore, mitigation of fiber nonlinearity is vital to enhance the capacity ensuring reliable transmission. Consequently, several nonlinear mitigation techniques have been proposed in the recent era, of which, digital backward propagation (DBP) [10], maximum likelihood sequence estimation (MLSE) based nonlinearity mitigation [17], spectral inversion [18–20], phase conjugated twin waves [21, 22] and perturbation based approaches [23] gained considerable attention. However, the real time implementation of these algorithms is extremely challenging owing to either the high required computational effort or the higher bandwidth consumption. Although computationally complex, DBP has drawn significant attention owing to its capability of mitigating linear and nonlinear impairments simultaneously, provided the channel characteristics are known and the step size is sufficiently small when solving the inverse nonlinear Schrodinger equation (NLSE). Several strategies have been proposed in the literature to reduce the number of required DBP steps and enhancing the performance either by including temporal correlations [27, 28] or optimizing the parameters [29]. Nevertheless, the performance of DBP significantly deteriorates in the presence of stochastic impairments like laser phase noise and NLPN [30–32]. Moreover, its applicability is limited to single channel systems [33, 34].

Apart from fiber linear and nonlinear impairment compensation, digital carrier synchronization has also become an essential component of the coherent receivers, for synchronizing the phase and frequency offsets between the transmitter laser and the local oscillator (LO), eliminating the necessity of a phase locked loop. Several carrier phase estimation (CPE) techniques have been proposed for suppressing the laser phase noise [35–44]. CPE being a low complex technique is also under wide investigation to compensate the nonlinear phase shift owing to Kerr effect, besides laser phase noise [30, 44–47]. Investigations were also carried on the combined performance of DBP and CPE, in order to reduce the number of DBP steps per span and there by its complexity [31, 48]. It was reported in [30, 45], that the considered CPE methods outperform the DBP technique implemented using asymmetric split step Fourier method (SSFM) with one step per span and without any parameter optimization. However, the accumulated ASE noise and the NLPN at high signal powers pose a challenging constraint on the conventional CPE limiting its nonlinear mitigation capability. Moreover, a phase unwrapping function [35] is typically required by CPE, which might increase the probability of cycle slips [49] and error propagation. Furthermore, the commonly employed CPE techniques have low tolerance towards the frequency offset (FO) between the transmitter laser and the LO. Therefore, a separate FO estimation module is necessary [50].

In the recent era, Kalman filtering has gained huge attention in the field of optical communication systems, owing to its potential capability to mitigate several optical transmission impairments simultaneously. The Kalman filter was first developed by R. E. Kalman in 1960. In [51], he presented a new approach to the linear filtering and prediction problems by introducing the state space notation, where the random processes/signals to be estimated are represented as the output of a linear dynamic system perturbed by uncorrelated noise. This approach facilitates recursive computation of the optimal solution and highly reduces the computational effort as compared to the conventional Wiener filter besides eliminating the memory problems. The so called Kalman filter computes the optimal solution recursively in the minimum mean square error (MMSE) sense. While the applicability of the Wiener filter is limited to stationary processes, the Kalman filters can be also applied to the non-stationary processes. An added advantage of the Kalman filter is its extended applicability also to the nonlinear systems through an approximate linearization and the so called filter is known as extended Kalman filter (EKF). This has attracted the Kalman filters for numerous real-time applications in the fields of navigation, radar, mobile communications, speech signal processing and so forth. Currently, EKF is under wide investigation in coherent optical communication systems for tracking and mitigating linear and nonlinear phase noise, amplitude noise, phase and frequency offsets as well as polarization de-multiplexing [12, 13, 31, 52–66].

Moreover, enhancing the performance obtained from EKF by incorporating with other existing techniques like DBP has also been studied in [31, 58, 59]. EKF requires only a few complex multiplications to recover one data symbol. Besides the advantage from low complexity, it further offers other benefits including faster convergence, joint tracking and compensation of fiber impairments. Therefore, it is worth discussing and reviewing the applications of Kalman filters for coherent optical communications in a nutshell.

This chapter is organized as follows: In Section 2, we discuss the principles of Kalman filter by describing the state space notation and the recursive equations. We further present some variations of the Kalman filter, namely, EKF and adaptive Kalman filter (AKF). Section 3 details the applications of EKF for coherent optical communications. We illustrate how to employ Kalman filtering for the joint tracking of several optical transmission impairments by formulating the state space model (SSM) and detailing the working principles. We also describe our numerical model and present the results to justify the theoretical findings. Finally, the chapter is concluded with a note on the key points, in Section 4.

2. The Kalman filter

A Kalman filter is an optimal recursive linear MMSE estimator that estimates the state of a linear dynamic perturbed by noise. Since the true state of the system is not observable, instead we obtain the measurements or observations that are corrupted by noise. Now, the goal of the Kalman filter is to obtain an optimal estimate of the unknown state from the noisy observations recursively. The stochastic process under estimation is modeled by a state space model (SSM) which facilitates the recursive nature of the Kalman filter. In the following, we present the general framework of the Kalman filtering and also discuss briefly the principles of the EKF and AKF.

2.1. Principles of Kalman filter

Consider a discrete-time, linear, time varying system in the state space notation, given by Eqs. (1) and (2). Eq. (1) describes how the true state of the system evolves over time and is known as the state or the process equation. Eq. (2) describes how the measurements are related to the states and is known as measurement or observation equation. Here, k denotes the time instant, $\mathbf{x}_k$ and $\mathbf{y}_k$ denote the state vector and the measurement vector, respectively. $\mathbf{F}_k$ denotes the state transition matrix that relates the states at the time instances k and $k-1$, in the absence of process noise $\mathbf{w}_k$. $\mathbf{H}_k$ denotes the measurement matrix that relates the states to the measurements in the absence of measurement noise $\boldsymbol{n}_k$. The process and measurement noise vectors $\mathbf{w}_k$ and $\boldsymbol{n}_k$ are assumed to be zero mean white Gaussian noise processes with co-variance matrices $\mathbf{Q}_k$ and $\mathbf{R}_k$, respectively. It is also assumed that the initial state $\mathbf{x}_0$ at time instant 0, is a Gaussian random vector. Given the SSM and these assumptions, the objective of the Kalman filter is to obtain a linear MMSE estimate of $\mathbf{x}_k$ based on the observations $\{\mathbf{y}_1, \mathbf{y}_2, \ldots \mathbf{y}_k\}$. The solution corresponds to the conditional mean [67] as given in Eq. (3). Here, E[·] denotes the expectation operator.

$$\mathbf{x}_k = \mathbf{F}_k\mathbf{x}_{k-1} + \mathbf{w}_k \tag{1}$$

$$\mathbf{y}_k = \mathbf{H}_k x_k + \boldsymbol{n}_k \tag{2}$$

$$\hat{\mathbf{x}}_k = \mathrm{E}[\mathbf{x}_k|\mathbf{y}_1, \mathbf{y}_2, \ldots, \mathbf{y}_k] \tag{3}$$

The Kalman filter computes the optimal state recursively, following a predictor-corrector structure, where a prediction is computed prior to the availability of the observation at current time instant k and updates the prediction when the observation at time instant k is available. Throughout this Chapter, we follow the typical notation convention for the Kalman filter equations: any variable with subscript $k|k-1$ denotes prediction or apriori estimate, and any variable with subscript $k|k$ or simply, denotes the updated or aposteriori estimate. During the prediction step, the Kalman filter makes the best guess about the system's state based on its dynamics, prior to the availability of the current observation. The state prediction denoted by $\hat{x}_{k|k-1}$ is given in Eq. (4). The uncertainty associated with the prediction is given by the apriori error covariance matrix $\mathbf{P}_{k|k-1}$, as in Eq. (5). Under the given assumptions and initial conditions, the conditional probability density function (pdf) $p(\mathbf{x}_k|\mathbf{y}_1,\mathbf{y}_2,\ldots,\mathbf{y}_{k-1})$ is also Gaussian, where the apriori state estimate $\hat{\mathbf{x}}_{k|k-1}$ and the apriori error covariance $\mathbf{P}_{k|k-1}$, reflects the mean and variance of the distribution, as given in Eq. (6). Here, N denotes normal or Gaussian distribution.

$$\hat{\mathbf{x}}_{k|k-1} = \mathrm{E}[\mathbf{x}_k|\mathbf{y}_1, \mathbf{y}_2, \ldots, \mathbf{y}_{k-1}] = \mathbf{F}_k\hat{\mathbf{x}}_{k-1|k-1} \tag{4}$$

$$\mathbf{P}_{k|k-1} = \mathrm{E}[(\mathbf{x}_k - \hat{\mathbf{x}}_{k|k-1})(\mathbf{x}_k - \hat{\mathbf{x}}_{k|k-1})^{\mathbf{H}}] = \mathbf{F}_k\mathbf{P}_k\mathbf{F}_k^{\mathbf{H}} + \mathbf{Q}_k \tag{5}$$

$$p(\mathbf{x}_k|\mathbf{y}_1, \mathbf{y}_2, \ldots, \mathbf{y}_{k-1}) \sim N(\hat{\mathbf{x}}_{k|k-1}, \mathbf{P}_{k|k-1}) \tag{6}$$

During the update step, when the new observation at time k is available, the optimal estimate is computed as a linear combination of the prediction and the new information available from the current measurement weighted by an optimal weighting matrix known as Kalman gain. The update equations can be summarized in Eqs. (7)–(10). The innovation denoted by $\mathbf{v}_k$, can be interpreted as the new information that is available in the observation $\mathbf{y}_k$ relative to all the past observations up to time instant $k-1$. It is computed as the difference between the actual and the predicted observation $\hat{\mathbf{y}}_{k|k-1}$, and is given in Eq. (7). The Kalman gain, denoted by $\mathbf{K}_k$, determines the extent up to which the innovation should be taken into account in updating the apriori state estimate and is computed according to Eq. (8). Here, $\mathbf{H}$ denotes the Hermitian operator. The updated or aposteriori state estimate $\hat{\mathbf{x}}_{k|k}$, and the aposteriori error covariance $\mathbf{P}_{k|k}$, are computed as given in Eqs. (9) and (10), respectively. The aposteriori pdf $p(\mathbf{x}_k|\mathbf{y}_1,\mathbf{y}_2,\ldots,\mathbf{y}_k)$ is also Gaussian distributed with mean and variance given by the aposteriori state estimate $\hat{\mathbf{x}}_{k|k}$ and the aposteriori error covariance $\mathbf{P}_{k|k}$, respectively, as given in Eq. (11). Thus, the Kalman filter propagates the first and second order moments of the state distribution recursively for computing the optimal state estimate.

$$\mathbf{v}_k = \mathbf{y}_k - \hat{\mathbf{y}}_{k|k-1} \tag{7}$$

$$\mathbf{K}_k = \mathbf{P}_{k|k-1}\mathbf{H}_k^{\mathrm{H}}(\mathbf{H}_k\mathbf{P}_{k|k-1}\mathbf{H}_k^{\mathrm{H}} + \mathbf{R}_k)^{-1} \tag{8}$$

$$\hat{\mathbf{x}}_{k|k} = \hat{\mathbf{x}}_{k|k-1} + \mathbf{K}_k\mathbf{v}_k \tag{9}$$

$$\mathbf{P}_{k|k} = \mathbf{P}_{k|k-1} - \mathbf{P}_{k|k-1}\mathbf{K}_k\mathbf{H}_k \tag{10}$$

$$p(\mathbf{x}_k|\mathbf{y}_1, \mathbf{y}_2, \ldots, \mathbf{y}_k) \sim N(\hat{\mathbf{x}}_{k|k}, \mathbf{P}_{k|k}) \tag{11}$$

2.2. Extended Kalman filtering

In Section 2.1, we addressed the problem of estimating the unknown state of a linear dynamic system from noisy observations. Now, we consider the filtering problem for nonlinear system dynamics (either the process or observation model or both being nonlinear). The Kalman filter solution can be adopted for the nonlinear dynamic systems through an approximate linearization procedure and the resulting filter is known as EKF. Consider a nonlinear dynamic system described by the SSM given in Eqs. (12) and (13). Here, $\mathbf{f}_k(\cdot)$ and $\mathbf{h}_k(\cdot)$ denote the nonlinear state transition function and the measurement function, respectively.

$$\mathbf{x}_k = \mathbf{f}_k(\mathbf{x}_{k-1}) + \mathbf{w}_k \tag{12}$$

$$\mathbf{y}_k = \mathbf{h}_k(\mathbf{x}_k) + \mathbf{n}_k \tag{13}$$

The nonlinear system dynamics can be linearized through a first order Taylor approximation at each time instant, around the most recent state estimate. This forms the basic idea of EKF. Let, $\mathbf{A}_k$ and $\mathbf{B}_k$ be the Jacobian matrices of $\mathbf{f}_k(\cdot)$ and $\mathbf{h}_k(\cdot)$, respectively, and are computed according to Eqs. (14) and (15). Under the given assumptions and, the initial conditions as discussed in the earlier section, the EKF recursive equations can be summarized in Eqs. (16)–(20).

$$\mathbf{A}_k = \frac{\partial \mathbf{f}_k(\mathbf{x})}{\partial \mathbf{x}} \ \textbf{at}\ \mathbf{x} = \hat{\mathbf{x}}_{k-1|k-1} \tag{14}$$

$$\mathbf{B}_k = \frac{\partial \mathbf{h}_k(\mathbf{x})}{\partial \mathbf{x}} \ \textbf{at}\ \mathbf{x} = \hat{\mathbf{x}}_{k|k-1} \tag{15}$$

$$\hat{\mathbf{x}}_{k|k-1} = \mathbf{f}_k(\hat{\mathbf{x}}_{k-1|k-1}) \tag{16}$$

$$\mathbf{P}_{k|k-1} = \mathbf{A}_k\mathbf{P}_k\mathbf{A}_k^{\mathbf{H}} + \mathbf{Q}_k \tag{17}$$

$$\mathbf{v}_k = \mathbf{y}_k - \hat{\mathbf{y}}_{k|k-1} = \mathbf{y}_k - \mathbf{h}_k(\hat{\mathbf{x}}_{k|k-1}) \tag{18}$$

$$\mathbf{K}_k = \mathbf{P}_{k|k-1}\mathbf{B}_k^{\mathbf{H}}(\mathbf{B}_k\mathbf{P}_{k|k-1}\mathbf{B}_k^{\mathbf{H}} + \mathbf{R}_k)^{-1} \tag{19}$$

$$\hat{\mathbf{x}}_{k|k} = \hat{\mathbf{x}}_{k|k-1} + \mathbf{K}_k\mathbf{v}_k \tag{20}$$

$$\mathbf{P}_{k|k} = \mathbf{P}_{k|k-1} - \mathbf{P}_{k|k-1}\mathbf{K}_k\mathbf{B}_k \tag{21}$$

2.3. Adaptive Kalman filtering

The Kalman filter computes the optimal solution, provided the process noise and measurement noise covariances, $\mathbf{Q}_k$ and $\mathbf{R}_k$, respectively, are known apriori. However, in practice, a precise knowledge about the noise statistics might not be available. The Kalman gain $\mathbf{K}_k$ takes into account the noise covariances, $\mathbf{Q}_k$ and $\mathbf{R}_k$, to determine the extent of reliability between the predicted state $\hat{\mathbf{x}}_{k|k-1}$ and the innovation $\mathbf{v}_k$. Therefore, a poor knowledge of the noise statistics might significantly degrade the filter performance and even leads to divergence. To overcome these difficulties, an adaptive approach can be followed to adaptively estimate the noise covariances from the noise samples, (for example, the innovation sequence) that are generated during the Kalman recursions at each time instant. This leads to the adaptive Kalman filtering. The different approaches for adaptive filtering are classified into four types: Bayesian, maximum likelihood, correlation and covariance matching methods [68]. Here, we discuss the approach based on covariance matching [68, 69] for adaptive estimation of noise statistics. The basic idea behind this approach lies on the fact that for an optimal filter, the theoretical covariance of the innovation $\mathbf{v}_k$, denoted by $\mathbf{S}_k$, given in Eq. (22) should be consistent with the empirically estimated covariance given in Eq. (23). Here, m denotes the window size to provide statistical smoothing.

$$\mathbf{S}_k = \mathbf{H}_k\mathbf{P}_{k|k-1}\mathbf{H}_{\mathbf{k}}^{\mathbf{H}} + \mathbf{R}_k \tag{22}$$

$$\mathrm{E}(\mathbf{v}_k\mathbf{v}_k^H) = \frac{1}{m}\sum_{i=0}^{m-1}\mathbf{v}_{k-i}\mathbf{v}_{k-i}^{\mathbf{H}} \tag{23}$$

$$\mathbf{H}_k\mathbf{P}_{k|k-1}\mathbf{H}_{\mathbf{k}}^{\mathbf{H}} + \mathbf{R}_k = \frac{1}{m}\sum_{i=0}^{m-1}\mathbf{v}_{k-i}\mathbf{v}_{k-\mathbf{i}}^{\mathbf{H}} \tag{24}$$

Since the Kalman gain $\mathbf{K}_k$ depends on the ratio of the process and measurement noise covariances $\mathbf{Q}_k/\mathbf{R}_k$, rather than on their individual values, if either of $\mathbf{Q}_k$ or $\mathbf{R}_k$, is known, the other can be adaptively estimated by satisfying the condition for covariance matching, given in Eq. (24). When $\mathbf{Q}_k$ is known, $\mathbf{R}_k$ can be directly estimated from Eq. (24). Alternatively, when $\mathbf{R}_k$ is given, $\mathbf{Q}_k$ can be estimated by a scaling procedure to improve the robustness of the filter. The basic idea behind this scaling method is that if the estimated covariance of $\mathbf{v}_k$, on the right hand side of Eq. (24), is much larger than the theoretical covariance, then $\mathbf{Q}_k$ (please note that $\mathbf{P}_{k|k-1} = \mathbf{F}_k\mathbf{P}_k\mathbf{F}_k^{\mathbf{H}} + \mathbf{Q}_k$) should be increased to bring the theoretical covariance closer to the estimated one and vice-versa. Therefore, $\mathbf{Q}_k$ can be adaptively updated in order to balance any deviations between the theoretical and estimated innovation covariance by considering a scaling factor α_k, as given in Eq. (25). The estimate of $\mathbf{Q}_k$, denoted by $\hat{\mathbf{Q}}_k$, is given in Eq. (26).

$$\alpha_k = \frac{\mathrm{trace}\left(\frac{1}{m}\sum_{i=0}^{m-1}\mathbf{v}_{k-i}\mathbf{v}_{k-i}^H - \mathbf{R}_k\right)}{\mathrm{trace}\left(\mathbf{H}_k\mathbf{P}_{k|k-1}\mathbf{H}_k^H\right)} \tag{25}$$

$$\hat{\mathbf{Q}}_k = \alpha_k\hat{\mathbf{Q}}_{k-1} \tag{26}$$

In the case of EKF, the same procedure can be followed for adaptive estimation of noise covariances, by replacing the measurement matrix with the Jacobian matrix.

3. Kalman filtering for coherent optical communications

3.1. Kalman filtering for carrier phase and amplitude noise estimation (CPANE)

Digital carrier phase estimation (CPE) has become an essential component of coherent optical receivers to recover the carrier phase perturbed by laser phase noise arising from the transmitter laser or LO [35–43]. Several CPE techniques have been developed in the literature based on feedback [39, 40] or feed forward loops [35–37]. Depending on how the data phase is wiped off, they can be further classified into decision directed (DD) [35, 40, 42, 46] or non-decision directed (NDD) methods [39, 41, 43]. NDD methods like Viterbi-Viterbi [41] CPE has gained high attention due to its ease of implementation. However, it employs m-th power scheme to remove the data modulation and therefore, are only better suited for QPSK systems. However, for higher QAM systems, DD-CPE methods exhibit better performance compared to NDD CPE methods [35, 42].

Apart from tracking the carrier phase, CPE being a low complex technique, can also be employed for compensating the nonlinear phase shift stemming from the Kerr nonlinear effects [30, 44–47]. However, the nonlinear mitigation performance of CPE is limited in the presence of ASE noise and at high launch powers. Moreover, a phase unwrapping function is typically required for CPE that might increase the probability of cycle slips [35, 49]. Addressing these problems, we have proposed a CPANE algorithm using EKF in [12, 53] for the joint mitigation of linear and nonlinear phase noise as well as ASE induced phase and amplitude distortions. Unlike CPE, EKF-CPANE estimates a complex quantity, and therefore, no argument function is required which eliminates the ambiguity associated with multiples of 2π and consequent cycle slips.

Kalman filter based CPE has been introduced and numerically verified in [52]. From the numerical results, it was reported that the Kalman based phase estimation combined with DD equalizer in a feedback configuration outperforms the conventional CMA based approach [52]. CPE based on EKF was demonstrated and verified experimentally for QPSK and 16-QAM systems in [57]. In [55], EKF has been investigated for characterizing the laser phase and amplitude noise. EKF based carrier synchronization has also been verified experimentally, in combination with expectation maximization (EM). A carrier recovery scheme based on block estimation process with Kalman filter has been demonstrated in [56]. This approach was verified experimentally for 16 and 64-QAM signals. However, these Kalman filter based approaches estimate an argument which involves sine and cosine functions, computation of the Jacobian matrix and also require matrix multiplications and inversions, which increases the computational complexity. The proposed method in [12, 53], estimates a complex quantity accounting also for the phase and amplitude distortions arising from the ASE noise in addition to the carrier phase. The variables in the SSM reduce to scalars and therefore, the vectors and matrices are reduced to scalars which will ease the computational effort. In the following, we first describe the general principles of CPE. Later, we explain our proposed CPANE algorithm illustrating its principles and implementation details using EKF.

3.1.1. Principles of CPE

Consider an m-ary QAM received signal on single polarization, which is sampled and compensated for linear impairments. Assuming perfect linear equalization, the k-th input signal to the CPE can be written as in Eq. (27). Here, r_k denotes the k-th input signal to CPE, a_k denotes the transmitted symbol, and n_k denotes the collective amplified spontaneous emission (ASE) noise which is assumed to be white Gaussian process. θ_k denotes the phase noise arising from the laser linewidth effects and fiber nonlinearity, which is typically modeled as a Wiener process and is given in Eq. (28). **Figure 1 (a)** describes the input signal model to CPE. It can be seen that after a_k is rotated by phase noise θ_k, n_k further adds additional phase noise n'_k and amplitude noise $\tilde{n}_k$. The objective of CPE is to estimate the phase noise θ_k, and derotate the received signal r_k, in order to recover the transmitted symbol a_k, as given in Eq. (29) and **Figure 1 (b)**. However, since the CPE targets at estimating an accurate $\hat{\theta}_k$, the recovered transmitted symbol $\hat{a}_k$, still suffers from the residual phase noise or amplitude noise or both. For more details, please refer [12].

$$r_k = a_k \mathbf{e}^{j\theta_k} + n_k \tag{27}$$

$$\theta_k = \theta_{k-1} + w_k \tag{28}$$

$$\hat{a}_k = r_k \mathbf{e}^{-j\hat{\theta}_k} \tag{29}$$

3.1.2. Principles of CPANE

The effects of n_k, as discussed in Section 3.1.1, can be taken into account by reformulating Eq. (27) as given in Eq. (30) which forms the input signal to CPANE. Here, r_k is modeled as the transmitted symbol a_k being rotated by a complex quantity ψ_k, that considers the effects of both phase and amplitude noise in its real and imaginary parts, respectively, as given in Eq. (30).

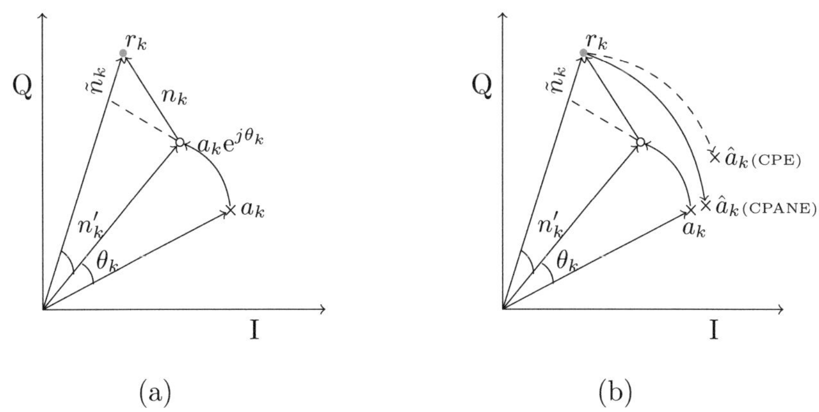

Figure 1. (a) Input signal model to CPE (b) recovered symbols using CPE and CPANE [54].

The objective of CPANE is to recover θ_k more accurately, by estimating the complex quantity ψ_k. The recovered transmitted symbol $\hat{a}_k$ is given in Eq. (32). Since ψ_k takes into account, both the phase and amplitude distortions, $\hat{a}_k$ can be recovered more accurately by employing CPANE compared to CPE, as depicted in **Figure 1 (b)**. Moreover, unlike CPE, CPANE eliminates the necessity of phase unwrapping function.

$$r_k = a_k e^{j\psi_k} \tag{30}$$

$$\psi_k = (\theta_k + n'_k) + j\tilde{n}_k \tag{31}$$

$$\hat{a}_k = r_k e^{-j\hat{\psi}_k} \tag{32}$$

3.1.3. EKF-CPANE for joint mitigation of phase and amplitude noise

As discussed earlier, CPANE algorithm can be employed for the joint mitigation of phase and amplitude noise. However, it requires a reliable tracking of the complex quantity ψ_k, which can be accomplished by an EKF. The required SSM for the EKF can be formulated using Eqs. (33) and (34). Eq. (33) represents the state or process equation that describes the time evolution of ψ_k. Eq. (34) represents the observation equation that describes the relation of the states ψ_k to the observations r_k. Eq. (34) is similar to Eq. (30), however, for consistency of the filter, the measurement noise m_k has been taken into account. Here, all the variables in the SSM are scalar quantities. Comparing to the standard SSM for EKF described in Section 2.2, it can be noted that the state transition is identity and the measurement matrix is the transmitted symbol, a_k, for simplicity, we call it measurement weight (MW), since it is a scalar. The EKF recursive equations can be derived analogously by relating the SSM to the standard SSM of EKF discussed in Section 2.2. Since the MW a_k is required to compute the update equations, which is not known apriori, EKF-CPANE is DD. The required decisions of a_k, denoted by d_k are obtained by de-rotating r_k with an average of the past updated estimates $\hat{\psi}_k$ over a window length of N, as given in Eq. (35). For more details on the prediction and update equations of EKF-CPANE, please refer [12]. **Figure 2** depicts the schematic of the EKF-CPANE algorithm, illustrating that the prediction $\hat{\psi}_{k|k-1}$ is the delayed version of the past updated estimate and the current updated state $\hat{\psi}_{k|k}$ is the linear combination of the prediction $\hat{\psi}_{k|k-1}$ and the innovation v_k weighted by the Kalman gain K_k. The process of making the required decisions for the update step has also been illustrated in **Figure 2**.

$$\psi_k = \psi_{k-1} + w_k \tag{33}$$

$$r_k = a_k e^{j\psi_k} + m_k \tag{34}$$

$$d_k = \text{decision}(t_k) \text{ where } t_k = r_k e^{-j\frac{1}{N}\sum_N \hat{\psi}_{k-N}} \tag{35}$$

3.1.4. Numerical analysis of EKF-CPANE

The performance of EKF-CPANE algorithm for mitigating the laser phase noise, fiber nonlinearity besides the ASE induced phase and amplitude distortions has been verified through numerical simulations on single channel systems in [12, 53] and multi-channel systems in [54]. Here, we briefly discuss the numerical model and present a few simulation results reproduced from [12], so that the flow of the readers is not interrupted. The numerical model of polarization multiplexed (PM) m-QAM coherent transmission system including a DSP module at the receiver, is depicted in **Figure 3**. Here, we consider the PM-m-QAM transmitter with m = 16 and 64, operated at 28 and 18.667 GBaud, respectively. These signals are transmitted through a standard single mode fiber (SSMF) link at different launch powers. The SSMF has the following parameters: attenuation coefficient (α) = 0.2 dB/km, dispersion coefficient (D) = 16 ps/nm-km, and nonlinearity coefficient (γ) = 1.2/W-km. The span length of SSMF is 80 km and a number of 12 and 6 spans have been considered for 16 and 64 QAM, respectively, yielding a total transmission distance of 960 and 480 km. The span losses are compensated by an EDFA with a gain of 16 dB and noise figure (NF) of 4 dB. For simplicity, PMD has been neglected in this study. At the receive end, we employ a dual polarization coherent receiver which is followed by a DSP module. The laser linewidth of the LO has been set to 100 kHz. After coherent detection, the signals are re-sampled to twice the symbol rate and are followed by linear compensation. Then the signals are further down sampled to the symbol rate and are further processed by the EKF-CPANE for mitigating linear and nonlinear phase noise besides amplitude noise. The performance of EKF-CPANE is compared to

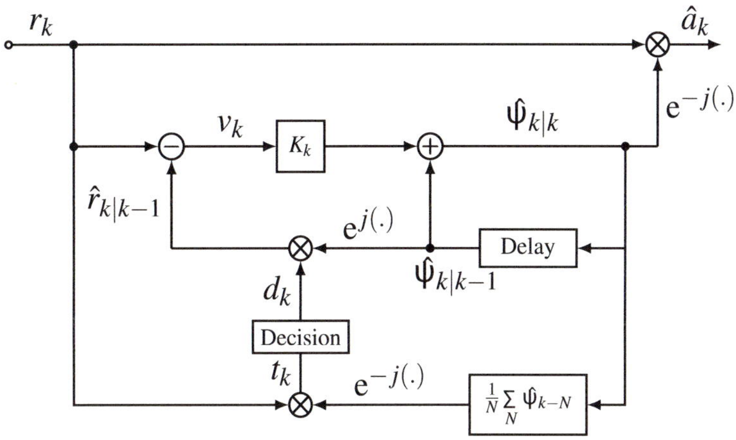

Figure 2. Block diagram of EKF-CPANE algorithm [54].

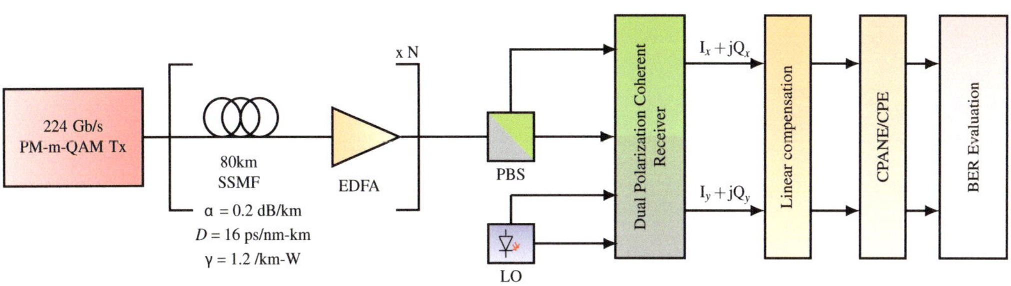

Figure 3. Numerical model of PM-m-QAM coherent transmission system with DSP module [12].

feedforward DD-CPE [46], feedback DD phase locked loop (DD-PLL) [36] and a NDD universal CPE (U-CPE) [39]. The noise covariances for EKF-CPANE, the tap length or step size for DD-CPE, DD-PLL and U-CPE are set to optimize the performance.

The bit error rate (BER) performance of the considered algorithms is evaluated and a Q-factor is computed as $20\log_{10}$ erfcinv(2*BER). The Q-factor vs. launch power curves for 16-QAM and 64-QAM are depicted in **Figure 4(a)** and **(b)**, respectively. It can be seen that EKF-CPANE exhibits better performance compared to DD-CPE, DD-PLL and U-CPE in both linear and nonlinear regimes. This performance enhancement is better visible compared to the DD-CPE method. For PM-64-QAM, it can also be seen that the DD-CPE experiences cycle slips occurring through the error propagation of wrong decisions which can be seen in **Figure 4(b)** at launch powers ranging from −2 to 1 dBm [12]. Since the performance of DD algorithms strongly depends on the pre-decisions made by the algorithm, we study the impact of ideal error free decisions on their performance by replacing the pre-decisions d_k with the true data symbols a_k. The algorithms with the ideal case are denoted by IEKF-CPANE, IDD-CPE and IDD-PLL. It can be seen from **Figure 4(a)** and **(b)**, that the IEKF-CPANE shows significant performance enhancement and better tolerance towards linear and nonlinear phase noise as well as amplitude noise, compared to IDD-CPE and IDD-PLL. Unlike EKF-CAPNE, no notable improvement can be obtained for the DD-CPE and DD-PLL between their practical and ideal cases. Although, the ideal case, where the true symbols a_k are already known, is not possible in practice, it should be noted that the performance of EKF-CPANE can be further improved by reducing the number of decision errors, which will be further discussed in the next Section 3.2.

3.2. EKF and DBP for fiber nonlinear mitigation

In Section 3.1, we have described how the EKF can be employed for the joint mitigation of phase and amplitude noise. From the numerical results discussed in Section 3.1.4, it can be concluded that the EKF-CAPNE algorithm shows promising results in mitigating the linear

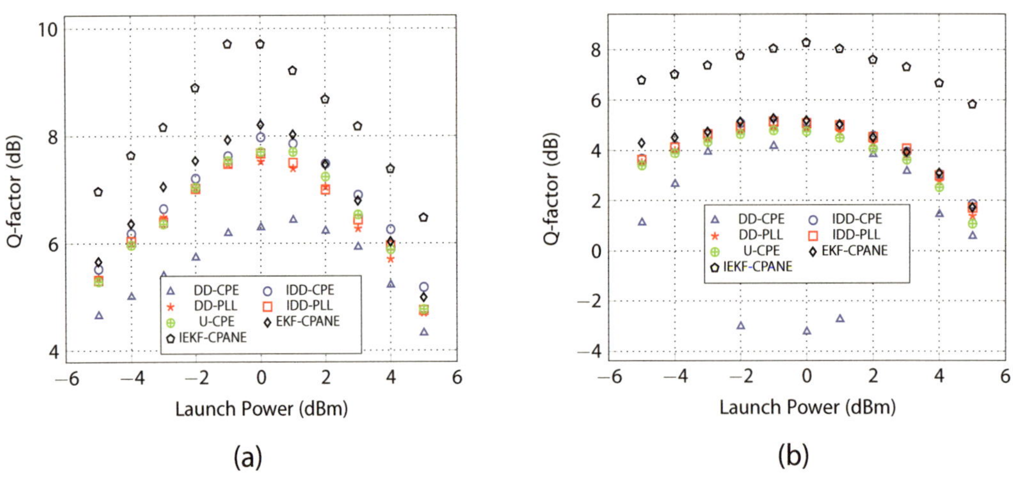

Figure 4. Q-factor vs. launch power curves for the considered algorithms (a) PM-16-QAM transmission over 960 km of SSMF transmission (b) PM-64-QAM transmission over 480 km of SSMF transmission [12].

and nonlinear phase noise as well as amplitude noise simultaneously besides less computational effort. Although, EKF-CPANE outperforms several other considered CPE methods, the effectiveness of EKF-CPANE in mitigating fiber nonlinear effects can be further enhanced by reducing the number of errors in the pre-decisions d_k. We have proposed a weighted innovation approach (WIA) in [12], where the innovation is computed as a weighted combination of the two nearest likely decisions. Although a gain of ≈ 0.3 dB in the Q-factor can be obtained compared to conventional EKF-CPANE, in the linear regime, no notable improvement can be seen in the nonlinear regime. On the other hand, DBP has emerged to be an effective technique in mitigating linear and nonlinear impairments simultaneously, provided the channel parameters are known a-priori and the step size is sufficiently small. However, DBP can compensate only the deterministic impairments of self-phase modulation and its performance deteriorates significantly in the presence of stochastic impairments like laser phase noise, ASE and NLPN. Moreover, the required huge computational effort keeps it far away from real-time implementation. Nevertheless, by employing a few DBP steps prior to EKF-CPANE would yield an enhanced tolerance towards nonlinearities since DBP is well capable of mitigating deterministic impairments and EKF takes into account the stochastic nature of ASE noise and NLPN. By partially compensating fiber nonlinear effects employing few DBP steps prior to EKF, would result in improved pre-decisions and thereby facilitates the residual compensation of nonlinearities along with amplitude and phase noise effectively. These theoretical findings are verified through numerical simulations on both single [31] and multichannel systems [58].

In [31], it was reported that the EKF-CPANE outperforms the asymmetric split step Fourier method (ASSFM) based one step per span (OSPS) DBP with optimized nonlinear co-efficient γ (ODBP), for single channel systems, for transmission on both SSFM and non-zero dispersion shifted fiber (NZ-DSF). A detailed investigation has also been carried out on the combined performance of DBP and EKF-CPANE with an analysis on the influence of the nonlinear coefficient and the step size of DBP when employed prior to EKF-CPANE. The numerical model employed in this study is similar to the one discussed in Section 3.1.4, with a few changes in the parameters of NF being 5 dB and the linewidth of LO being 500 kHz. The influence of DBP step size on the combined performance of DBP and EKF-CPANE for both SSMF as well as NZ-DSF transmission is illustrated in **Figure 5(a)** [31]. Here, OCDBP denotes the optimized DBP which has a nonlinear coefficient different from ODBP when employed prior to EKF. A worth noting result is that at a launch power of 3 dBm and a transmission distance of 960 km, a gain of 1 dB in the Q-factor can be obtained by employing 0.3 DBP steps per span prior to EKF-CPANE, for both SSMF and NZ-DSF transmission. At the expense of additional computational effort, the deployment of a few DBP steps prior to EKF-CPANE further enhances its performance trading off to complexity.

For the case of multi-channel systems, also, a detailed analysis has been performed in [58], on the combined performance of DBP and EKF for mitigation of inter and intra channel nonlinearities besides phase and amplitude noise. Here, the DBP is employed by considering the temporal correlations between the neighboring signal samples and is termed as correlated DBP (CDBP) [27, 28]. This approach will improve the accuracy in computing the nonlinear phase shift and there by enhances the nonlinear mitigation performance. Since the optimization of nonlinear coefficient plays a vital role on the performance of DBP, we proposed an amplitude

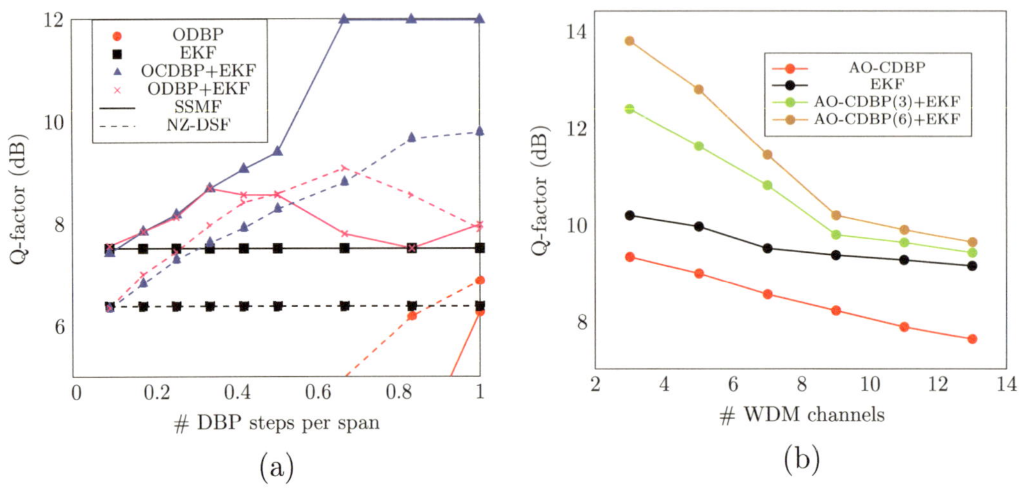

Figure 5. (a) Q-factor vs. DBP step size for the combined performance of EKF and DBP after 960 km of SSMF/NZ-DSF transmission at a launch power of 3 dBm [31] (b) Q-factor vs. number of WDM channels after 960 km of SSMF transmission at launch power of 3 dBm [58].

dependent optimization (AO) [58] of the nonlinear coefficient, according to the discrete amplitude levels present in the higher order modulation formats like 16-QAM. The combined performance of AO-CDBP and EKF-CPANE for WDM systems with varying number of channels has been investigated in [58]. Analogous to the single channel systems, the combined performance of AO-CDBP and EKF yields an improved performance also for the WDM case. However, with increasing impact of the cross phase modulation (XPM) as the number of channels increase, the gain obtained from their combined performance starts vanishing which can be observed in **Figure 5(b)**.

3.3. EKF for mitigation of nonlinearities in dispersion managed links

Since the advent of coherent detection and DSP for coherent optical receivers, CD can be effectively compensated by digital equalization in the electric domain ad thereby, eliminating the need for dispersion compensating fibers (DCF). However, nonlinear mitigation in the dispersion managed (DM) links is also vital in order to upgrade existing links. Although, the computational complexity of DBP is quite high, for DM links, the DBP algorithm can be simplified by assuming that the nonlinear behavior repeats itself every span and therefore, the total nonlinearity after N spans of transmission can be approximated to N times the nonlinearity from a single span [70]. This is termed as distance folded DBP [70] and it reduces the complexity by a factor of N assuming the step size of DBP is equal to the span length and the span length is assumed to be constant. Assuming the dispersion is fully compensated in each span, only the nonlinear term in the nonlinear Schrödinger equation (NLSE) can be solved in the time domain avoiding the Fourier and inverse Fourier transformation (FFT/IFFT) pairs which reduces the computational cost of DBP drastically. We call this approach single step nonlinearity mitigation (SSNL). Similar to the unmanaged links as discussed in the earlier section of 3.2, we investigated the combined performance of SSNL and EKF-CPANE for mitigating the fiber nonlinearity in DM links [59].

The numerical model of PM-16-QAM coherent transmission system over DM link [59] is depicted in **Figure 6**. Here, a fully compensated periodical DM link with several spans has been considered. Each span consists of 80 km of SSMF and 17 km of dispersion compensating fiber (DCF). The SSMF has the following parameters: $\alpha = 0.2$ dB/km, $D = 17$ dB/nm-km, $\gamma = 1.2$/W-km. The parameters of DCF are given by: $\alpha = 0.5$ dB/km, $D = -80$ dB/nm-km and $\gamma = 5$/W-km. In this study, the input power to DCF was set to half of the input power to SSMF. Therefore, the gains of EDFA1 and EDFA2 are adjusted accordingly, to compensate the span losses. The NF of both the EDFAs are set to 4 dB. As described earlier, after coherent detection, the signals are further processed by the SSNL and EKF-CPANE algorithms for mitigating fiber nonlinearities. It has been reported in [59], that the combined performance of SSNL and EKF yields an improved tolerance towards nonlinearities of up to 2 dB for a transmission distance of 1200 km and at a BER of $2*10^{-2}$. Further, their combined performance increases the transmission reach by ≈ 250 km at a launch power of 3 dBm and at a BER of $2*10^{-2}$ as depicted in **Figure 7**.

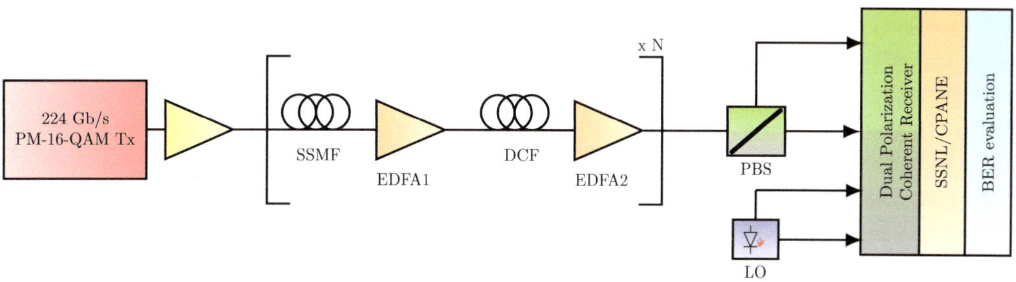

Figure 6. Simulation model of PM-16-QAM coherent transmission over DM link with DSP module [59].

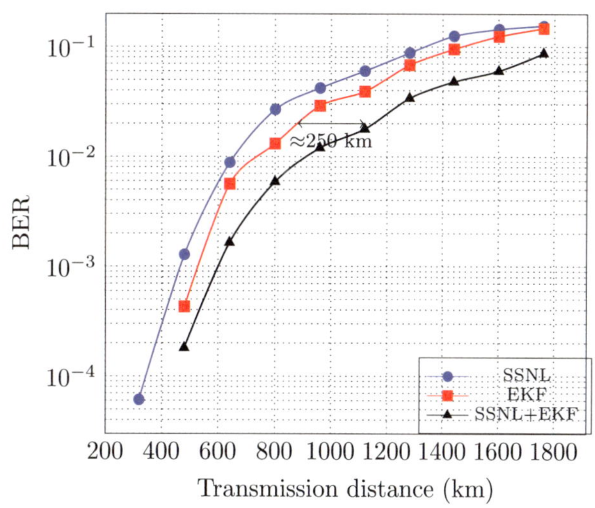

Figure 7. BER vs. transmission distance curves for PM-16-QAM coherent transmission system over DM link [59].

3.4. Kalman filtering for polarization de-multiplexing

An effective way to double the transmission capacity is to employ PDM which allows the transmission of two information signals simultaneously on the orthogonal polarization states of the same optical carrier wave. However, due to fiber birefringence, the state of polarization is not preserved during the propagation on the fiber that leads to crosstalk upon the receipt of the signal. In coherent receivers, CMA [15] or MMA [16] is commonly employed in order to align the polarization states and recover the transmitted signal fully. However, CMA or MMA suffer from the drawbacks of low convergence speed and singularity problem [71]. Moreover, a separate phase estimation scheme is required to track the laser phase noise. Since the Kalman filter allows simultaneous tracking of several state variables provided a precise SSM, the Kalman filter and its variations including radius directed linear Kalman filter (RD-LKF), EKF and UKF are widely investigated for tracking the complex elements of the Jones matrix along with the carrier phase [61–63].

3.4.1. RD-LKF, EKF and UKF for joint tracking polarization state and phase noise

An EKF has been proposed in [61] for joint tracking of the polarization and phase noise. It has also been reported that the EKF shows faster convergence than the conventional approach based on CMA and VV-CPE [61]. However, the variables in the state vector are restricted to real values, which would lead to singularity problems or divergence of the filter [63], besides increasing the dimensions of the vectors and matrices in the Kalman recursive equations. A polarization state tracking scheme using Kalman filter, which is immune to phase/frequency offset, has been introduced in [62], and is termed as RD-LKF. Although, it shows faster convergence compared to CMA, this method needs significant modifications for applying to higher order QAM. Moreover, it is not possible to track the carrier phase simultaneously with the polarization state. The joint tracking of polarization state and carrier phase using EKF has been experimentally verified in [57]. A reduced SSM using UKF has been introduced in [63], which facilitates the joint tracking of polarization state and phase noise. Here, the variables of the state vector are considered to be complex valued. This approach exhibits better performance compared to EKF at high OSNRs at the expense of additional computational effort.

3.4.2. Adaptive cascaded Kalman filtering (A-CKF) for polarization de-multiplexing with simultaneous tracking of phase and amplitude distortions

A cascaded Kalman filtering (CKF), a series of EKF and linear Kalman filtering (LKF) for joint tracking of phase and amplitude distortions besides polarization state, has been proposed in [13]. By splitting up the conventional SSM into linear and nonlinear SSM, the inaccuracies in the linearization of the SSM as a whole can be reduced and thus CKF exhibits enhanced performance besides no increased computational cost compared to the approaches like UKF [63] and radius directed (RD) LKF [62]. Since the optimal performance of the Kalman filter depends on the noise covariances, we proposed an adaptive CKF (A-CKF) [13] to adapt the process noise covariance recursively using the covariance matching method as described in Section 2.3.

3.4.2.1. *Principles of A-CKF*

The transmitted and received signal in the presence of phase noise and polarization rotation can be related as given in Eq. (36). Here, $\mathbf{t}_k$, $\mathbf{r}_k$ and $\mathbf{n}_k$ denote the transmitted and received signal and ASE noise in dual polarization, respectively. $\mathbf{J}_k$ denotes the Jones matrix, θ_k denotes the phase noise and α denotes the loss factor. Assuming negligible, the inverse of the Jones matrix can be described as in Eq. (37) and the elements of the Jones matrix satisfy $J_{yy} = J_{xx}^*$ and $J_{yx} = -J_{xy}^*$ [15]. From now on, for simplified notation, we omit the time variable k in this section. The observation model in Eq. (36) can be rewritten in dual polarization as given in Eq. (38). Here, the subscripts x and y denote the x and y polarizations, respectively. The conventional approach to track the phase and the polarization effects using EKF, the state vector consists of the parameters, a, b, c, and d. However, we reduce the dimensions of the state vector and also the other matrices in the SSM by considering the complex elements in the state vector given by $S(k) = [J_{xx} J_{xy} \;\; \varphi]$. Moreover, we also split up the nonlinear observation model given in Eq. (38), into a nonlinear and linear observation model, where we employ an EKF-CPANE for the joint tracking of phase and amplitude distortions and an LKF for tracking of the complex elements in the Jones matrix. The process noise covariance has been adaptively updated by employing the covariance matching method as described in Section 2. For more details on the A-CKF algorithm, please refer to [13].

$$\mathbf{t}_k = \mathbf{e}^{-j\theta_k}(\alpha\mathbf{J}_k)^{-1}\mathbf{r}_k + \mathbf{n}_k \tag{36}$$

$$\mathbf{J}^{-1} = \begin{bmatrix} J_{xx} & J_{xy} \\ J_{yx} & J_{yy} \end{bmatrix} = \begin{bmatrix} a+jb & c+jd \\ -c+jd & a-jb \end{bmatrix} \tag{37}$$

$$\begin{pmatrix} t_x \\ t_y \end{pmatrix} = \mathrm{e}^{-j\theta_k}(\alpha\mathbf{J}_k)^{-1}\begin{pmatrix} r_x \\ r_y \end{pmatrix} + \begin{pmatrix} n_x \\ n_y \end{pmatrix} \tag{38}$$

Numerical investigations on both back-to-back (BTB) and transmission scenarios, have been carried out in [13], on the variations of the Kalman filter including EKF, UKF, CKF and A-CKF, for tracking the polarization state and phase noise jointly and are compared to the conventional MMA algorithm. Since the MMA can track only the polarization state, it is accompanied by a DD-CPE algorithm for the phase noise mitigation. It can be concluded from [13] that the CKF and A-CKF outperform the rest of the considered algorithms with a better tolerance towards polarization rotations, phase and amplitude noise. This can be attributed to the decrement in the inaccuracies through the linearization of the whole SSM in CKF/A-CKF, compared to EKF and UKF. The benefit from the adaptive computation of process noise covariance compared to the CKF can be observed at rotation angular frequencies of 400 Mrad/s and higher in the BTB case and at higher launch powers of 5 dBm in the transmission case [13].

3.5. Kalman filtering for joint compensation of phase and frequency offset

Apart from digital equalization, carrier synchronization is also vital to mitigate the phase and frequency offsets between the transmitter laser and free running LO. Since the CPE methods have low tolerance towards FO, which may go as high as ±5 GHz, a separate FO estimation (FOE) is required. Consequently, several FOE algorithms have been proposed in the literature that are either based on the phase increments between adjacent symbols [72] or spectrum based methods [73]. These methods are either not accurate for higher order QAM systems or computationally complex.

3.5.1. LKF and EKF for FO estimation

A novel FOE algorithm using Kalman filtering have been proposed and numerically verified in [60]. The simulation results in [60] concludes that the Kalman filter can achieve faster convergence and outperforms the conventional FO estimation at low OSNR. In [64], FOE schemes based on blind and training data, using LKF and EKF have been proposed for QPSK systems. These Kalman based FOE algorithms are evaluated both numerically and experimentally, and are compared to FFT based FOE methods. The investigations in [64] report that the training data based Kalman FOE methods show better accuracy in estimating the FO in case of fewer symbols and high OSNR, compared to FFT based methods. However, a separate phase estimation has to be carried out after FO compensation.

3.5.2. Two stage EKF for joint compensation of FO, phase and amplitude noise

The Kalman based FOE algorithms proposed in [60, 64] can compensate only for the FO and therefore, the carrier phase has to be recovered separately after FO compensation. In [65], a two stage EKF method based on training data has been proposed for joint tracking of FO, phase and amplitude noise. In the first stage, a coarse estimate of FO is obtained using a set of training data symbols following the training data scheme proposed in [64]. In the second stage, CPANE algorithm has been employed to jointly compensate for the residual FO, phase and amplitude noise.

3.5.2.1. Principles of two stage EKF

After linear equalization, the received signal on single polarization, with frequency and phase offset can be represented as given in Eq. (39). Here, r_k and a_k denote the received and transmitted symbol, respectively, at the time instant. w denotes the FO between the transmitter laser and the LO. T_s denotes the symbol duration. $\O_k$ and n_k denote the phase noise and ASE noise, respectively. In order to obtain the measurement for FO, the first step is to wipe off the data phase which is performed by employing training data. Then the phase difference between the adjacent symbols [64] is computed, which gives the measurement of FO denoted by m_k, in Eq. (40). Here, v_k is given by $\O_k - \O_{k-1}$, and follows a Gaussian distribution. By considering the observation model given in Eq. (40) for EKF, a coarse FO estimation is performed in the first stage using a set of training data sequence. The input signal $\hat{r}_k$ to the second stage after coarse

FO estimation is given in Eq. (41). Here, the CPANE algorithm is employed to compensate the residual FO, phase noise and ASE induced phase and amplitude distortions. **Figure 8** illustrates the basic structure of this two stage EKF [65]. A similar two stage model using LKF has also been evaluated in [65] and compared to EKF.

$$r_k = a_k \mathbf{e}^{j(wkT_s + \varnothing_k)} + n_k \tag{39}$$

$$m_k = \mathbf{e}^{j(w+v_k)} + \vartheta_k \tag{40}$$

$$\hat{r}_k = a_k \mathbf{e}^{j(\Delta w_k + \varnothing_k)} + n_k \tag{41}$$

The BER vs OSNR curves for LKF and EKF after the 2nd stage, using 200 and 500 training data symbols, for a FO of 1 GHz, are depicted in **Figure 9** [65]. It can be concluded from [65], that both

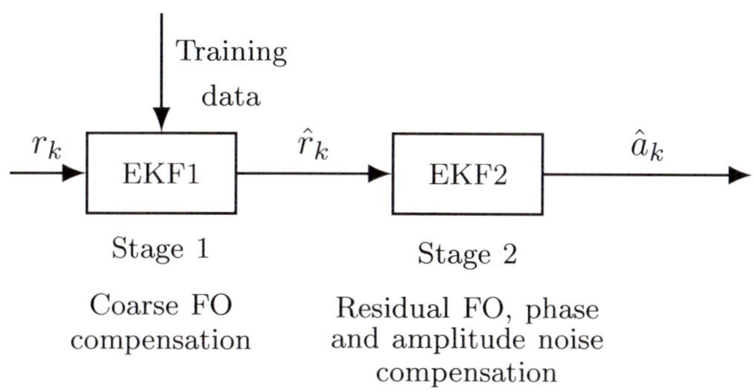

Figure 8. Block diagram of two stage EKF for the joint compensation of FO, phase and amplitude noise [65].

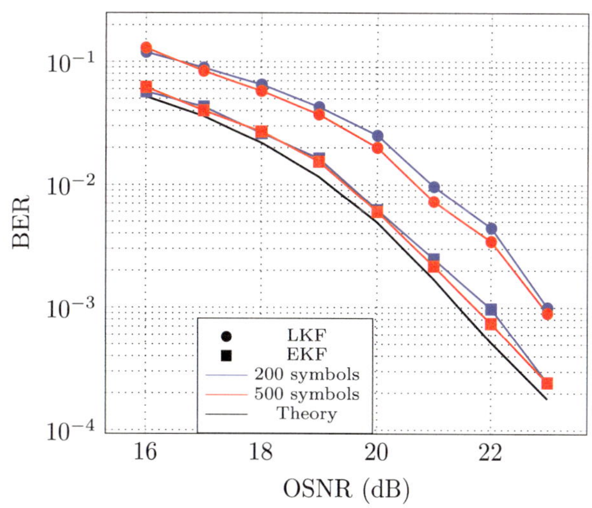

Figure 9. BER vs. OSNR curves for LKF and EKF after residual FO compensation for a FO of 1 GHz [65].

LKF and EKF show faster convergence irrespective of the number of training data symbols utilized in the first stage. However, since the EKF estimates a complex quantity, it facilitates in compensating also for the amplitude noise and therefore, outperforms LKF. Moreover, as discussed earlier, EKF does not require any angle operations unlike LKF, and thereby the additional few computations required by the EKF compared to LKF can be sought to be compensated with the additional benefit of better tracking capability.

This two stage EKF model has been extended in [66] to compensate also for the fiber nonlinearity in addition to FO, phase and amplitude noise. The first stage is similar and compensates FO coarsely, as discussed earlier. In the second stage, the total phase noise to be estimated comprises of both laser phase noise and fiber nonlinearities. The EKF-CPANE algorithm is employed for tracking the residual FO and the total phase noise in addition to amplitude noise. From the numerical analysis, it was reported in [66] that compared to LKF, the maximum possible transmission reach can be increased by an additional 500 km using EKF, at a BER of $2.4*10^{-2}$.

4. Conclusions

We have discussed in detail on how to exploit the potential of Kalman filters for the joint mitigation of several fiber optical transmission impairments in coherent optical transmission systems. Various Kalman based approaches for tracking carrier phase and frequency offset, polarization state have been reviewed. The CPANE algorithm and its implementation details using EKF for joint mitigation of linear and nonlinear phase noise as well as amplitude noise have been illustrated in detail. It is also verified that the combination of DBP and EKF enhances the nonlinear mitigation performance, at the expense of few DBP steps. A cascaded structure using LKF and EKF is illustrated for tracking the polarization state and carrier phase besides amplitude noise, simultaneously. A two stage EKF model for simultaneous tracking of FO, phase and amplitude noise is also discussed. From the discussed numerical verifications, it can be concluded that the Kalman filter based approaches for tracking the optical transmission impairments outperforms the conventional methods in coherent optical communication systems, with faster convergence, better tracking ability and more tolerance towards the optical transmission impairments. Since the Kalman filter is an optimal recursive MMSE estimator, with its attractive properties of hardware efficient implementation feasibility, less computational effort as well as memory requirements, it seems to be an essential component of future coherent optical receivers.

Acknowledgements

The authors gratefully acknowledge the support of the Erlangen Graduate School in Advanced Optical Technologies (SAOT) funded by the German National Science Foundation (DFG) in the framework of the excellence initiative.

Author details

Lalitha Pakala* and Bernhard Schmauss

*Address all correspondence to: lalitha.pakala@fau.de

Institute of Microwaves and Photonics (LHFT) and Erlangen Graduate School for Advanced Optical Technologies (SAOT), University of Erlangen-Nuremberg, Erlangen, Germany

References

[1] Zhu Y, Li A, Peng W, Kan C, Li Z, Chowdary S, Cui Y, Bai Y. Spectrally efficient single carrier 400G transmission enabled by probabilistic shaping. In: Optical Fiber Communications (OFC); 19-23 March 2017; Los Angeles, CA, USA. 2017. pp. 1-3

[2] Yu Y, Lu Y, Liu L, Huang Y, Wang X, Li L. Experimental demonstration of single carrier 400G/500G in 50-GHz grid for 1000-km transmission. In: Optical Fiber Communications; 19-23 March 2017; Los Angeles, CA, USA. 2017. pp. 1-3

[3] Chien H, Yu J. On single-carrier 400G line side optics using PM-256 QAM. In: European Conference on Optical Communications; 18-22 September 2016; Dusseldorf, Germany. 2016. pp. 1-3

[4] Maeda H, Kotanigawa T, Saito K, Yokota M, Hamaoka F, Yoshida M, et al. Field trial of simultaneous 100-Gbps and 400-Gbps transmission using advanced digital coherent technologies. In: Optical Fiber Communications (OFC); 20-24 March 2016; Anaheim, CA, USA. 2016. pp. 1-3

[5] Lau APT, Gao Y, Sui Q, Wang D, Zhuge Q, Morsy-Osman MH, et al. Advanced DSP techniques enabling high spectral efficiency and flexible transmissions: Toward elastic optical networks. IEEE Signal Processing Magazine. 2014;**31**(2):82-92. DOI: 10.1109/MSP.2013.2287021

[6] Ip E, Lau APT, Barros DJF, Kahn JM. Coherent detection in optical fiber systems. Optics Express. 2008;**16**(2):753-791. DOI: https://doi.org/10.1364/OE.16.000753

[7] Kikuchi K. Coherent optical communications: Historical perspectives and future directions. In: Optical and Fiber Communications Reports. Berlin, Heidelberg, Germany: Springer; 2010. pp. 11-49. DOI: 10.1007/978-3-642-10419-0_2

[8] Ip EM, Kahn JM. Fiber impairment compensation using coherent detection and digital signal processing. Journal of Lightwave Technology. 2010;**28**(4):502-519. DOI: 10.1109/JLT.2009.2028245

[9] Ip E, Kahn JM. Digital equalization of chromatic dispersion and polarization mode dispersion. Journal of Lightwave Technology. 2007;**25**(8):2033-2043. DOI: 10.1109/JLT.2007.900889

[10] Ip E, Kahn JM. Compensation of dispersion and nonlinear impairments using digital backpropagation. Journal of Lightwave Technology. 2008;**26**(20):3416-3425. DOI: 10.1109/JLT.2008.927791

[11] Du LB, Rafique D, Napoli A, Spinnler B, Ellis AD, Kuschnerov M, et al. Digital fiber nonlinearity compensation: Toward 1-Tb/s transport. IEEE Signal Processing Magazine. 2014;**31**(2):46-56. DOI: 10.1109/MSP.2013.2288110

[12] Pakala L, Schmauss B. Extended Kalman filtering for joint mitigation of phase and amplitude noise in coherent QAM systems. Optics Express. 2016;**24**(6):6391-6401. DOI: 10.1364/OE.24.006391

[13] Pakala L, Schmauss B. Joint tracking of polarization state and phase noise using adaptive cascaded Kalman filtering. IEEE Photonics Technology Letters. 2017;**29**(16):1297-1300. DOI: 10.1109/LPT.2017.2719279

[14] Zhou X, Nelson L. Advanced DSP for 400 Gb/s and beyond optical networks. Journal of Lightwave Technology. 2014;**32**(16):2716-2725. DOI: 10.1109/JLT.2014.2321135

[15] Kikuchi K. Polarization-demultiplexing algorithm in the digital coherent receiver. In: Proceedings of Digest LEOS Summer Topical Meetings; 21-23 July; Acapulco, Mexico. 2008. pp. 101-102. DOI: 10.1109/LEOSST.2008.4590509

[16] Fatadin I, Ives D, Savory SJ. Blind equalization and carrier phase recovery in a 16-QAM optical coherent system. Journal of Lightwave Technology. 2009;**27**(15):3042-3049. DOI: 10.1109/JLT.2009.2021961

[17] Stojanovic N, Huang Y, Hauske FN, Fang Y, Chen M, Xie C, et al. MLSE-based nonlinearity mitigation for WDM 112 Gbit/s PDM-QPSK transmissions with digital coherent receiver. In: Optical Fiber Communication Conference and Exposition (OFC/NFOEC), 2011 and the National Fiber Optic Engineers Conference; 6-11 March; Los Angeles, CA, USA. 2011. pp. 1-3

[18] Marcenac DD, Nesset D, Kelly AE, Brierley M, Ellis AD, Moodie DG, et al. 40 Gbit/s transmission over 406 km of NDSF using mid-span spectral inversion by four-wave-mixing in a 2 mm long semiconductor optical amplifier. Electronics Letters. 1997;**33** (10):879-880. DOI: 10.1049/el:19970583

[19] Morshed M, Du LB, Lowery AJ. Mid-span spectral inversion for coherent optical OFDM systems: Fundamental limits to performance. Journal of Lightwave Technology. 2012;**31** (1):58-66. DOI: 10.1109/JLT.2012.2227942

[20] Sackey I, Ros FD, Fischer JK, Richter T, Jazayerifar M, Peucheret C. Kerr nonlinearity mitigation: Mid-link spectral inversion versus digital backpropagation in 5x28-GBd PDM 16-QAM signal transmission. Journal of Lightwave Technology. 2015;**33**(9):1821-1827. DOI: 10.1109/JLT.2015.2393152

[21] Liu X, Chandrasekhar S, Winzer PJ, Tkach RW, Chraplyvy AR. Fiber-nonlinearity-tolerant superchannel transmission via nonlinear noise squeezing and generalized phase

conjugated twin waves. Journal of Lightwave Technology. 2014;**32**(4):766-775. DOI: 10.1109/JLT.2013.2280998

[22] Tao Z, Dou L, Yan W, Li L, Hoshida T, Rasmussen JC. Multiplier-free intrachannel nonlinearity compensating algorithm operating at symbol rate. Journal of Lightwave Technology. 2011;**29**(17):2570-2576. DOI: 10.1109/JLT.2011.2160933

[23] Gao Y, Cartledge JC, Karar AS, Yam SS-H, O'Sullivan M, Laperle C, et al. Reducing the complexity of perturbation based nonlinearity precompensation using symmetric EDC and pulse shaping. Optics Express. 2014;**22**(2):1209-1219. DOI: 10.1364/OE.22.001209

[24] Essiambre RJ, Kramer G, Winzer PJ, Foschini GJ, Goebel B. Capacity limits of optical fiber networks. Journal of Lightwave Technology. 2010;**28**(4):662-701. DOI: 10.1109/JLT.2009.2039464

[25] Ellis AD, Zhao J, Cotter D. Approaching the non-linear Shannon limit. Journal of Lightwave Technology. 2010;**28**(4):423-433. DOI: 10.1109/JLT.2009.2030693

[26] Wu M, Way WI. Fiber nonlinearity limitations in ultra-dense WDM systems. Journal of Lightwave Technology. 2004;**22**(6):1483-1498. DOI: 10.1109/JLT.2004.829222

[27] Li L, Tao Z, Dou L, Yan W, Oda S, Tanimura T, et al. Implementation efficient equalizer based on correlated backpropagation. In: Optical Fiber Communication Conference and Exposition (OFC/NFOEC), 2011 and the National Fiber Optics Engineers Conference; 6-10 March; Los Angeles, CA, USA. 2011. pp. 1-3

[28] Rafique D, Mussolin M, Forzati M, Martensson J, Chugtai MN, Ellis AD. Compensation of intra-channel nonlinear fibre impairments using simplified digital back-propagation algorithm. Optics Express. 2011;**19**(10):9453-9460. DOI: 10.1364/OE.19.009453

[29] Lin C, Holtmannspoetter M, Asif MR, Schmauss B. Compensation of transmission impairments by digital backward propagation for different link designs. In: 36th European Conference and Exhibition on Optical Communication; 19-23 September; Torino, Italy. 2010. pp. 1-3. DOI: 10.1109/ECOC.2010.5621413

[30] Pakala L, Schmauss B. Performance evaluation of modulation format independent carrier phase estimation and decision directed carrier phase estimation for fiber nonlinearity mitigation. In: Proc. SPIE 9288, Photonics North 2014; 28 May; Montreal, Canada. 2014. DOI: 10.1107/12.2074897

[31] Pakala L, Schmauss B. Enhanced transmission performance using digital back propagation and extended Kalman filtering for dispersion unmanaged links. In: Advanced Photonics 2016; 18-20 July; Vancouver, Canada. OSA Technical Digest (Optical Society of America); 2016. DOI: 10.1364/SPPCOM.2016.SpW1G.3

[32] Rafique D, Zhao J, Ellis AD. Fundamental limitations of digital back-propagation in coherent transmission systems. In: Transparent Optical Networks (ICTON), 2011 13th International Conference on; 26-30 June; Stockholm, Sweden. 2011. pp. 1-4. DOI: 10.1109/ICTON.2011.5971101

[33] Secondini M, Rommel S, Meloni G, Fresi F, Forestieri E, Poti L. Single-step digital backpropagation for nonlinearity mitigation. Photonics Network Communications. 2016;**31**(3): 493-502. DOI: 10.1007/s11107-015-0586-z

[34] Rafique D, Ellis AD. Nonlinear and ROADM induced penalties in 28 GBaud dynamic optical mesh networks employing electronic signal processing. Optics Express. 2011;**19**(18): 16739-16748. DOI: 10.1364/OE.19.016739

[35] Ip E, Kahn JM. Feedforward carrier recovery for coherent optical communications. Journal of Lightwave Technology. 2007;**25**(9):2675-2692. DOI: 10.1109/JLT.2007.902118

[36] Pfau T, Hoffmann S, Noe R. Hardware-efficient coherent digital receiver concept with feedforward carrier recovery for M-QAM constellations. Journal of Lightwave Technology. 2009;**27**(8):989-999. DOI: 10.1109/JLT.2008.2010511

[37] Li J, Li L, Tao Z, Hoshida T, Rasmussen JC. Laser-linewidth-tolerant feed-forward carrier phase estimator with reduced complexity for QAM. Journal of Lightwave Technology. 2011;**29**(16):2358-2364. DOI: 10.1109/JLT.2011.2159580

[38] Taylor MG. Phase estimation methods for optical coherent detection using digital signal processing. Journal of Lightwave Technology. 2009;**27**(7):901-914. DOI: 10.1109/JLT.2008.927778

[39] Gao Y, Lau APT, Lu C. Modulation-format-independent carrier phase estimation for square M-QAM systems. IEEE Photonics Technology Letters. 2013;**25**(11):1073-1076. DOI: 10.1109/LPT.2013.2259226

[40] Pakala L, Schmauss B. Improved decision directed carrier phase estimation for nonlinearity mitigation in 16-QAM systems. In: 16th International Conference on Transparent Optical Networks (ICTON); 6-10 July; Graz, Austria. 2014. pp. 1-4. DOI: 10.1109/ICTON.2014.6876047

[41] Viterbi A. Nonlinear estimation of PSK-modulated carrier phase with application to burst digital transmission. IEEE Transactions on Information Theory. 1983;**29**(4):543-551. DOI: 10.1109/TIT.1983.1056713

[42] Zhang S, Kam PY, Yu C, Chen J. Decision-aided carrier phase estimation for coherent optical communications. Journal of Lightwave Technology. 2010;**28**(11):1597-1607. DOI: 10.1109/JLT.2010.2048198

[43] Rice F, Rice M, Cowley B. A new algorithm for 16QAM carrier phase estimation using QPSK partitioning. Digital Signal Processing. 2002;**12**(1):77-86

[44] Li L, Tao Z, Liu L, Yan W, Oda S, Hoshida T, et al. XPM tolerant adaptive carrier phase recovery for coherent receiver based on phase noise statistics monitoring. In: 35th European Conference on Optical Communication 2009 (ECOC'09); 20-24 September; Vienna, Austria. 2009. pp. 1-3

[45] Pakala L, Schmauss B. Nonlinearity and phase noise mitigation using feed-forward carrier phase estimation and digital backward propagation in coherent QAM transmission. Physics Procedia. 2014;**56**:1353-1357. DOI: 10.1016/j.phpro.2014.08.062

[46] Piyawanno K, Kuschnerov M, Spinnler B, Lanki B. Nonlinearity mitigation with carrier phase estimation for coherent receivers with higher-order modulation formats. In: LEOS Annual Meeting Conference Proceedings, 2009 (LEOS'09); 4-8 October; Belek-Antalya, Turkey. 2009. pp. 1-2. DOI: 10.1109/LEOS.2009.5343204

[47] Pakala L, Schmauss B. Non-linear mitigation using carrier phase estimation and K-Means clustering. In: Photonic Networks; 16. ITG Symposium; Proceedings of; 7-8 May; Leipzig, Germany. 2015

[48] Lin C, Asif R, Holtmannspoetter M, Schmauss B. Nonlinear mitigation using carrier phase estimation and digital backward propagation in coherent QAM transmission. Optics Express. 2012;**20**(26):405-412. DOI: 10.1364/OE.20.00B405

[49] Bisplinghoff A, Vogel C, Kupfer T, Langenbach S, Schmauss B. Slip-reduced carrier phase estimation for coherent transmission in the presence of non-linear phase noise. In: Optical Fiber Communication Conference and Exposition and the National Fiber Optic Engineers Conference (OFC/NFOEC), 2013; 17-21 March; Anaheim, CA, USA. 2013. pp. 1-3. DOI: 10.1364/OFC.2013.OTu3I.1

[50] Savory SJ. Digital coherent optical receivers: Algorithms and subsystems. IEEE Journal of Selected Topics in Quantum Electronics. 2010;**16**(5):1164-1179. DOI: 10.1109/JSTQE.2010.2044751

[51] Kalman RE. A new approach to linear filtering and prediction problems. Transaction of the ASME-Journal of Basic Engineering. 1960;**82**:35-45

[52] Pessoa LM, Salgado HM, Darwazeh I. Efficient implementation of a phase estimation algorithm in coherent optical systems. In: IEEE LEOS, STM; 2008. pp. 1-3

[53] Pakala L, Schmauss B. Joint compensation of phase and amplitude noise using extended Kalman filter in coherent QAM systems. In: European Conference on Optical Communication (ECOC), 2014; 21-25 September; Cannes, France. 2014. pp. 1-3. DOI: 10.1109/ECOC.2014.6964165

[54] Pakala L, Schmauss B. Extended Kalman filtering for simultaneous phase and amplitude noise mitigation in WDM systems. In: 17th International Conference on Transparent Optical Networks (ICTON); 5-9 July 2015; Budapest, Hungary. 2015. pp. 1-4. DOI: 10.1109/ICTON.2015.7193301

[55] Zibar D, Hecker de Carvalho LH, Piels M, Doberstein A, Diniz J, Nebendahl B, et al. Application of machine learning techniques for amplitude and phase noise characterization. Journal of Lightwave Technology. 2015;**33**(7):1333-1343. DOI: 10.1109/JLT.2015.2394808

[56] Inoue T, Namiki S. Carrier recovery for M-QAM signals based on a block estimation process with Kalman filter. Optics Express. 2014;**22**(13):15376-15387. DOI: 10.1364/OE.22.015376

[57] Szafraniec B, Marshall TS, Nebendahl B. Performance monitoring and measurement techniques for coherent optical systems. Journal of Lightwave Technology. 2013;**31**(4):648-663. DOI: 10.1109/JLT.2012.2212234

[58] Pakala L, Schmauss B. Evaluation of correlated digital back propagation and extended Kalman filtering for nonlinear mitigation in PM-16-QAM WDM systems. In: Proc. SPIE 10130, Next-Generation Optical Communication: Components, Sub-systems, and Systems VI, 101300K; 28 January; San Francisco, CA, USA; 2017. pp. 1-9. DOI: 10.1107/12.2250488

[59] Pakala L, Schmauss B. Effective mitigation of non-linearities using extended Kalman filtering for dispersion managed links. In: Proceedings of Photonic Networks; 17. ITG-Symposium; 12-13 May; Leipzig, Germany. 2016

[60] Zhang S, Kam PY, Yu C, Chen J. Frequency offset estimation using a Kalman filter in coherent optical phase-shift keying systems. In: Conference on Lasers and Electro-Optics (CLEO) and Quantum Electronics and Laser Science Conference (QELS); 16-21 May; San Jose, CA, USA. 2010. pp. 1-2

[61] Marshall T, Szafraniec B, Nebendahl B. Kalman filter carrier and polarization-state tracking. Optics Letters. 2010;**35**(13):2203-2205. DOI: 10.1364/OL.35.002203

[62] Yang Y, Cao G, Zhong K, Zhou X, Yao Y, Pak A, et al. Fast polarization-state tracking scheme based on radius-directed linear Kalman filter. Optics Express. 2015;**23**(15):19673-19680. DOI: 10.1364/OE.23.019673

[63] Jignesh J, Corcoran B, Zhu C, Lowery A. Unscented Kalman filters for polarization state tracking and phase noise mitigation. Optics Express. 2016;**24**(19):22282-22295. DOI: 10.1364/OE.24.022282

[64] Jiang W, Yang Y, Zhang Q, Sun Y, Zhong K, Zhou X, et al. Application of Kalman filter in frequency offset estimation for coherent optical quadrature phase-shift keying communication system. Optical Engineering. 2016;**55**(9):096-102. DOI: 10.1117/1.OE.55.9.096102

[65] Pakala L, Schmauss B. Joint compensation of frequency offset, phase and amplitude noise using two stage extended Kalman filtering. In: Proceedings of Photonic Networks; 18. ITG-Symposium; 11-12 May; Leipzig, Germany. 2017. pp. 1-4

[66] Pakala L, Schmauss B. Two stage extended Kalman filtering for joint compensation of frequency offset, linear and nonlinear phase noise and amplitude noise in coherent QAM systems. In: 19th International Conference on Transparent Optical Networks (ICTON); 2-6 July; Girona, Spain; 2017. pp. 1-4

[67] Bar-Shalom Y, Rong Li X, Kirubarajan T. Estimation with Applications to Tracking and Navigation: Theory, Algorithms and Software. New York: Wiley; 2001. 584 p. DOI: 10.1002/0471221279

[68] Mehra R. Approaches to adaptive filtering. IEEE Transactions on Automatic Control. 1972;**17**(5):693-698. DOI: 10.1109/TAC.1972.1100100

[69] Ding W, Wang J, Rizos C. Improving adaptive Kalman estimation in GPS/INS integration. The Journal of Navigation. 2007;**60**:517-529. DOI: 10.1017/S0373463307004316

[70] Zhu L, Li G. Folded digital backward propagation for dispersion managed fiber-optic transmission. Optics Express. 2011;**19**(7):5953-5959. DOI: 10.1364/OE.19.005953

[71] Kikuchi K. Performance analyses of polarization demultiplexing based on constant-modulus algorithm in digital coherent optical receivers. Optics Express. 2011;**19**(2):9868-9880. DOI: 10.1364/OE.19.009868

[72] Leven A, Kaneda N, Koc U, Chen Y. Frequency estimation in intradyne reception. IEEE Photonics Technology Letters. 2007;**19**(6):366-368. DOI: 10.1109/LPT.2007.891893

[73] Selmi M, Jaouen Y, Ciblat P. Accurate digital frequency offset estimator for coherent PolMux QAM transmission systems. In: 35th European Conference on Optical Communications (ECOC' 2009); 20-24 September; Vienna, Austria. 2009. pp. 1-3

Kalman Filter for Moving Object Tracking: Performance Analysis and Filter Design

Kenshi Saho

Additional information is available at the end of the chapter

http://dx.doi.org/10.5772/intechopen.71731

Abstract

This chapter presents Kalman filters for tracking moving objects and their efficient design strategy based on steady-state performance analysis. First, a dynamic/measurement model is defined for the tracking systems, assuming both position-only and position-velocity measurements. Then, problems with the Kalman filter design in tracking systems are summarized, and an efficient steady-state performance index proposed by the author [termed the root-mean-squared error index (the RMS index)] is introduced to resolve these concerns. The analytical relationship between the proposed RMS index and the covariance matrix of the process noise is shown, leading to a proposed design strategy that is based on this relationship. Theoretical performance analysis is conducted using the performance indices to show the optimality of the design strategy. Numerical simulations show the validity of the theoretical analyses and effectiveness of the proposed strategy in realistic situations. In addition, the optimal performance of the position-only-measured and position-velocity-measured systems is analyzed and compared. This comparison shows that the position-velocity-measured Kalman filter tracking is accurate when compared with the position-only-measured filter.

Keywords: Kalman tracking filter, moving object tracking, steady-state analysis, performance index, filter design, process noise

1. Introduction

Remote monitoring systems for cars and robots require accurate tracking of moving objects. Representative tracking algorithms include the Kalman filter [1–5] and its variants, such as the extended/unscented Kalman [6–9] and particle filters [10–12]. These can accurately track movement based on adaptive filtering by using a state-space model.

To use the Kalman filter for the tracking of moving objects, it is necessary to design a dynamic model of target motion. The most common dynamic model is a constant velocity (CV) model

[1, 10], which assumes that the velocity is constant during a sampling interval. This model has been used in many applications because of its versatility, effectiveness, and simplicity. However, in almost conventional tracking systems, the selection of process noise (zero-mean white noise in the dynamic model) is conducted empirically [4, 6, 8]. This is because conventional studies tend to assume that process noise takes one of a limited number of forms, which is known as appropriate selections. Thus, despite the large number of investigations into Kalman filter trackers, the optimal selection of a process noise model has not been discussed. The general problems of model selection for Kalman filter trackers were discussed by Ekstrand in 2012 [1]. In the years since, further research on these issues has been conducted, but no satisfactory solutions to the abovementioned problems have been presented. Crouse [13] described a general solution for optimal trackers in a steady state. However, this method also requires an empirical selection of the dynamic models. A detailed analysis of the Kalman filter has been provided for various applications, including global navigation satellite systems [14] and video trackers [15]. However, only limited systems have yet been considered, and no definitive parameter-setting procedure for the Kalman tracking filter has been provided. Although various criteria have been proposed and investigated for the design of Kalman filters and its variants to achieve better tracking accuracy, robustness, and real-time capability, relationship between these performance indices and the model parameters such as the process noise variance is not discussed even in recent studies [16].

Another significant problem in a measurement model of the conventional Kalman tracking filter is that most studies consider only position measurements and therefore cannot make full use of modern sensors that are able to measure velocity, such as ultrawideband Doppler radar [17, 18]. Moreover, sensor fusion based on Internet of Things technology also enables the simultaneous measurement of position and velocity (e.g., sensor data fusion based on the communication between radars/lasers/sonars and speedometers embedded in targets). Consequently, Kalman filters for such systems have become an important area of research [19–24]. In Ref. [24], the extended Kalman filter for radar measurements is modified for range (position) and range-rate (velocity) measurements, and its effectiveness in realistic radar applications is verified. However, concrete design criterion is not shown. The number of conventional studies on position-velocity-measured (PVM) Kalman filters is smaller than those on the more common position-only-measured (POM) Kalman filters, and the performance and design of PVM Kalman filters are not sufficiently considered.

To resolve the two problems described above concerning the process noise selection and PVM systems, our previous work clarified the fundamental properties of PVM tracking filters [25, 26] and generated an efficient performance index to design an optimal process noise matrix [3, 5]. In the studies of PVM tracking filters [25, 26], fixed-gain PVM filter properties were analytically clarified, but there was no optimization of the PVM Kalman filters. In our work on the process noise matrix [3], an optimal POM Kalman filter, with respect to position prediction, was presented. In this chapter, an appropriate process noise design strategy, based on our proposed efficient steady-state performance index (introduced in Section 3), and its applicability are verified. Our previous work highlighted the following issues, which we address in this chapter:

i. Analysis of the performance of a PVM Kalman filter with a CV model, based on the proposed index.

ii. Application of the proposed process noise design strategy to a PVM Kalman filter.

iii. Comparison of the performance of optimal POM and PVM Kalman filters.

This chapter presents the theoretical analyses and simulations required to tackle these issues. The remainder of this chapter is organized as follows: Section 2 defines the tracking filtering problem dealt in this chapter and explains the existing concerns and models for POM and PVM Kalman filter design. Section 3 introduces our proposed efficient performance indices with their mathematical formulations. Section 4 presents the proposed process noise design strategy based on the performance index. Section 5 shows the theoretical analysis of the optimal POM and PVM Kalman filter performance in a steady state. The effectiveness of the PVM Kalman filter is proven by the comparison with the POM filter. Section 6 shows realistic maneuvering-target-tracking application examples. Section 7 concludes this chapter and proposes future tasks.

2. Problem statement

This section introduces the Kalman filter for moving object tracking and defines the model assumed in this chapter.

2.1. Dynamic model

The Kalman filter for tracking moving objects estimates a state vector comprising the parameters of the target, such as position and velocity, based on a dynamic/measurement model. For simplicity, this chapter deals with a typical second-order one-dimensional Kalman filter tracker whose true state vector is defined as

$$\mathbf{x}_{\mathrm{t}} = (x_{\mathrm{t}} \;\; v_{\mathrm{t}})^{\mathrm{T}}, \tag{1}$$

where x_{t} and v_{t} are the true position and velocity of the target moving object, respectively, and ^{T} denotes the transpose. The assumed dynamic model is a CV model, which is a simple and popular model for tracking moving objects. The CV model assumes that the velocity is constant during the sampling interval, which is expressed as

$$\mathbf{x}_{\mathrm{t}k} = \mathbf{\Phi}\mathbf{x}_{\mathrm{t}k-1} + \mathbf{w}_k, \tag{2}$$

where $\mathbf{x}_{\mathrm{t}k}$ denotes the true state at time kT, T is the sampling interval, $\mathbf{w}_k$ is the process noise with covariance matrix $\mathbf{Q}$, and $\mathbf{\Phi}$ is the transition matrix from kT to $(k+1)T$, which is expressed as

$$\mathbf{\Phi} = \begin{pmatrix} 1 & T \\ 0 & 1 \end{pmatrix}, \tag{3}$$

The Kalman filter predicts the target state based on this dynamic model.

2.2. Measurement model

The measurements are simply modeled as

$$\mathbf{z}_k = \mathbf{H}\mathbf{x}_{tk} + \mathbf{v}_k, \tag{4}$$

where $\mathbf{z}_k$ denotes the measurement vector, $\mathbf{H}$ denotes the measurement matrix, and $\mathbf{v}_k$ is the measurement noise with covariance matrix $\mathbf{R}$. This chapter considers two types of measurement systems, which are discussed as follows.

2.2.1. Position-only-measured system

The POM system assumes that the sensors (such as radar, laser, and sonar) can measure only the position of the target. This is a general assumption in the moving object tracking. $\mathbf{H}$ and $\mathbf{R}$ of this model are expressed as

$$\mathbf{H} = \begin{pmatrix} 1 & 0 \end{pmatrix}, \tag{5}$$

$$\mathbf{R} = (B_\mathrm{x}), \tag{6}$$

where $B_\mathrm{x}$is the variance of the position measurement errors.

2.2.2. Position-velocity-measured system

The PVM system assumes that the sensor system can measure position and velocity simultaneously. One example of the PVM model system is a pulse Doppler radar. Sensor fusion systems using communications of position/velocity sensors can also be expressed by the PVM model. $\mathbf{H}$ of this model is expressed as

$$\mathbf{H} = \begin{pmatrix} 1 & 0 \\ 0 & 1 \end{pmatrix}. \tag{7}$$

We now assume that the noises of position and velocity measurements are uncorrelated, and $\mathbf{R}$ of PVM systems under this assumption is defined as

$$\mathbf{R} = \begin{pmatrix} B_\mathrm{x} & 0 \\ 0 & B_\mathrm{v} \end{pmatrix}. \tag{8}$$

where $B_\mathrm{v}$is the variance of the velocity measurement errors.

2.3. Kalman filter tracking

The Kalman filter tracker based on the abovementioned models sequentially estimates state vectors via the Kalman filter equations. The prediction and estimation are calculated as

$$\tilde{\mathbf{x}}_k = \boldsymbol{\Phi}\hat{\mathbf{x}}_{k-1}, \tag{9}$$

$$\hat{\mathbf{x}}_k = \tilde{\mathbf{x}}_k + \mathbf{K}_k(\mathbf{z}_k - \mathbf{H}\tilde{\mathbf{x}}_k), \tag{10}$$

where predicts and estimates are denoted by ˜ and ^, respectively, and $\mathbf{K}_k$ denotes the Kalman gain that minimizes the errors in the estimated position and velocity. $\mathbf{K}_k$ is calculated as

$$\mathbf{K}_k = \tilde{\mathbf{P}}_k\mathbf{H}^{\mathrm{T}}\left(\mathbf{H}\tilde{\mathbf{P}}_k\mathbf{H}^{\mathrm{T}} + \mathbf{R}\right). \tag{11}$$

where $\mathbf{P}_k$ is the covariance matrix of errors determined from

$$\tilde{\mathbf{P}}_k = \boldsymbol{\Phi}\hat{\mathbf{P}}_{k-1}\boldsymbol{\Phi}^{\mathrm{T}} + \mathbf{Q}. \tag{12}$$

$$\hat{\mathbf{P}}_k = \tilde{\mathbf{P}}_k - \mathbf{K}_k\mathbf{H}\tilde{\mathbf{P}}_k. \tag{13}$$

2.4. Aspects of tracking filter design

Moving object tracking obtains accurate and sequential estimation of the target position and velocity by using Eqs. (9)–(13). As indicated in Eqs. (1)–(13), the design parameters of the Kalman filter tracker are elements of the covariance matrix of the process noise $\mathbf{Q}$. We must set $\mathbf{Q}$ to achieve tracking errors that are as small as possible. Thus, we must know how to design an appropriate $\mathbf{Q}$. Moreover, we must be able to define the evaluation index of the filter performance. However, these issues have not been sufficiently deliberated because the selection of $\mathbf{Q}$ has not been sufficiently addressed in previous studies. Here, the design of $\mathbf{Q}$ is empirically carried out.

In the conventional tracking systems, the most commonly used random acceleration (RA) process noise is often selected because it has a better performance. Its $\mathbf{Q}$ is

$$\mathbf{Q}_{\mathrm{ra}} = \begin{pmatrix} T^4/4 & T^3/2 \\ T^3/2 & T^2 \end{pmatrix}\sigma_{\mathrm{q}}^2. \tag{14}$$

The appropriate selection of σ_{q} is important because σ_{q} (and sensor noise variance $\mathbf{R}$) directly determines the performance of the tracking filter with the CV model. However, in conventional studies, process noises and their parameters are empirically selected, and the validity of the selection is discussed only casually [1, 16]. Many conventional tracking systems select the RA process noises ($\mathbf{Q}_{\mathrm{ra}}$), with variance σ_{q} set based on the assumed target motion. However, no definitive method of determining σ_{q} has been established. Although tracking index defined by Kalata [27] is known as an effective design parameter, its empirical selection is still required. Moreover, the validity in selecting the RA process noise is also questionable. Various other forms of $\mathbf{Q}$ are known and have been used for different target motions [12]. For example, random velocity model [2] and the diagonal $\mathbf{Q}$, which do not include correlations in process noise [7], are also frequently used. However, for the reasons discussed earlier, the differences in performance between the various process noise models are not known.

3. The efficient steady-state performance index (RMS index)

The process noise selection problems discussed in Section 2.3 must be solved to effectively design Kalman tracking filters. Thus, we must properly evaluate the performance of the filter. The effective steady-state performance index was derived [3] and is termed root-mean-squared error index (an RMS index). This section introduces the RMS index for POM and PVM systems and shows the analytical relationships between the RMS index and **Q**.

3.1. Definition of RMS index

In tracking filtering, the following two functions are required:

- Function 1. Reduces random errors caused by measurement noises.
- Function 2. Tracks targets with complicated motions (e.g., accurate tracking of an accelerating target is required for the CV model).

The RMS index is proposed for the comprehensive evaluation of the performance of these two functions and is defined as

$$\varepsilon_{\mathrm{p}} \equiv \sqrt{\mathrm{E}\left[(x_{\mathrm{ta}k} - \tilde{x}_k)^2\right]} \tag{15}$$

where $\tilde{x}_k$ is the predicted target position (second element of $\tilde{\mathbf{x}}_k$), E[] indicates the mean with respect to k, and $x_{\mathrm{ta}k}$ is the true position of a constant acceleration target which is

$$x_{\mathrm{ta}k} = a_{\mathrm{c}}(kT)^2/2 \tag{16}$$

where a_{c} is constant acceleration of the target. In the Kalman filter tracker using the CV model, it is assumed that the target velocity is constant during the sampling interval. Thus, for the constant acceleration target, a steady-state bias error occurs because of the difference between the target motion and the assumed dynamic model. Moreover, $\tilde{x}_k$ includes random errors due to measurement noise. Thus, the RMS index ε_{p} expresses both bias errors and random errors. With the steady-state bias error due to the model/motion difference of e_{ac} and the steady-state standard deviation of the random errors in $\tilde{x}_k$ of σ_{p}, ε_{p} is expressed as

$$\varepsilon_{\mathrm{p}} = \sqrt{e_{\mathrm{ac}}^2 + \sigma_{\mathrm{p}}^2} \tag{17}$$

σ_{p} expresses the performance corresponding to Function 1 and e_{ac} expresses the performance corresponding to Function 2. The smaller these errors are, the better is the tracking filter. Thus, the minimum ε_{p} achieves the best tracking filter in a steady state.

3.2. RMS index of a POM system

One important advantage of the RMS index is that it can be expressed in closed form. The closed form of ε_{p} for the POM system was derived in Ref. [3]. This subsection introduces the RMS index and its relationship to the design parameter **Q** in the POM system.

First, the arbitrary **Q** is defined as

$$\mathbf{Q}_{\mathrm{gen}} = \begin{pmatrix} a & b \\ b & c \end{pmatrix} \tag{18}$$

where $a > 0$, $b > 0$, and $c > 0$, and the dimensions of a, b, and c are [m^2], [m^2/s], and [m^2/s^2], respectively. For example, substituting $(a, b, c) = \left(\sigma_{\mathrm{q}}^2 T^4/4, \sigma_{\mathrm{q}}^2 T^3/2, \sigma_{\mathrm{q}}^2 T^2\right)$ into Eq. (18) gives the $\mathbf{Q}_{\mathrm{ra}}$ of Eq. (14) and $b = 0$ leads to the diagonal **Q**. The analytical relationship between $\mathbf{Q}_{\mathrm{gen}}$ and ε_{p} is expressed by the following closed form.

$$\varepsilon_{\mathrm{p,\,pom}}^2 = \frac{{a_{\mathrm{c}}}^2 T^4}{\beta^2} + \frac{2\alpha^2 + 2\beta + \alpha\beta}{\alpha(4 - 2\alpha - \beta)} B_{\mathrm{x}} \tag{19}$$

where α and β are components of the steady-state Kalman gain $\mathbf{K}_\infty = (\alpha, \beta/T)^{\mathrm{T}}$ calculated from (a, b, c) using the following equations:

$$\beta = \frac{C + \sqrt{C(16 + 4A - 4B + C)}}{4} - \sqrt{\frac{C^2(16 + 4A - 4B + C)}{8\sqrt{C(16 + 4\mathrm{A} - 4B + C)}} + \frac{C(2A - 2B + C}{8}} \tag{20}$$

$$A = a/B_{\mathrm{x}}, \quad B = bT/B_{\mathrm{x}}, \quad C = cT^2/B_{\mathrm{x}} \tag{21}$$

$$\alpha = 1 - \beta^2/C \tag{22}$$

The derivation process of these equations is shown in Ref. [3]. As shown in Eqs. (19)–(22), the optimal (a, b, c) is designed minimizing ε_{p}.

3.3. RMS index of a PVM system

In a similar manner to the treatment of the POM system, this subsection introduces the RMS index of a PVM system and its relationship to $\mathbf{Q}_{\mathrm{gen}}$. The RMS index of the PVM system is

$$\varepsilon_{\mathrm{p,\,pvm}}^2 = \left(\frac{2 - 2\eta - \theta}{2(\beta + \alpha\theta - \beta\eta)}\right)^2 a_{\mathrm{c}}^2 T^4 + \frac{g_2(\alpha, \beta, \eta, \theta) B_{\mathrm{x}} + g_3(\alpha, \beta, \eta, \theta) T^2 B_{\mathrm{v}}}{g_1(\alpha, \beta, \eta, \theta)} \tag{23}$$

where $\alpha, \beta, \eta,$ and θ are components of the steady-state Kalman gain:

$$\mathbf{K}_\infty = \begin{pmatrix} \alpha & T\eta \\ \beta/T & \theta \end{pmatrix} \tag{24}$$

and,

$$g_1(\alpha, \beta, \eta, \theta) = (\beta\eta - \alpha\theta - \beta)(\alpha\theta - \beta\eta - \alpha - \theta)(4 - 2\alpha - \beta - 2\theta + \alpha\theta - \beta\eta) \tag{25}$$

$$g_2(\alpha,\beta,\eta,\theta) = \alpha^3\theta(\theta-2)(\theta-1) + \alpha^2\beta(2-2\theta-3\eta\theta^2+6\eta\theta-2\eta) + \alpha^2\theta^2(2-\theta) + 2\alpha\beta\theta(\eta-2)(\theta-2) + \alpha\beta^2(1+2\eta-\theta-3\eta^2+3\eta^2\theta) + \beta^3\eta(1+\eta)(1-\eta) + \beta^2\eta(2-\eta)(\theta-2) + \beta^2(2-\theta) \tag{26}$$

$$g_3(\alpha,\beta,\eta,\theta) = \alpha\theta(2\eta^2+2\eta\theta+\theta^2-\theta) + \eta\beta(2\eta+2\theta-2\eta^2-\theta^2-2\eta\theta) + \theta^2(2-\theta) \tag{27}$$

Eq. (22) is obtained from σ_p^2 and e_{ac} of the steady-state PVM Kalman filter ($\alpha-\beta-\eta-\theta$ filter) by using Eq. (17). The derivation processes for these are shown in Ref. [25]. The relationship between the steady-state Kalman gains and $\mathbf{Q}_{gen}$ is derived as follows:

$$a = \frac{T^2B_v}{1-R_{xv}\beta^2-(1-\theta)\alpha-\theta}\{(\alpha\beta^2+2\beta^3+\beta^2+(\theta-1)\theta\beta)R_{xv} + \alpha^2R_{xv}((1-\theta)(1+2\alpha\beta)+\beta^2\theta+\beta(3\theta-2-\theta^2)) + \theta(1-\theta)(\alpha-1)\} \tag{28}$$

$$b = \frac{\beta^3R_{xv}^2 + R_{xv}(\alpha(\beta(1-\theta)-\theta^2+\theta)+\beta^2\theta+\beta\theta+\theta^2-\theta)}{1-R_{xv}\beta^2-(1-\theta)\alpha-\theta}TB_v \tag{29}$$

$$c = \frac{R_{xv}(\alpha\beta\theta+\beta^2(\theta+1)-\beta\theta)-\alpha\theta(\beta+\theta)+\beta\theta+\theta^2}{1-R_{xv}\beta^2-(1-\theta)\alpha-\theta}B_v \tag{30}$$

where

$$R_{xv} = B_x/T^2B_v \tag{31}$$

$$\eta = R_{xv}\beta \tag{32}$$

The derivation of these is given in the Appendix. Note that the dimensionless parameter R_{xv} corresponds to the ratio of the measurement accuracies in position and velocity and directly affects the tracking accuracy in PVM tracking systems. From these results, we also obtain the closed form of the RMS index for PVM systems and can design optimal $\mathbf{Q}$ using Eqs. (22)–(32).

4. Filter design strategy based on the RMS index

Using the RMS index introduced in the previous section, we can design the Kalman filter parameters (i.e., $\mathbf{Q}$) to achieve optimal tracking. This section defines the optimization problems for POM and PVM systems with a $\mathbf{Q}$ that minimizes the RMS index ε_p.

4.1. RMS-index minimization problem

4.1.1. POM system optimization

The evaluating function to determine optimal $\mathbf{Q}$ is $\varepsilon_{p,pom}$ normalized by B_x, which is defined as

$$\mu_{\text{pom}} = \varepsilon_{\text{p,pom}}^2 / B_{\text{x}} = \frac{a_{\text{D}}^2}{\beta^2} + \frac{2\alpha^2 + 2\beta + \alpha\beta}{\alpha(4 - 2\alpha - \beta)} \tag{33}$$

where

$$a_{\text{D}}^2 = a_{\text{c}}^2 T^4 / B_{\text{x}} \tag{34}$$

is the preset parameter for the proposed strategy. Substituting Eqs. (20)–(22) into (33), we obtain $\mu_{\text{pom}}(a, b, c, a_{\text{D}})$. Using this, the optimal (a, b, c) for the POM system is determined by solving

$$\begin{aligned} &\underset{a,b,c}{\arg\min} && \mu_{\text{pom}}(a, b, c, a_{\text{D}}) \\ &\text{sub. to.} && a > 0, b > 0, c > 0, \text{and } a_{\text{D}} = \text{Const.} \end{aligned} \tag{35}$$

4.1.2. PVM system optimization

Like the POM system, a normalized RMS index can be used for the design of the PVM system. Normalizing Eq. (22) by B_{x} and substituting Eqs. (31) and (32) into this, the evaluating function for the PVM system is given by

$$\mu_{\text{pvm}} = \varepsilon_{\text{p,pvm}}^2 / B_{\text{x}} = \left(\frac{2 - 2\beta R_{\text{xv}} - \theta}{2(\beta + \alpha\theta - \beta^2 R_{\text{xv}})} \right)^2 a_{\text{D}}^2 + \frac{g_2(\alpha, \beta, \theta, R_{\text{xv}}) + g_3(\alpha, \beta, \theta, R_{\text{xv}}) / R_{\text{xv}}}{g_1(\alpha, \beta, \theta, R_{\text{xv}})} \tag{36}$$

To design optimal (a, b, c) for the PVM system, the optimal steady-state Kalman gains are calculated by solving the following minimization problem.

$$\begin{aligned} &\underset{\alpha,\beta,\theta}{\arg\min} && \mu_{\text{pvm}}(\alpha, \beta, \theta, R_{\text{xv}}) \\ &\text{sub. to.} && \text{Stability conditions are satisfied, and } a_{\text{D}} = \text{Const.} \end{aligned} \tag{37}$$

where the stability conditions with respect to Kalman gains are easily derived by the well-known Jury's test as

$$(1 - \eta)\beta < \alpha\theta \text{ and } 4 - 2\alpha - \beta - 2\theta + \alpha\theta - \eta\beta > 0 \text{ and } |\alpha\theta - \eta\beta - \alpha - \theta + 1| < 1 \tag{38}$$

Substituting the optimal (α, β, θ) calculated by Eq. (37) into Eqs. (28)–(30), we obtain an optimal (a, b, c) for the PVM Kalman filter.

4.2. Procedure and notes of the proposed design strategy

The procedure of the proposed strategy for each system is summarized in this section.

4.2.1. Design procedure for a POM system

1. Set B_{x} from the sensor performance.

2. Preset a_D based on the approximate target acceleration.
3. Determine (a, b, c) by solving Eq. (35).

The methodology of presetting a_D is discussed in the simulation section.

4.2.2. Design procedure for a PVM system

1. Set B_x and B_v from the sensor performance.
2. Preset a_D based on the approximate target acceleration.
3. Determine $(\alpha, \beta, \theta)$by solving Eq. (37).
4. Determine (a, b, c) from (α, β, θ) using Eqs. (28)–(30).

4.2.3. Notes on computation in the proposed strategy

With respect to the proposed strategy, note that:

- Eqs. (35) and (37) can be solved by gradient descent with several initial values. This is because that the parameter searching range is narrow due to the stability conditions.
- The proposed design process is only carried out once before using the Kalman filter. Although the computational cost of the above optimization process is not small, it does not affect the Kalman filtering process.

4.3. Discussion on preset parameter a_D

Here, the appropriate presetting for a_D in practical use is discussed. The covariance matrix of process noise **Q** determined by the proposed strategy is only optimal when a_D is matched to the target acceleration and the target is moving with constant acceleration corresponding to a_D. However, using the proposed strategy, the tracking accuracy is always better than when using conventional models as verified in Ref. [3]. Consequently, the proposed method achieves sufficient accuracy, even if a_D is not matched to the true target acceleration. This means that the relatively small difference between the true and preset acceleration is acceptable. Thus, in practical use, we estimate an approximate or a typical value for the acceleration (e.g., mean and maximum) in advance based on the assumed motion of the target and then set a_D by using this estimated value. The example application presented in Section 6 assumes the approximate maximum acceleration of the target is known and is used for the Kalman filter design.

Thus, target acceleration information is required for accurate Kalman filter tracking by using the proposed strategy. As a method to obtain an approximated acceleration, communications between the tracking systems and the accelerometers embedded in targets can be considered. Many sensing targets have acceleration sensors; for example, robots and vehicles have inertial sensors, and humans have accelerometers embedded in smartphones. Soon, Internet of Things technology will make data communications between robots, smartphones, and radar possible. Thus, we can obtain approximated acceleration based on this novel technology.

5. Theoretical steady-state performance analysis

This section presents theoretical performance analyses of the Kalman tracking filters by using the proposed design strategy. With respect to POM systems, our previous study [3] verified the effectiveness of the proposed strategy by comparison with a conventional random acceleration model based filter design. Thus, the RMS indices for the following filters are compared:

- Optimal POM filter: the Kalman filter for the POM system designed using the strategy mentioned in Section 4.2.1.
- Optimal PVM filter: the Kalman filter for the PVM system designed using the strategy mentioned in Section 4.2.2.
- RA filter: the Kalman filter for the PVM system with the RA process noises by using optimal σ_q with respect to the RMS index.

The comparison of the optimal PVM filter with the RA filter indicates the effectiveness of the proposed strategy (i.e., considering the arbitrary covariance matrix of the process noise $\mathbf{Q}_{gen}$) and the comparison of the optimal POM and PVM filters illustrates the enhancement of tracking accuracy by using the velocity measurements in the proposed strategy. This section assumes that B_x and T are normalized to 1.

Figure 1 shows the relationship between the design parameter a_D and the minimum RMS index $\varepsilon_{p,opt}$ for R_{xv} = 1 (**Figure 1** left) and R_{xv} = 10 (**Figure 1** right). It can be seen that the optimal PVM filter achieves the best performance. This result verifies that the proposed strategy determines steady-state gains corresponding to a better covariance matrix of process noise than the RA model. The optimal PVM filter also achieves better performance compared with the optimal POM filter even for $R_{xv} = 1$, which means that the measurement accuracy of the

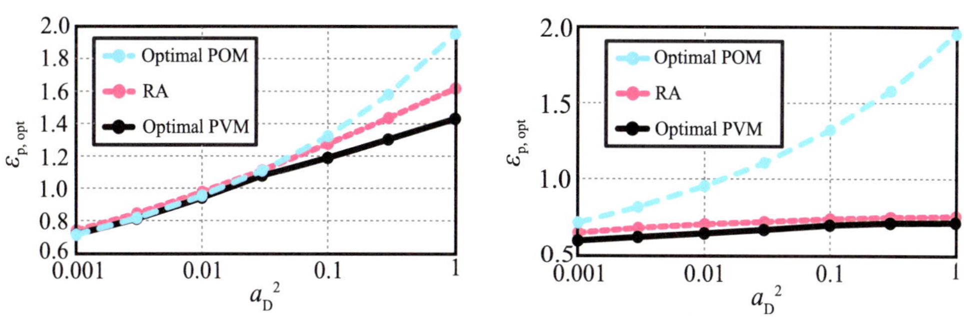

Figure 1. Analytical relationship between a_D and $\varepsilon_{p,opt}(R_{xv} = 1$ (left), $R_{xv} = 10$ (right)).

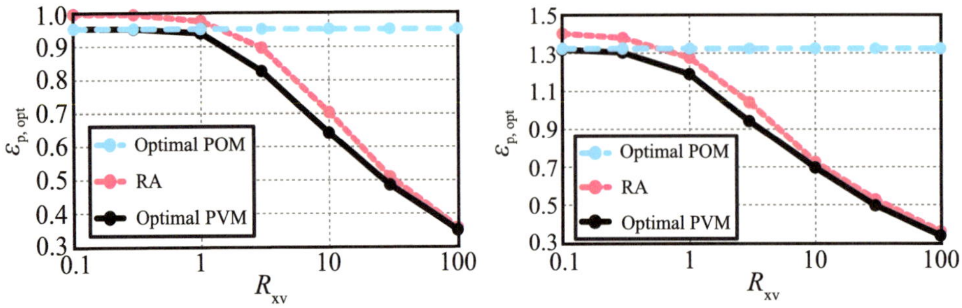

Figure 2. Analytical relationship between R_{xv} and $\varepsilon_{p,opt}$ ($a_D = 0.01$ (left), $a_D = 0.1$ (right)).

position and velocity is the same. The addition of the velocity measurements effectively enhances the tracking accuracy. Furthermore, when the velocity measurement accuracy is high, the optimal PVM filter achieves greater accuracy than the POM filter.

Figure 2 shows the relationship between R_{xv} and $\varepsilon_{p,opt}$ for $a_D^2 = 0.01$ (left) and 0.1 (right). Both cases exhibit the same trend. For both optimal PVM and RA filters, better performance is achieved with better velocity measurement accuracy. The performance of the optimal PVM filter is better than that of the optimal POM filter including relatively small R_{xv} (the velocity measurement accuracy is low). In contrast, the performance of the RA filter is worse than that of the optimal POM filter for small R_{xv} because the covariance matrix is limited to Eq. (14). Moreover, by comparing the two insets of **Figure 2**, we see the greater effectiveness of the proposed strategy for relatively large a_D.

6. Application to radar tracking simulation

Finally, this section provides an example of the Kalman filter tracker designed with the proposed strategy in a realistic application, namely, pulse Doppler radar tracking.

6.1. Simulation setup

We simulated the pulse Doppler radar tracking of a maneuvering target and compared the tracking errors of the filters assumed in the previous section. **Figure 3** shows the simulation scenario and the true target acceleration. The true target position is $(x_{tk}, y_{tk}) = \left((kT)^2, 20 + (kT)^{1.5}\cos(\pi kT/5)\right)$. Two-dimensional tracking in the x-y plane of the point target is assumed. We consider two pulse Doppler radars located at $(x, y) = (0.5$ m, 0) and (1.0 m, 0). The sampling interval T is 100 ms, and the observation time is 4 s. The transmitted signal is a pulse with central frequency of 60 GHz and bandwidth of 500 MHz. The received radar signals are calculated using ray tracing with the addition of the Gaussian white noise. The radar measurement parameter depends on the system under consideration: the POM system assumes the measurement of the position by using ranging results, and the PVM system assumes the position and velocity measurements where the position measurement is the same as the POM

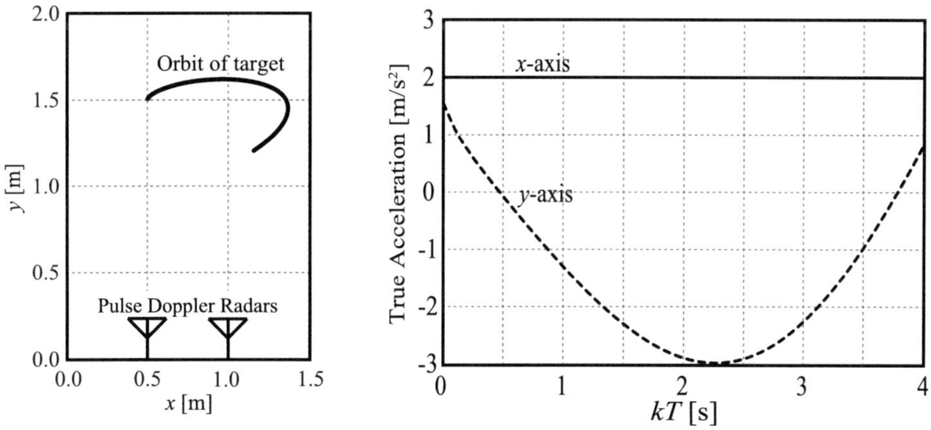

Figure 3. Simulation setting (simulation scenario (left), true target acceleration (right)).

system, and the velocity measurement is based on the Doppler shift with the method presented in Ref. [18]. We determine a variance for this noise to set $B_x = 9 \times 10^{-4}$ m^2 and $B_v = 0.09$ m^2/s^2. In these settings, $R_{xv} = 1$. These values are the averages along the two axes. Using the RMS prediction error calculated from 1000 Monte Carlo simulations, the performance is defined as

$$\varepsilon_k = \sqrt{\frac{1}{1000}\sum_{m=1}^{1000}\left\{\left(x_{tk} - x_{pmk}\right)^2 + \left(y_{tk} - y_{pmk}\right)^2\right\}} \tag{39}$$

where x_{pmk} and y_{pmk} are the predicted positions in the mth Monte Carlo simulation.

6.2. Implementation of Kalman filter

First, the implementation of the Kalman filters for two-dimensional system is presented. The implementation of a two-dimensional optimal POM filter is as follows:

$$\mathbf{x}_t = \begin{pmatrix} x_t & v_{xt} & y_t & v_{yt} \end{pmatrix}^T \tag{40}$$

$$\mathbf{\Phi} = \begin{pmatrix} 1 & T & 0 & 0 \\ 0 & 1 & 0 & 0 \\ 0 & 0 & 1 & T \\ 0 & 0 & 0 & 1 \end{pmatrix} \tag{41}$$

$$\mathbf{H} = \begin{pmatrix} 1 & 0 & 0 & 0 \\ 0 & 0 & 1 & 0 \end{pmatrix} \tag{42}$$

$$\mathbf{R} = \begin{pmatrix} B_x & 0 \\ 0 & B_x \end{pmatrix} \tag{43}$$

$$\mathbf{Q} = \begin{pmatrix} a_{\text{opt}} & b_{\text{opt}} & 0 & 0 \\ b_{\text{opt}} & c_{\text{opt}} & 0 & 0 \\ 0 & 0 & a_{\text{opt}} & b_{\text{opt}} \\ 0 & 0 & b_{\text{opt}} & c_{\text{opt}} \end{pmatrix} \tag{44}$$

where v_{yt} is the true velocity in the y-axis and $(a_{\text{opt}}, b_{\text{opt}}, c_{\text{opt}})$is optimized (a, b, c), calculated using the procedure in Section 4.2.1. $\mathbf{x}_{\text{t}}$ and Φ of a two-dimensional PVM filter are the same as for a POM filter. **H** and **R** are

$$\mathbf{H} = \begin{pmatrix} 1 & 0 & 0 & 0 \\ 0 & 1 & 0 & 0 \\ 0 & 0 & 1 & 0 \\ 0 & 0 & 0 & 1 \end{pmatrix} \tag{45}$$

$$\mathbf{R} = \begin{pmatrix} B_{\text{x}} & 0 & 0 & 0 \\ 0 & B_{\text{v}} & 0 & 0 \\ 0 & 0 & B_{\text{x}} & 0 \\ 0 & 0 & 0 & B_{\text{v}} \end{pmatrix} \tag{46}$$

In addition, the formulation of **Q** is the same as in Eq. (43) and $(a_{\text{opt}}, b_{\text{opt}}, c_{\text{opt}})$ is calculated using the procedure in Section 4.2.2. A two-dimensional RA filter is the same as the optimal PVM filter, with the exception of **Q**. **Q** of the RA filter is

$$\mathbf{Q}_{\text{ra}} = \begin{pmatrix} T^4/4 & T^3/2 & 0 & 0 \\ T^3/2 & T^2 & 0 & 0 \\ 0 & 0 & T^4/4 & T^3/2 \\ 0 & 0 & T^3/2 & T^2 \end{pmatrix} \sigma_{\text{q}}^2 \tag{47}$$

Next, the design for an appropriate a_{D} is presented. We presume an approximate prediction of accelerations. For instance, when the maximum acceleration of the target in **Figure 3** is predicted to be approximately $a_{\text{c}} = 3\ \text{m/s}^2$, a_{D} is then 1.0, from Eq. (34). Using this $a_{\text{D}}$and the radar settings described in the previous section, we have $(a_{\text{opt}}, b_{\text{opt}}, c_{\text{opt}})$ for each filter.

6.3. Results and discussion

Figure 4 shows the simulation results. Clearly, the filters using velocity measurements achieve greater accuracy than the optimal POM filter. The mean steady-state prediction RMS errors ($\text{E}[\varepsilon_k]$ in 2 s < kT) of the optimal POM, RA, and optimal PVM filters are 0.59, 0.46, and 0.19 m, respectively. These results indicate that the proposed strategy achieves greater accuracy than the conventional RA filter even in realistic situations. The mean RMS error of the optimal PVM filter is 41% of that of the RA filter. This is because the RA model cannot track the abrupt motion of the high-maneuvering target because of limitations in expressing the process noise. In contrast, the optimal PVM filter can set gains corresponding to the appropriate process

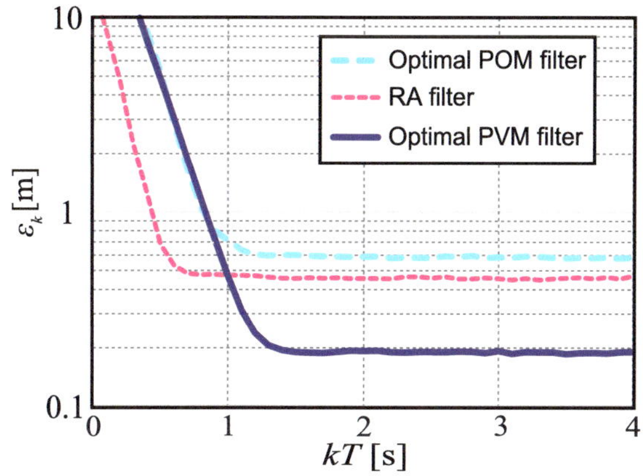

Figure 4. Simulation results.

noise to accurately track high-maneuvering target. Moreover, the mean RMS error of the optimal PVM filter is 32% of the error in optimal POM filter, and this clearly indicates the effectiveness of the velocity measurement, even when $R_{xv} = 1$ (when the measurement reliability of the position and velocity are the same). These simulation results are consistent with the theoretical analyses presented in **Figure 1**.

7. Final remarks

7.1. Conclusions

In this chapter, the efficient steady-state performance index, known as the RMS index, was introduced for both POM and PVM Kalman filters for systems that track moving objects. Automatic design (preset) of the covariance matrix of the process noise **Q**, to realize optimal position prediction, was achieved using the analytical relationship between **Q** and the RMS index. The validity of the proposed design strategy was shown via analyses and simulations. These results verified that the proposed index attained accurate tracking when compared with the conventional RA-model-based Kalman filter design. A simulation of a realistic situation indicated that the optimal performance given by the proposed strategy is 41% better than that given by the conventional design procedure for a PVM system. Moreover, the optimal performance of the optimal POM and PVM Kalman filters was compared showing that the optimal PVM Kalman filter is accurate when compared with the POM filter in a steady state.

7.2. Future works

The most important future objective is the extension of the RMS index-based design strategy to the third-order (and higher order) Kalman filters that are widely used for real applications. In third-order tracking, an acceleration is added to the state vector, becoming one of the input parameters of the Kalman filter. Performance analysis and the establishment of a design strategy for such systems (i.e., position/acceleration and position/velocity/acceleration measured Kalman filters) are important considerations for advanced sensor fusion systems under development. Moreover, considerations of other dynamic models (e.g., the constant turn model) should also be probed for use in many applications including pedestrian tracking.

A. Appendix

A.1. Derivation of Eqs. (28)–(30)

Because we assume a steady state, the index k of all parameters and matrices is omitted in the following calculations. The ith row and jth column of a matrix $\mathbf{P}$ are denoted as $P^{i,j}$.

Eq. (11) is also written using $\widehat{\mathbf{P}}$ as

$$\mathbf{K} = \widehat{\mathbf{P}}\mathbf{H}^{\mathrm{T}}\mathbf{R}^{-1} \tag{48}$$

As indicated in Eq. (7), $\mathbf{H}$ of the PVM Kalman filter is the identity matrix. Thus, from Eq. (48), the relationship between the Kalman gains and the error covariance matrix in the estimated state $\widehat{\mathbf{P}}$ is calculated using Eqs. (8) and (24) as

$$\widehat{\mathbf{P}} = \mathbf{KR} = \begin{pmatrix} \alpha B_{\mathrm{x}} & T\eta B_{\mathrm{v}} \\ \beta B_{\mathrm{x}}/T & \theta B_{\mathrm{v}} \end{pmatrix} \tag{49}$$

With $P^{1,2} = P^{2,1}$ and Eq. (31), we have the following relationship:

$$\eta = \beta \frac{B_{\mathrm{x}}}{T^2 B_{\mathrm{v}}} = \beta R_{\mathrm{xv}} \tag{50}$$

Eq. (50) is equal to Eq. (32), showing that this relationship is satisfied in the assumed PVM Kalman filter without depending on the process noise. $\widehat{\mathbf{P}}$ is also calculated using Eq. (13) by substituting Eqs. (7) and (24) as

$$\widehat{\mathbf{P}} = \begin{pmatrix} (1-\alpha)\tilde{P}^{1,1} - T\eta\tilde{P}^{1,2} & (1-\alpha)\tilde{P}^{1,2} - T\eta\tilde{P}^{2,2} \\ (1-\theta)\tilde{P}^{1,2} - (\beta/\mathrm{T})\tilde{P}^{1,1} & (1-\theta)\tilde{P}^{2,2} - (\beta/\mathrm{T})\tilde{P}^{1,2} \end{pmatrix} \tag{51}$$

Elements of $\tilde{\mathbf{P}}$ are required to calculate Eq. (51) and are calculated using Eqs. (3), (12), and (18) as

$$\tilde{\mathbf{P}} = \begin{pmatrix} P^{1,1} + 2TP^{1,2} + T^2P^{2,2} + a & P^{1,2} + TP^{2,2} + b \\ P^{1,2} + TP^{2,2} + b & P^{2,2} + c \end{pmatrix} \tag{52}$$

Substituting Eq. (52) into Eq. (51), and comparing elements of Eq. (49), we have the following linear system:

$$\alpha B_x = (1-\alpha)\left(\alpha B_x + 2\eta T^2 B_v + \theta T^2 B_v + a\right) - \eta T(\eta T B_v + \theta T B_v + b) \tag{53}$$

$$\beta B_x/T = (1-\alpha)(\eta T B_v + \theta T B_v + b) - \eta T(\theta B_v + c) \tag{54}$$

$$\theta B_v = (1-\theta)(\theta B_v + c) - \left(\beta/T\right)(\eta T B_v + \theta T B_v + b) \tag{55}$$

Solving this linear system with respect to (*a*, *b*, *c*) and substituting Eq. (50) into the solutions, we arrive at Eqs. (28)–(30).

Author details

Kenshi Saho

Address all correspondence to: saho@pu-toyama.ac.jp

Department of Intelligent Systems Design Engineering, Toyama Prefectural University, Imizu, Toyama, Japan

References

[1] Ekstrand B. Some aspects on filter design for target tracking. Journal of Control Science and Engineering. 2012;**2012**:870890. DOI: 10.1155/2012/870890

[2] Bar-Shalom Y, Li XR, Kirubarajan T. Estimation and Tracking: Principles, Techniques, and Software. Boston, MA, United States: Artech House Publishers; 1998

[3] Saho K, Masugi M. Automatic parameter setting method for an accurate Kalman filter tracker using an analytical steady-state performance index. IEEE Access. 2015;**3**:1919-1930. DOI: 10.1109/ACCESS.2015.2486766

[4] Hashlamon I, Erbatur K. An improved real-time adaptive Kalman filter with recursive noise covariance updating rules. Turkish Journal of Electrical Engineering & Computer Sciences. 2016;**24**:524-540. DOI: 10.3906/elk-1309-60

[5] Basso G, Amorim TG, Brito AV, Nascimento TP. Kalman filter with dynamical setting of optimal process noise. IEEE Access. 2017;**5**:8385-8393. DOI: 10.1109/ACCESS.2017.2697072

[6] Fan Y, Lu F, Zhu W, Bai G, Yan L. A hybrid model algorithm for hypersonic glide vehicle Maneuver tracking based on the aerodynamic model. Applied Sciences. 2017;**7**:159. DOI: 10.3390/app7020159

[7] Li W, Sun S, Jia Y, Du J. Robust unscented Kalman filter with adaptation of process and measurement noise covariances. Digital Signal Processing. 2016;**48**:93-103. DOI: 10.1016/j.dsp.2015.09.004

[8] Vivone F, Braca P, Horstmann J. Knowledge-based multitarget ship tracking for HF surface wave radar systems. IEEE Transactions on Geoscience and Remote Sensing. 2015;**53**:3931-3949. DOI: 10.1109/TGRS.2014.2388355

[9] Du G, Zhang P. A Markerless human robot Interface using particle filter and Kalman filter for dual robots. IEEE Transactions on Industrial Electronics. 2015;**62**:2257-2264. DOI: 10.1109/TIE.2014.2362095

[10] Martino L, Read J, Elvira V, Louzada F. Cooperative parallel particle filters for on-line model selection and applications to urban mobility. Digital Signal Processing. 2017;**60**: 172-185. DOI: 10.1016/j.dsp.2016.09.011

[11] Drovandi CC, McGree J, Pettitt AN. A sequential Monte Carlo algorithm to incorporate model uncertainty in Bayesian sequential design. Journal of Computational and Graphical Statistics. 2014;**23**:324. DOI: 10.1080/10618600.2012.730083

[12] Li XR, Jilkov VP. Survey of maneuvering target tracking. Part I: Dynamic models. IEEE Transactions on Aerospace and Electronic Systems. 2003;**39**:1333-1364 DOI: 10.1109/TAES.2003.1261132

[13] Crouse DF. A general solution to optimal fixed-gain (alpha-beta-gamma etc.) filters. IEEE Signal Processing Letters. 2015;**22**:901-904. DOI: 10.1109/LSP.2014.2376876

[14] Tang X, Falco G, Falletti E, Presti LL. Complexity reduction of the Kalman filter-based tracking loops in GNSS receivers. GPS Solutions. 2017;**21**:685-699. DOI: 10.1007/s10291-016-0557-6

[15] Jatoth RK, Gopisetty S, Hussai M. Performance analysis of alpha beta filter, Kalman filter and meanshift for object tracking in video sequences. International Journal of Image, Graphics and Signal Processing. 2015;**3**:24-30. DOI: 10.5815/ijigsp.2015.03.04

[16] Li Z, Wu H. A survey of Maneuvering target tracking using Kalman filter. Proc. 4th Int. Conf. Mechatronics Mater. Chem. Comput. Eng. (ICMMCCE). 2015;**39**:542-545. DOI: 10.2991/icmmcce-15.2015.109

[17] Saho K, Homma H, Sakamoto T, Sato T, Inoue K, Fukuda T. Accurate image separation method for two closely spaced pedestrians using UWB Doppler imaging radar and supervised learning. IEICE Transactions on Communications. 2014;**97**:1223-1233. DOI: 10.1587/transcom.E97.B.1223

[18] Anabuki M, Okumura S, Sato T, Sakamoto T, Saho K, Yoshioka M, Inoue K, Fukuda T, Sakai H. Ultrawideband radar imaging using adaptive Array and Doppler separation. IEEE Transactions on Aerospace and Electronic Systems. 2017;**53**:190-200. DOI: 10.1109/TAES.2017.2649798

[19] Cossaboom M, Georgy J, Karamat T, Noureldin A. Augmented kalman filter and map matching for 3D RISS/GPS integration for land vehicles. International Journal of Navigation and Observation. 2012;**2012**:576807. DOI: 10.1155/2012/576807

[20] Afia AB, Escher AC, Macabiau C, Roche S. A GNSS/IMU/WSS/VSLAM hybridization using an extended Kalman filter. In: Proc. of ION 2015 Pacific PNT Meeting. 2015. pp. 719-732

[21] Kellner D, Barjenbruch M, Klappstein J, Dickmann J, Dietmayer K. Tracking of extended objects with high-resolution Doppler radar. IEEE Transaction on Intelligent Transportation Systems. 2016;**17**:1341-1353. DOI: 10.1109/TITS.2015.2501759

[22] Carrolla P, Domresea K, Zhoub H, Zhoua S, Willetta P. Doppler-aided localization of mobile nodes in an underwater distributed antenna system. Physical Communication. 2016;**18**:49-59. DOI: 10.1016/j.phycom.2015.08.008

[23] Zhoua G, Wub L, Xiea J, Denga W, Quan T. Constant turn model for statically fused converted measurement Kalman filters. Signal Processing. 2015;**108**:400-411. DOI: 10.1016/j.sigpro.2014.09.022

[24] Cho MH, Tahk MJ, Kim MC, Lee D, Yoon J. Modified gain pseudo-measurement filter design for radar target tracking with range rate measurement. In: Proc. IEEE 2017 25th Mediterranean Conf. on Control and Automation (MED). 2017. pp. 1195-1200. DOI: 10.1109/MED.2017.7984280

[25] Saho K, Masugi M. Performance analysis of fixed-gain position-velocity-measured tracking filter. In: 2016 2nd International Conference on Control, Automation and Robotics (ICCAR); 28-30 April 2016; Hong Kong, China. IEEE; 2016. DOI: 10.1109/ICCAR.2016.7486760

[26] Saho K, Masugi M. Performance analysis of alpha-beta-gamma tracking filters using position and velocity measurements. EURASIP Journal on Advances in Signal Processing. 2015;**2015**:35. DOI: 10.1186/s13634-015-0220-3

[27] Kalata PR. The tracking index: A generalized parameter for alpha-beta and alpha-beta-gamma target trackers. IEEE Transactions on Aerospace and Electronic Systems. 1984; **AES-20**:174-182. DOI: 10.1109/TAES.1984.310438

Chapter 13

Distributed Kalman Filter

Felix Govaers

Additional information is available at the end of the chapter

http://dx.doi.org/10.5772/intechopen.71941

Abstract

The continuing trend toward connected sensors ("internet of things" and" ubiquitous computing") drives a demand for powerful distributed estimation methodologies. In tracking applications, the distributed Kalman filter (DKF) provides an optimal solution under Kalman filter conditions. The optimal solution in terms of the estimation accuracy is also achieved by a centralized fusion algorithm, which receives all associated measurements. However, the centralized approach requires full communication of all measurements at each time step, whereas the DKF works at arbitrary communication rates since the calculation is fully distributed. A more recent methodology is based on "accumulated state density" (ASD), which augments the states from multiple time instants to overcome spatial cross-correlations. This chapter explains the challenges in distributed tracking. Then, possible solutions are derived, which include the DKF and ASD approach.

Keywords: distributed Kalman filter, target tracking, multisensor fusion, information fusion, accumulated state density

1. Introduction

Modern tracking and surveillance system development is increasingly driving technological trends and algorithm developments toward networks of dislocated sensors. Besides classical target tracking, many other applications can be found, for instance, in robotics, manufacturing, health care, and financial economics. A multisensor network can exploit spatial diversity to compensate for the weak attributes of a single sensor such as limited field of view or high measurement noise. Also, heterogeneous sensors can reveal a more complete situational awareness and a more precise estimate of the underlying phenomena. Additionally, a sensor network is more robust against a single point of failure in comparison to a standalone system, if its architecture is chosen carefully.

Despite its unquestioned advantages, a multisensor network must cope with design-specific challenges, for instance, a full transmission of all the measurements to a fusion center can be hindered, when the communication links are unreliable or constrained in bandwidth or costs. A well-known approach to cope with the limited bandwidth data links is to apply data processing on the sensor sites to generate local track parameters that are transmitted to the fusion center. The latter then reconstructs the global track parameters by an application of a *Track-to-Track Fusion* (T2TF) scheme. Depending on the scenario constraints, this is a nontrivial task, since the local tracks are mutually correlated due to the common process noise. The first known solution in the literature to the T2TF problem proposed to apply an information filter-based multisensor fusion algorithm in [1], which later became famous as the "tracklet fusion." However, the tracklet fusion also requires a transmission from each sensor after each time step in order to reconstruct the optimal solution.

This chapter presents the theory and the derivation of the *distributed Kalman filter* (DKF), which is an optimal solution of the T2TF problem under Kalman filter assumptions with respect to the *mean squared error* (MSE). Assuming Kalman conditions means that linear Gaussian models are provided for the motion model and all measurement models of the sensors. Moreover, it is assumed that measurement-to-track (at the sensors) and track-to-track (at the fusion center) association has been solved. The DKF approach requires, however, all remote sensor models to be known at each local sensor site, which is infeasible in most practical scenarios. Therefore, this chapter also presents a solution based on the *accumulated state density* (ASD), which is closely related to the DKF but does not require the measurement models to be known. Surveys that reflect the history of research in distributed estimation can be found, for instance, in [2, 3].

This chapter is structured as follows: Section 2 summarizes the problem formulation. A basic approach to T2TF is given in Section 3, where we present the least squares solution. Section 4 presents a simple fusion methodology, which is easy to compute and provides practical results for various applications. The reason why this approach is suboptimal is investigated in Section 5 by means of a recursive computation of the cross-covariances of the local tracks. In Section 6, the product representation of Gaussian probability densities is introduced, which is the basis for the derivation of the distributed Kalman filter in Section 7. An alternative derivation in terms of information parameters is provided in Section 8. Since the local measurement models are often unknown in practical applications, the distributed accumulated state density filter is introduced in Section 9. The chapter concludes with Section 10.

2. Problem formulation

Throughout this chapter, it is assumed that all S sensors produce their measurements $z_k^s \in \mathbb{R}^m$ at each time step t_k of the same target with its true state $x_k \in \mathbb{R}^n$ in a synchronized way. The extension to the unsynchronized case is straightforward by means of standard Kalman filter predictions, and is therefore omitted for the sake of notational simplicity. The measurement equation for sensor $s \in \{1, \ldots, S\}$ is given by

$$z_k^s = H_k^s x_k + v_k^s \tag{1}$$

where $v_k^s \sim \mathrm{N}(v_k^s; 0, R_k^s)$ is the Gaussian distributed, zero-mean random variable, which models the noise of the sensing process. Therefore, the likelihood for a single measurement is fully described by the Gaussian

$$p(z_k^s|x_k) = \mathrm{N}(z_k^s; H_k^s x_k, R_k^s) \tag{2}$$

Since the measurement processes across all sensors $Z_k = \{z_k^1, \ldots, z_k^S\}$ are mutually independent, the joint likelihood of all sensor data produced at time t_k factorizes:

$$\begin{aligned} p(Z_k|x_k) &= \prod_{s=1}^{S} p(z_k^s|x_k) \\ &= \prod_{s=1}^{S} \mathrm{N}(z_k^s; H_k^s x_k, R_k^s) \end{aligned} \tag{3}$$

The true state of the target itself is modeled as a time-variant stochastic process, where the transition from time t_{k-1} to time t_k is given by the following motion equation:

$$x_k = F_{k|k-1} x_{k-1} + w_k \tag{4}$$

where $w_k \sim \mathrm{N}(w_k; 0, Q_k)$ is the Gaussian distributed, zero-mean random variable to model the process noise of the system. Analogously to the likelihood, this provides the probability density function for the *transition model*:

$$p(x_k|x_{k-1}) = \mathrm{N}(x_k; F_{k|k-1} x_{k-1}, Q_k) \tag{5}$$

Based on the local processors, each sensor node produces a *track* at time t_k in terms of an estimate $x_{k|k}^s$ and a corresponding estimation error covariance $P_{k|k}^s$. In a T2TF scheme, these parameters are the only information, which is transmitted to a *fusion center* (FC). It should be noted that the FC may also not be centralized, distinguished instance in the architecture, but each and every processing node can act as a FC. The introduction of a distinguished FC is only for clarification of different computation layers. An excellent overview of pros and cons of various data fusion layers can be found in [4].

The T2TF problem can now be stated as follows: compute a fused estimate $x_{k|k}$ of the state x_k and a consistent error covariance $P_{k|k}$ by means of the local tracks and knowledge on their models:

$$x_{k|k} \leftarrow \left\{x_{k|k}^1, \ldots, x_{k|k}^S\right\} \tag{6}$$

3. Least squares estimate

In order to compute an estimate as a well-suited combination of the local tracks, it is useful to consider the joint likelihood function given by the following Gaussian:

$$p\left(x_{k|k}^{1}, \ldots, x_{k|k}^{S} | x_k\right) = \mathrm{N}\left(\begin{pmatrix} x_{k|k}^{1} \\ \vdots \\ x_{k|k}^{S} \end{pmatrix}; \begin{pmatrix} x_k \\ \vdots \\ x_k \end{pmatrix}, \begin{pmatrix} P_{k|k}^{1,1} & \cdots & P_{k|k}^{1,S} \\ \vdots & \ddots & \vdots \\ P_{k|k}^{S,1} & \cdots & P_{k|k}^{S,S} \end{pmatrix}\right), \tag{7}$$

where $P_{k|k}^{s,s} = P_{k|k}^{s}$ are the track covariances on the block-diagonal entries and $P_{k|k}^{i,j} \triangleq \mathrm{cov}\left[x_{k|k}^{i}, x_{k|k}^{j} | x_k\right]$ $= P_{k|k}^{j,i\,T}$ are the cross-covariances on the off-diagonal entries of the joint error covariance.

Since the joint likelihood from above is proportional to an exponential function:

$$p\left(x_{k|k}^{1}, \ldots, x_{k|k}^{S} | x_k\right) \propto \exp\left\{-\frac{1}{2}\left(\begin{pmatrix} x_{k|k}^{1} \\ \vdots \\ x_{k|k}^{S} \end{pmatrix} - \begin{pmatrix} x_k \\ \vdots \\ x_k \end{pmatrix}\right)^{T} \begin{pmatrix} P_{k|k}^{1,1} & \cdots & P_{k|k}^{1,S} \\ \vdots & \ddots & \vdots \\ P_{k|k}^{S,1} & \cdots & P_{k|k}^{S,S} \end{pmatrix}^{-1} \left(\begin{pmatrix} x_{k|k}^{1} \\ \vdots \\ x_{k|k}^{S} \end{pmatrix} - \begin{pmatrix} x_k \\ \vdots \\ x_k \end{pmatrix}\right)\right\} \tag{8}$$

the *maximum likelihood* (ML) estimate can be computed in terms of the least squares:

$$x_{k|k} = \mathrm{argmin}_{x_k}\left\{\left(x_{k|k}^{1:S} - I^{1:S} x_k\right)^{T} \left(P_{k|k}^{1:S}\right)^{-1} \left(x_{k|k}^{1:S} - I^{1:S} x_k\right)\right\}, \tag{9}$$

where $x_{k|k}^{1:S} = \left(x_{k|k}^{1\,T}, \ldots, x_{k|k}^{S\,T}\right)^{T}$, $I^{1:S} = (I, \ldots, I)^{T}$, and $P_{k|k}^{1:S}$ for the joint error covariance have been introduced for notational simplicity. A closed form solution of the ML estimates can be obtained by setting the gradient with respect to the state to zero:

$$0 = \nabla_{x_k}\left(x_{k|k}^{1:S} - I^{1:S} x_k\right)^{T} \left(P_{k|k}^{1:S}\right)^{-1} \left(x_{k|k}^{1:S} - I^{1:S} x_k\right) = 2\left(I^{1:S}\right)^{T} \left(P_{k|k}^{1:S}\right)^{-1} \left(x_{k|k}^{1:S} - I^{1:S} x_k\right), \tag{10}$$

Therefore, the ML estimate is given by:

$$x_{k|k} = \left(\left(I^{1:S}\right)^{T} \left(P_{k|k}^{1:S}\right)^{-1} \left(I^{1:S}\right)\right)^{-1} \left(I^{1:S}\right)^{T} \left(P_{k|k}^{1:S}\right)^{-1} x_{k|k}^{1:S}. \tag{11}$$

For information fusion applications, it is also important to have a consistent estimate of the squared error, in other words, we need to compute the corresponding error covariance:

$$\begin{aligned}
\mathrm{cov}\left[x_{k|k} | x_k\right] &= \mathrm{E}\left[\left(x_k - x_{k|k}\right)\left(x_k - x_{k|k}\right)^{\mathrm{T}}\right] \\
&\triangleq \mathrm{E}\left[\left(x_k - x_{k|k}\right)^{2}\right] \\
&= \mathrm{E}\left[\left(x_k - \left(\left(I^{1:S}\right)^{\mathrm{T}} \left(P_{k|k}^{1:S}\right)^{-1} \left(I^{1:S}\right)\right)^{-1} \left(I^{1:S}\right)^{\mathrm{T}} \left(P_{k|k}^{1:S}\right)^{-1} x_{k|k}^{1:S}\right)^{2}\right] \\
&= \mathrm{E}\left[\left(\left(\left(I^{1:S}\right)^{\mathrm{T}} \left(P_{k|k}^{1:S}\right)^{-1} \left(I^{1:S}\right)\right)^{-1} \left(\left(\left(I^{1:S}\right)^{\mathrm{T}} \left(P_{(k|k)}^{1:S}\right)^{-1} \left(I^{1:S}\right)\right) x_k - \left(I^{1:S}\right)^{\mathrm{T}} \left(P_{k|k}^{1:S}\right)^{-1} x_{k|k}^{1:S}\right)\right)^{2}\right]
\end{aligned} \tag{12}$$

$$= \left(\left(I^{1:S}\right)^{\mathrm{T}} \left(P^{1:S}_{k|k}\right)^{-1} \left(I^{1:S}\right) \right)^{-1} \left(I^{1:S}\right)^{\mathrm{T}} \left(P^{1:S}_{k|k}\right)^{-1} \mathrm{E}\left[\left(\left(I^{1:S}\right) x_k - x^{1:S}_{k|k} \right)^2 \right] \left(P^{1:S}_{k|k}\right)^{-1} \left(I^{1:S}\right) \left(\left(I^{1:S}\right)^{\mathrm{T}} \left(P^{1:S}_{k|k}\right)^{-1} \left(I^{1:S}\right) \right)^{-1}$$

$$= \left(\left(I^{1:S}\right)^{\mathrm{T}} \left(P^{1:S}_{k|k}\right)^{-1} \left(I^{1:S}\right) \right)^{-1}.$$

The last equation holds due to the fact that $\mathrm{E}\left[\left(\left(I^{1:S}\right) x_k - x^{1:S}_{k|k} \right)^2 \right] = P^{1:S}_{k|k}$ is the joint covariance. Concluding the derivations from above equation, one can obtain:

$$x_{k|k} = P_{k|k} \left(I^{1:S}\right)^{\mathrm{T}} \left(P^{1:S}_{k|k}\right)^{-1} x^{1:S}_{k|k},$$
$$P_{k|k} = \left(\left(I^{1:S}\right)^{\mathrm{T}} \left(P^{1:S}_{k|k}\right)^{-1} \left(I^{1:S}\right) \right)^{-1} \tag{13}$$

4. Naïve fusion

It is obvious that for the ML estimate, it is assumed that the cross-covariances $P^{i,j}_{k|k}, i, j \in \{1, \dots, S\}$ are known. Since this might not be given in practical scenarios, a simple approximation is to assume them to be zero. This approach is called *Naïve fusion*. It implies that the joint error covariance is given in block-diagonal form:

$$P^{1:S}_{k|k} = \begin{pmatrix} P^{1,1}_{k|k} & & \\ & \ddots & \\ & & P^{S,S}_{k|k} \end{pmatrix} \tag{14}$$

As a direct consequence of the *matrix inversion lemma* (see Appendix 12.1), the inverse can be obtained in closed form solution:

$$\left(P^{1:S}_{k|k}\right)^{-1} = \begin{pmatrix} \left(P^{1,1}_{k|k}\right)^{-1} & & \\ & \ddots & \\ & & \left(P^{S,S}_{k|k}\right)^{-1} \end{pmatrix} \tag{15}$$

Filling into the maximum likelihood formulas directly yields.

$$x_{k|k} = P_{k|k} \sum_{s=1}^{S} \left(P^{s}_{k|k}\right)^{-1} x^{s}_{k|k},$$
$$P_{k|k} = \left(\sum_{s=1}^{S} \left(P^{s}_{k|k}\right)^{-1} \right)^{-1} \tag{16}$$

Thus, by means of a simple approximation of the ML estimate, we have obtained a first practical fusion rule for the FC, which we call *convex combination* due to its structure for further

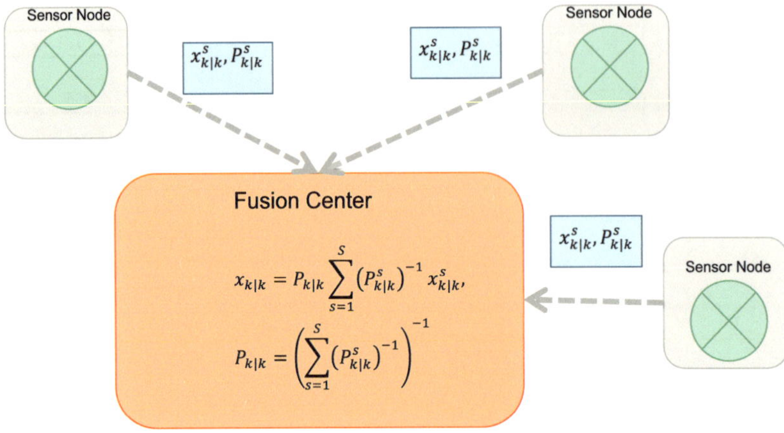

Figure 1. Fusion scheme of the Naive Fusion approach.

usage. The fusion scheme as a whole can be outlined schematically as in the flowing **Figure 1**. Each sensor node s processes its produced sensor data by means of a local filter, which results in a track in terms of an estimate together with an error covariance. These parameters are transmitted to the fusion center, which applies the convex combination to compute the fused result.

5. What makes the Naïve fusion naïve?

For the Naïve fusion, we have assumed that the cross-covariances vanish. It is worth to be aware of the structure of the cross-covariances to see the conditions whether this holds or does not hold. This can be achieved by a recursive computation of the posterior cross-covariance $P^{i,j}_{k|k}$ of two sensors with indices i and j, which process their data by means of local Kalman filters. At the beginning of the estimation process at t_0 tracks are not yet correlated, that is, $P^{i,j}_{0|0} = 0$, due to the fact that initial measurements are mutually uncorrelated. A recursive computation can be achieved by a *prediction-filtering* cycle.

5.1. Cross-covariance prediction

For the prediction step, it is assumed that a previous posterior cross-covariance $P^{i,j}_{k|k-1}$ has been computed. The prior parameters are obtained by means of the motion model:

$$\begin{aligned} P^{i,j}_{k|k-1} &= \mathrm{E}\left[\left(x_k - x^i_{k|k-1}\right)\left(x_k - x^j_{k|k-1}\right)^{\mathrm{T}}\right] \\ &= \mathrm{E}\left[\left(F_{k|k-1}x_{k-1} + w_k - F_{k|k-1}x^i_{k-1|k-1}\right)\left(F_{k|k-1}x_{k-1} + w_k - F_{k|k-1}x^j_{k-1|k-1}\right)^{\mathrm{T}}\right] \\ &= F_{k|k-1}\mathrm{E}\left[\left(x_{k-1} - x^i_{k-1|k-1}\right)\left(x_{k-1} - x^j_{k-1|k-1}\right)^{\mathrm{T}}\right]F^{\mathrm{T}}_{k|k-1} \end{aligned} \tag{17}$$

$$
\begin{aligned}
&- F_{k|k-1} \mathrm{E}\left[\left(x_{k-1} - x^i_{k-1|k-1}\right) w_k{}^{\mathrm{T}}\right] - \mathrm{E}\left[w_k \left(x_{k-1} - x^j_{k-1|k-1}\right)^{\mathrm{T}}\right] F^{\mathrm{T}}_{k|k-1} + \mathrm{E}\left[w_k w_k{}^{\mathrm{T}}\right] \\
&= F_{k|k-1} P^{i,j}_{k-1|k-1} F^{\mathrm{T}}_{k|k-1} + Q_k,
\end{aligned}
$$

where the last equality holds due to the fact that the estimation errors at time t_{k-1} of both sensor processors are uncorrelated to the process noise w_k.

5.2. Cross-covariance filtering

In the filtering step, both sensors compute their posterior parameters based on the produced measurements z^i_k and z^j_k, respectively. It is assumed that the local processors have applied local Kalman filter update steps with Kalman gains $W^i_{k|k-1}$ and $W^j_{k|k-1}$. The cross-covariance posterior matrix can be obtained by a straightforward computation:

$$
\begin{aligned}
P^{i,j}_{k|k} &= \mathrm{E}\left[\left(x_k - x^i_{k|k}\right)\left(x_k - x^j_{k|k}\right)^{\mathrm{T}}\right] \\
&= \mathrm{E}\left[\left(x_k - x^i_{k|k-1} - W^i_{k|k-1}\left(z^i_k - H^i_k x^i_{k|k-1}\right)\right)\left(x_k - x^j_{k|k-1} - W^j_{k|k-1}\left(z^j_k - H^j_k x^j_{k|k-1}\right)\right)^{\mathrm{T}}\right] \\
&= \mathrm{E}\left[\left(x_k - x^i_{k|k-1} - W^i_{k|k-1}\left(H^i_k x_k + v^i_k - H^i_k x^i_{k|k-1}\right)\right)\left(x_k - x^j_{k|k-1} - W^j_{k|k-1}\left(H^j_k x_k + v^j_k - H^j_k x^j_{k|k-1}\right)\right)^{\mathrm{T}}\right] \\
&= \mathrm{E}\left[\left(\left(I - W^i_{k|k-1} H^i_k\right)\left(x_k - x^i_{k|k-1}\right) - W^i_{k|k-1} v^i_k\right)\left(\left(I - W^j_{k|k-1} H^j_k\right)\left(x_k - x^j_{k|k-1}\right) - W^j_{k|k-1} v^j_k\right)^{\mathrm{T}}\right] \\
&= \left(I - W^i_{k|k-1} H^i_k\right) \mathrm{E}\left[\left(x_k - x^i_{k|k-1}\right)\left(x_k - x^j_{k|k-1}\right)^{\mathrm{T}}\right]\left(I - W^j_{k|k-1} H^i_k\right)^{\mathrm{T}} + W^i_{k|k-1} \mathrm{E}\left[v^i_k v^j_k \mathrm{T}\right] W^{j\,\mathrm{T}}_{k|k-1} \\
&= \left(I - W^i_{k|k-1} H^i_k\right) P^{i,j}_{k|k-1} \left(I - W^i_{k|k-1} H^i_k\right)^{\mathrm{T}}
\end{aligned}
\tag{18}
$$

For these equations, we have used the fact that the prior estimate error $\left(x_k - x^i_{k|k-1}\right)$ is independent of the measurement noise v^i_k, and that v^i_k and v^j_k are mutually independent.

Concluding the calculations from this section, we have obtained a recursive solution for the cross-covariances:

$$
\begin{aligned}
P^{i,j}_{0|0} &= 0 \\
P^{i,j}_{k|k-1} &= F_{k|k-1} P^{i,j}_{k-1|k-1} F^{\mathrm{T}}_{k|k-1} + Q_k \\
P^{i,j}_{k|k} &= \left(I - W^i_{k|k-1} H^i_k\right) P^{i,j}_{k|k-1} \left(I - W^i_{k|k-1} H^i_k\right)^{\mathrm{T}}
\end{aligned}
\tag{19}
$$

One can see that the cross-covariances are zero, if and only if the process noise covariance Q_k vanishes. In other words, if the tracks refer to a deterministically moving target and all sensors do local Kalman filtering, then the convex combination equations yield the optimal fusion result. If this is not the case, as maybe in most practical applications, then the Naïve fusion method is an approximation and its degree of approximation depends primarily on the level of the process noise.

6. Gaussian product representation

The basic concept of the distributed Kalman filter is to make the local parameters stochastically independent, even if process noise is present. This is achieved by a *product representation*, which directly follows from the fact that.

$$\exp\left\{-\frac{1}{2}\left(\begin{pmatrix} x_{k|k}^1 \\ \vdots \\ x_{k|k}^S \end{pmatrix} - \begin{pmatrix} x_k \\ \vdots \\ x_k \end{pmatrix}\right)^{\mathrm{T}} \begin{pmatrix} \left(P_{k|k}^1\right)^{-1} & & \\ & \ddots & \\ & & \left(P_{k|k}^S\right)^{-1} \end{pmatrix} \left(\begin{pmatrix} x_{k|k}^1 \\ \vdots \\ x_{k|k}^S \end{pmatrix} - \begin{pmatrix} x_k \\ \vdots \\ x_k \end{pmatrix}\right)\right\} \tag{20}$$

$$= \exp\left\{-\frac{1}{2}\sum_{s=1}^{S}\left(x_{k|k}^s - x_k\right)^{\mathrm{T}}\left(P_{k|k}^s\right)^{-1}\left(x_{k|k}^s - x_k\right)\right\} \propto \prod_{s=1}^{S} N\left(x_k; x_{k|k}^s, P_{k|k}^s\right)$$

Thus, the Gaussian product representation is equivalent to uncorrelated track parameters for each processing node. It should be noted that the product representation is not normalized, that is, the integral for $S > 1$ is not unity. This, however, is not of practical relevance since the fused estimate density is a Gaussian and the parameters of which are provided by the convex combination.

7. Derivation of the distributed Kalman filter

For the DKF, we are going to modify the local processing scheme for each sensor in order to have the product representation hold at each instant of time. Then, when the fusion center receives the parameters from all sensors, the convex combination can be applied to compute the optimal global estimate. Note that the convex combination does not consider a local prior of the fusion center; therefore, the result will be independent from previous transmissions. This can be of great benefit, if communication channels with unreliable links have to be considered, since the full information on the target state is distributed in the sensor network. However, for completeness, it should also be noted that the modified local parameters are not optimal anymore in a local sense. One could say that local optimality is given up for the sake of global optimality [5].

In the following sections, the derivation of a prediction-filtering recursion of the DKF is discussed.

7.1. DKF prediction

For the prediction, it is assumed that the previous posterior is given in product representation:

$$p\left(x_{k-1}|Z^{k-1}\right) \propto \prod_{s=1}^{S} N\left(x_{k-1}; x^{s}_{k-1|k-1}, P^{s}_{k-1|k-1}\right) \tag{21}$$

For notational simplicity, we have conditioned the posterior on the full data set $Z^{k-1} = \{Z_1^{k-1},\ldots, Z_s^{k-1}\}$, where $Z_s^{k-1}\{Z_1^s, \ldots, Z_{k-1}^s\}$ are from all sensors and all time steps up to time t_{k-1}, of which the local tracks are sufficient statistics. At the initialization phase, that is, $k-1=0$, the product representation is trivial, since the initial estimates can be based on first measurements, which are mutually independent.

To derive a closed form solution for the prediction of product representation, it is required to *globalize* the covariances $P^s_{k-1|k-1}$ of the local processing nodes at first. This process changes the local parameters such that the same previous posterior density is factorized; however, the local covariances will be unified to a single $\widetilde{P}_{k-1|k-1}$. Rigorously speaking, this matrix does not represent a meaningful covariance in the sense of an expected estimation error squared anymore. Still, the fused result will be optimal since the global density is not changed during this process. Thus, if we set

$$p\left(x_{k-1}|Z^{k-1}\right) \propto \prod_{s=1}^{S} N\left(x_{k-1}; \widetilde{x}^{s}_{k-1|k-1}, \widetilde{P}_{k-1|k-1}\right) \tag{22}$$

where

$$\begin{aligned} \widetilde{x}^s_{k-1|k-1} &= SP_{k-1|k-1}\left(P^s_{k-1|k-1}\right)^{-1} x^s_{k-1|k-1} \\ \widetilde{P}_{k-1|k-1} &= SP_{k-1|k-1} \\ P_{k-1|k-1} &= \left(\sum_{s=1}^{S}\left(P^s_{k-1|k-1}\right)^{-1}\right)^{-1} \end{aligned} \tag{23}$$

then the same fused density will be obtained, which is easily verified by means of the convex combination. It should be noted that the remote error covariances $P^s_{k-1|k-1}$ are required to compute the globalized covariance matrix $\widetilde{P}_{k-1|k-1}$. Since Kalman filter conditions are assumed, they can be computed without data transmission, as they do not depend on the local sensor measurements. Therefore, it is sufficient to be aware of the remote sensor models.

The prediction formulas can now be obtained by a marginalization of the joint density of the current and the last time step:

$$
\begin{aligned}
p(x_k|Z^{k-1}) &= \int \mathrm{d}x_{k-1}\, p(x_k, x_{k-1}|Z^{k-1}) \\
&= \int \mathrm{d}x_{k-1}\, p(x_k|x_{k-1}, Z^{k-1})\, p(x_{k-1}|Z^{k-1}) \\
&= \int \mathrm{d}x_{k-1}\, p(x_k|x_{k-1})\, p(x_{k-1}|Z^{k-1})
\end{aligned}
\tag{24}
$$

The last equality holds due to the Markov property of the system. Filling in our linear Gaussian transition model and the previous posterior yields

$$
p(x_k|Z^{k-1}) \propto \int dx_{k-1} \mathrm{N}(x_k; F_{k|k-1}x_{k-1}, Q_k) \prod_{s=1}^{S} N\left(x_{k-1}; \widetilde{x}^s_{k-1|k-1}, \widetilde{P}_{k-1|k-1}\right)
\tag{25}
$$

By means of a simple algebraic manipulation, it is possible to factorize the transition kernel Gaussian up to proportionality:

$$
\begin{aligned}
\mathrm{N}(x_k; F_{k|k-1}x_{k-1}, Q_k) &\propto \exp\left\{-\frac{1}{2}(F_{k|k-1}x_{k-1} - x_k)^{\mathrm{T}}(Q_k)^{-1}(F_{k|k-1}x_{k-1} - x_k)\right\} \\
&= \exp\left\{-\frac{1}{2}(F_{k|k-1}x_{k-1} - x_k)^{\mathrm{T}} S(SQ_k)^{-1}(F_{k|k-1}x_{k-1} - x_k)\right\} \\
&\propto \left[\mathrm{N}(x_k; F_{k|k-1}x_{k-1}, SQ_k)\right]^S
\end{aligned}
\tag{26}
$$

Thus, we can factorize the integration term of the global prior completely:

$$
p(x_k|Z^{k-1}) \propto \int dx_{k-1} \prod_{s=1}^{S} N(x_k; F_{k|k-1}x_{k-1}, SQ_k) N\left(x_{k-1}; \widetilde{x}^s_{k-1|k-1}, \widetilde{P}_{k-1|k-1}\right)
\tag{27}
$$

An application of the product formula (Section 12.2 in the appendix) yields:

$$
p(x_k|Z^{k-1}) \propto \prod_{s=1}^{S} \mathrm{N}\left(x_k; x^s_{k|k-1}, \widetilde{P}^s_{k|k-1}\right) \int dx_{k-1} \prod_{s=1}^{S} N(x_{k-1}; y^s, Y),
\tag{28}
$$

where

$$
\begin{aligned}
x^s_{k|k-1} &= F_{k|k-1}\widetilde{x}^s_{k-1|k-1} \\
\widetilde{P}^s_{k|k-1} &= F_{k|k-1}\widetilde{P}_{k-1|k-1}F_{k|k-1}{}^{\mathrm{T}} + SQ_k \\
y^s &= Y\left(\left(\widetilde{P}^s_{k|k-1}\right)^{-1} x^s_{k|k-1} + F_{k|k-1}{}^{\mathrm{T}}(SQ_k)^{-1}x_k\right) \\
Y &= \left(\left(\widetilde{P}^s_{k|k-1}\right)^{-1} + F_{k|k-1}{}^{\mathrm{T}}(SQ_k)^{-1}F_{k|k-1}\right)^{-1}.
\end{aligned}
\tag{29}
$$

At this point, we have derived factorized prediction formulas for the DKF prediction, and to our knowledge, the remaining integral is part of the normalization constant. This, however, is not trivial, since the parameter y^s depends on x_k. A successive application of the product formula yields the desired result:

$$\begin{aligned} &\int \mathrm{d}x_{k-1}\prod_{s=1}^{S} N\left(x_{k-1} - YF_{k|k-1}{}^{\mathrm{T}}(SQ_k)^{-1}x_k; Y\left(\widetilde{P}^{s}_{k|k-1}\right)^{-1}x^{s}_{k|k-1}, Y\right) \\ &\quad \propto \int \mathrm{d}x_{k-1} N\left(x_{k-1} - YF_{k|k-1}{}^{\mathrm{T}}(SQ_k)^{-1}x_k; \overline{\overline{y}}, \overline{\overline{Y}}\right) = 1, \end{aligned} \tag{30}$$

for some auxiliary variables $\overline{\overline{y}}$ and $\overline{\overline{Y}}$. All factors, which are independent of x_k have been omitted. This proves that the integral is independent of the state variable x_k. Concluding the derivations from above, we have derived the prediction formulas of the local estimation parameters as:

$$\begin{aligned} \widetilde{x}^{s}_{k-1|k-1} &= SP_{k-1|k-1}\left(P^{s}_{k-1|k-1}\right)^{-1}x^{s}_{k-1|k-1} \\ \widetilde{P}_{k-1|k-1} &= SP_{k-1|k-1} \\ x^{s}_{k|k-1} &= F_{k|k-1}\widetilde{x}^{s}_{k-1|k-1} \\ \widetilde{P}^{s}_{k|k-1} &= F_{k|k-1}\widetilde{P}_{k-1|k-1}F_{k|k-1}{}^{\mathrm{T}} + SQ_k. \end{aligned}$$

7.2. DKF filtering

Let Z_k denotes the set of measurements produced by all sensors at time t_k. The posterior density can be inferred by using the Bayes theorem:

$$p\left(x_k|Z^k\right) = \frac{p(Z_k|x_k)p\left(x_k|Z^{k-1}\right)}{p\left(Z_k|Z^{k-1}\right)} \tag{31}$$

Due to the mutual independence of the measurement noises, the joint likelihood function is given by:

$$p(Z_k|x_k) = \prod_{s=1}^{S} p\left(Z^{s}_{k}|x_k\right) \tag{32}$$

This is particularly useful for the structure of the product representation used for the DKF. Filling in the linear Gaussian models and neglecting the normalization constant in the denominator directly yields:

$$p\left(x_k|Z^k\right) \propto \prod_{s=1}^{S} \mathrm{N}\left(Z_k; H^{s}_{k}x_k, R^{s}_{k}\right)\mathrm{N}\left(x_k; x^{s}_{k|k-1}, \widetilde{P}^{s}_{k|k-1}\right) \tag{33}$$

Thus, the product formula again can be applied to compute the posterior parameters:

$$p\left(x_k|Z^k\right) \propto \prod_{s=1}^{S} \mathrm{N}\left(x_k; x^{s}_{k|k}, P^{s}_{k|k}\right) \tag{34}$$

where

$$\begin{aligned}
x^s_{k|k} &= x^s_{k|k-1} + W^s_{k|k-1} v^s_k \\
P^s_{k|k} &= P^s_{k|k-1} - W^s_{k|k-1} S^s_k W^{s\,\mathrm{T}}_{k|k-1} \\
v^s_k &= z^s_k - H^s_k x^s_{k|k-1} \\
W^s_{k|k-1} &= P^s_{k|k-1} H^{s\,\mathrm{T}}_k \left(S^s_k\right)^{-1} \\
S^s_k &= H^s_k P^s_{k|k-1} H^{s\,\mathrm{T}}_k + R^s_k .
\end{aligned} \tag{35}$$

Again, we have omitted the factors, which are independent of x_k.

8. Information filter formulation of the DKF

In [6], an elegant derivation of the DKF formulas was published based on the *information filter* (IF). The IF uses information matrices, which are inverted covariances, and information states, which are information matrices multiplied with states. The optimal, centralized update formulas for S sensors based on the predicted information parameters $\left(P_{k|k-1}\right)^{-1}$ and $\left(P_{k|k-1}\right)^{-1} x_{k|k-1}$ are given by:

$$\begin{aligned}
\left(P_{k|k}\right)^{-1} x_{k|k} &= \left(P_{k|k-1}\right)^{-1} x_{k|k-1} + \sum_{s=1}^{S} i^s_k \\
\left(P_{k|k}\right)^{-1} &= \left(P_{k|k-1}\right)^{-1} + \sum_{s=1}^{S} I^s_k
\end{aligned} \tag{36}$$

where $i^s_k = H^{s\,\mathrm{T}}_k \left(R^s_k\right)^{-1} z^s_k$ and $I^s_k = H^{s\,\mathrm{T}}_k \left(R^s_k\right)^{-1} H^s_k$ are the local information contribution from the current measurements at time t_k, which were received by the FC. If we want to distribute the computation to S nodes, we will have them uncorrelated as in the DKF previously. Since the fused estimate is obtained via the convex combination, the local information parameters are summed up:

$$\begin{aligned}
\left(P_{k|k}\right)^{-1} x_{k|k} &= \sum_{s=1}^{S} \left(P^s_{k|k}\right)^{-1} x^s_{k|k} \\
\left(P_{k|k}\right)^{-1} &= \sum_{s=1}^{S} \left(P^s_{k|k}\right)^{-1}
\end{aligned} \tag{37}$$

This summation structure can be used to provide a closed prediction-filtering cycle.

8.1. Information DKF prediction

The prediction of the state is easier than a direct transition of the information parameters. Based on the fused estimate, one can obtain.

http://dx.doi.org/10.5772/intechopen.71941

$$\begin{aligned} x_{k|k-1} &= F_{k|k-1} x_{k-1|k-1} \\ &= F_{k|k-1} P_{k-1|k-1} \sum_{s=1}^{S} \left(P^s_{k-1|k-1}\right)^{-1} x^s_{k-1|k-1} \\ &= \sum_{s=1}^{S} F_{k|k-1} P_{k-1|k-1} \left(P^s_{k-1|k-1}\right)^{-1} x^s_{k-1|k-1}. \end{aligned} \tag{38}$$

Thus, we have given the local predicted state parameters as:

$$x^s_{k|k-1} = F_{k|k-1} P_{k-1|k-1} \left(P^s_{k|k}\right)^{-1} x^s_{k|k} \tag{39}$$

Analogously, one obtains for the prior covariance:

$$\begin{aligned} P_{k|k-1} &= F_{k|k-1} P_{k-1|k-1} F^{\mathrm{T}}_{k|k-1} + Q_k \\ &= \frac{1}{S} F_{k|k-1} S P_{k-1|k-1} F^{\mathrm{T}}_{k|k-1} + S Q_k \end{aligned} \tag{40}$$

Thus, if we set $P^s_{k|k-1} = SP_{k|k-1} = F_{k|k-1} S P_{k-1|k-1} F^{\mathrm{T}}_{k|k-1} + SQ_k$, then the convex combination yields the exact global fused covariance.

8.2. Information DKF filtering

For the filtering, it is assumed that each sensor has computed its local information contribution parameter i^s_k and I^s_k from its own sensor model and, in addition, the information matrix contributions I^l_k from all remote sensors l by using the individual sensor models which are again assumed to be known. Then, the information state is updated via.

$$\begin{aligned} \left(P_{k|k}\right)^{-1} x_{k|k} &= \left(P_{k|k-1}\right)^{-1} x_{k|k-1} + \sum_{s=1}^{S} i^s_k \\ &= \sum_{s=1}^{S} \left(P^s_{k|k-1}\right)^{-1} x^s_{k|k-1} + i^s_k. \end{aligned} \tag{41}$$

As a direct consequence, the updated parameters of the local processors follow the standard IF filtering equations:

$$\left(P^s_{k|k}\right)^{-1} x^s_{k|k} = \left(P^s_{k|k-1}\right)^{-1} x^s_{k|k-1} + i^s_k. \tag{42}$$

For the globalized information matrix, the remote information parameters from the sensor models are used:

$$\begin{aligned}\left(P_{k|k}\right)^{-1} &= \left(P_{k|k-1}\right)^{-1} + \sum_{s=1}^{S} I_k^s \\ &= \sum_{s=1}^{S} \left(P_{k|k-1}^s\right)^{-1} + I_k^s, \\ \left(P_{k|k}^s\right)^{-1} &= \left(P_{k|k-1}^s\right)^{-1} + I_k^s\end{aligned} \tag{43}$$

It is important to note that the local processing nodes compute both the local pseudo information matrix $\left(P_{k|k}^s\right)^{-1}$ and the global fused error covariance $P_{k|k}$, which is required for the prediction to the next time step.

9. Distributed accumulated state density filter

The DKF from Sections 7 and 8 can be considered a big step toward distributed state estimation, tracking, and information inference. However, in practical applications, the exact solution is often hindered by the fact that the exact remote sensor model parameters are unknown and can only be approximated based on local state estimates. The good news is that there is another exact solution based on the *accumulated state density* (ASD). The distributed ASD equations turn the spatial correlations into temporal correlations of successive states. Originally, the ASD equations were introduced to solve the out-of-sequence problem, which handles delayed transmissions of measurements into an ongoing fusion process in an optimal manner. Therefore, the temporal correlations can well be coped with the ASD approach.

At first, let us introduce the ASD state $x_{k:n}$ as

$$x_{k:n} = \left(x_k^T, \ldots, x_n^T\right)^T \tag{44}$$

where t_k refers to the current time of the filtering process and t_n refers to the initialization time. The ASD approach now considers the conditional joint density $p\left(x_{k:n}|z^k\right)$, that is, the posterior of the full trajectory of the target between t_n and t_k. In particular, the individual state densities of a single instant of time can be obtained via marginalization. Also, it should be noted that the *Rauch-Tung-Striebel* (RTS) smoothing equations are inherently integrated in the ASD posterior, since all states are conditioned on the full set of measurements up to time t_k [7].

A recursive computation of the ASD posterior can be achieved by using the Bayes theorem:

$$p\left(x_{k:n}|Z^k\right) = \frac{p(Z_k|x_{k:n})p\left(x_{k:n}|Z^{k-1}\right)}{p\left(Z_k|Z^{k-1}\right)} \tag{45}$$

Since the measurements conditioned on the whole trajectory only depend on the state at time t_k, the joint likelihood function is given by:

$$p(Z_k|x_{k:n}) = p(Z_k|x_k) = \prod_{s=1}^{S} p\left(z_k^s|x_k\right) \tag{46}$$

The second factor can be reformulated as follows:

$$\begin{aligned} p\left(x_{k:n}|Z^{k-1}\right) &= p\left(x_k|x_{k-1:n}, Z^{k-1}\right)p\left(x_{k-1:n}|Z^{k-1}\right) \\ &= p(x_k|x_{k-1})p\left(x_{k-1:n}|Z^{k-1}\right) \end{aligned} \tag{47}$$

where we have used the Markov property of the system in the last equation. This recursive representation can now be repeated on the term $p\left(x_{k-1:n}|Z^{k-1}\right)$. A successive application of this procedure yields

$$\begin{aligned} p\left(x_{k:n}|Z^k\right) \propto &\prod_{s=1}^{S}\left\{p\left(z_k^s|x_k\right)\right\}p(x_k|x_{k-1})\prod_{s=1}^{S}\left\{p\left(z_{k-1}^s|x_{k-1}\right)\right\}p(x_{k-1}|x_{k-2})\cdots \\ &\prod_{s=1}^{S}\left\{p\left(z_{n+1}^s|x_{n+1}\right)\right\}p(x_{n+1}|x_n)\cdot p(x_n|Z^n) \end{aligned} \tag{48}$$

where we have neglected the normalization constant in the denominator. Filling in our Gaussian models and using the factorization of the transition model from above equation yields

$$p\left(x_{k:n}|Z^k\right) \propto \prod_{l=n+1}^{k}\prod_{s=1}^{S}\left\{N\left(z_l^s; H_l^s x_l, R_l^s\right)N\left(x_l; F_{l|l-1}x_{l-1}, SQ_l\right)\right\}\cdot p(x_n|Z^n) \tag{49}$$

Since the initial density usually is based on a first measurement, we can assume that it factorizes into independent local track starts:

$$p(x_n|Z^n) \propto \prod_{s=1}^{S} N\left(x_n; x_{n|n}^s, P_{n|n}^s\right) \tag{50}$$

When the posterior is fully factorized in the number of sensors and in the time steps, each processing node can compute the resulting ASD Gaussian with mean $x_{k:n|k}^s$ and covariance matrix $P_{k:n|k}^s$ [7]:

$$p\left(x_{k:n}|z^k\right) \propto \prod_{s=1}^{S} N\left(x_{k:n}; x_{k:n|k}^s, P_{k:n|k}^s\right) \tag{51}$$

where the parameters are given by:

$$x^s_{k:n|k} = \left(x^{s\,\mathrm{T}}_{k|k}, \ldots, x^{s\,\mathrm{T}}_{n|k}\right)^{\mathrm{T}},$$

$$P^s_{k:n|k} = \begin{pmatrix} P^s_{k|k} & P^s_{k|k}W^{s\,\mathrm{T}}_{k-1|k} & P^s_{k|k}W^{s\,\mathrm{T}}_{k-2|k} & \cdots & P^s_{k|k}W^{s\,\mathrm{T}}_{n|k} \\ W^s_{k-1|k}P^s_{k|k} & P^s_{k-1|k} & P^s_{k-1|k}W^{s\,\mathrm{T}}_{k-2|k-1} & * & P^s_{k-1|k}W^{s\,\mathrm{T}}_{n|k-1} \\ W^s_{k-2|k}P^s_{k|k} & W^s_{k-2|k-1}P^s_{k-1|k} & P^s_{k-2|k} & * & P^s_{k-2|k}W^{s\,\mathrm{T}}_{n|k-2} \\ \vdots & * & * & * & \vdots \\ W^s_{n|k}P^s_{k|k} & W^s_{n|k-1}P^s_{k-1|k} & \cdots & W^s_{n|+1}P^s_{n+1|k} & P^s_{n|k} \end{pmatrix}$$

We have used a short notation such that $x_{l|k} = \mathrm{E}\left[x_l|z^k\right]$ are the smoothed state estimates and $P_{l|k} = \mathrm{cov}\left[x_l|z^k\right]$ are the covariances, respectively, which result from the Rauch-Tung-Striebel equations. Also, the combined retrodiction gain matrices are known from the RTS smoother:

$$\begin{aligned} W_{l|k} &= \prod_{i=l}^{k-1} W_{i|i+1} \\ W_{i|i+1} &= P_{i|i}F^{\mathrm{T}}_{i+1|i}\left(P_{i+1|i}\right)^{-1} \end{aligned} \tag{52}$$

Thus, when the FC receives the local ASD parameters, the optimal fused estimate can be obtained via the convex combination:

$$\begin{aligned} x_{k:n|k} &= P_{k:n|k}\sum_{s=1}^{S}\left(P^s_{k:n|k}\right)^{-1} x^s_{k:n|k}, \\ P_{k:n|k} &= \left(\sum_{s=1}^{S}\left(P^s_{k:n|k}\right)^{-1}\right)^{-1} \end{aligned} \tag{53}$$

For a continuous state estimation process, it is convenient to formulate the distributed ASD solution in terms of a prediction-filtering cycle.

9.1. Distributed ASD prediction

For the prediction step, it is assumed that the local processing node s has computed the previous filtering parameters $x^s_{k-1:n|k-1}$ and $P^s_{k-1:n|k-1}$, which refer to time t_{k-1}. Then, the prior parameters are given by:

$$\begin{aligned} x^s_{k:n|k-1} &= \left(x^{s\,\mathrm{T}}_{k|k-1}, x^{s\,\mathrm{T}}_{k-1:n|k-1}\right)^{\mathrm{T}} \\ P^s_{k:n|k-1} &= \begin{pmatrix} P^s_{k|k-1} & P^s_{k-1:n|k-1}W^{s\,\mathrm{T}}_{k-1:n} \\ W^s_{k-1:n}P^s_{k-1:n|k-1} & P^s_{k-1:n|k-1} \end{pmatrix} \\ x^s_{k|k-1} &= F_{k|k-1}x^s_{k-1|k-1} \end{aligned} \tag{54}$$

http://dx.doi.org/10.5772/intechopen.71941

$$P^s_{k|k-1} = F_{k|k-1} P^s_{k-1|k-1} F^{\mathrm{T}}_{k|k-1} + SQ_k$$

$$W^s_{k-1:n} = \begin{pmatrix} W^s_{k-1|k} \\ W^s_{k-2|k} \\ \vdots \\ W^s_{n|k} \end{pmatrix}$$

9.2. Distributed ASD filtering

Since the prior is factorized in form of a product representation and the current measurements from time t_k are mutually uncorrelated, local Kalman filters can be applied to obtain the posterior parameters:

$$x^s_{k:n|k} = x^s_{k:n|k-1} + W^s_{k:n|k}\left(z^s_k - H^s_k \Pi_k x^s_{k:n|k-1}\right)$$

$$P^s_{k:n|k} = P^s_{k:n|k-1} - W^s_{k:n|k} S^s_k W^{s\ \mathrm{T}}_{k:n|k}$$

$$S^s_k = H^s_k \Pi_k P^s_{k:n|k-1} \Pi^{\mathrm{T}}_k H^{s\ \mathrm{T}}_k + R^s_k$$

$$W^s_{k:n|k} = P^s_{k:n|k-1} \Pi^T_k H^{s\ T}_k \left(S^s_k\right)^{-1} \tag{55}$$

10. Conclusion

In this chapter, we have introduced the least squares solution to the track-to-track fusion problem, where cross-covariances of the track estimation errors are required. Neglecting the cross-covariances has led us to the Naïve fusion, a simple but powerful fusion algorithm for practical applications. By recursive computation of the cross-covariances, we have seen that they primarily depend on the process noise of the state transition kernel. Since a centralized computation of the cross-covariances is infeasible in practical applications, more sophisticated solutions are required for optimal fusion results. The distributed Kalman filter, which uses the product representation to keep the local parameters decorrelated achieved this. However, this approach only works, if the local processors know all measurement models at each time step. Then, the distributed accumulated state density filter uses the temporal correlations to factorize the global posterior density. This approach does not require remote sensor models and is therefore, well suited for extensions with measurement ambiguity or nonlinear measurement functions.

In the study, one can find more extensions based on the distributed Kalman filter to overcome the lack of knowledge on the remote sensor models. In [8] and the references therein, a debiasing matrix is introduced to compensate for globally biased gain matrices of the local filters. An application of the tracklet fusion based on the distributed accumulated state density filter can be found in [9]. Then, in [10], the information filter formulation of the distributed Kalman filter also was extended to scenarios with input information on the transition process.

A. Appendix

A.1. Matrix inversion lemma

Let A, B, C, and D be matrices of a block matrix such that A and D are invertable, and also such that the Schur-complements $D - CA^{-1}B$ and $A - BD^{-1}C$ have full rank. Then, the inversion of the block matrix is given by:

$$\begin{pmatrix} A & B \\ C & D \end{pmatrix}^{-1} = \begin{pmatrix} (A - BD^{-1}C)^{-1} & -(A - BD^{-1}C)^{-1}BD^{-1} \\ -D^{-1}C(A - BD^{-1}C)^{-1} & D^{-1} + D^{-1}C(A - BD^{-1}C)^{-1}BD^{-1} \end{pmatrix}$$

$$= \begin{pmatrix} A^{-1} + A^{-1}B(D - CA^{-1}B)^{-1}CA^{-1} & -A^{-1}B(D - CA^{-1}B)^{-1} \\ -(D - CA^{-1}B)^{-1}CA^{-1} & (D - CA^{-1}B)^{-1} \end{pmatrix}. \tag{56}$$

In particular, it holds that

$$\begin{gathered} (A - BD^{-1}C)^{-1} = A^{-1} + A^{-1}B(D - CA^{-1}B)^{-1}CA^{-1} - (A - BD^{-1}C)^{-1}BD^{-1} = \\ -A^{-1}B(D - CA^{-1}B)^{-1} - D^{-1}C(A - BD^{-1}C)^{-1} = -(D - CA^{-1}B)^{-1}CA^{-1}D^{-1} \\ +D^{-1}C(A - BD^{-1}C)^{-1}BD^{-1} = (D - CA^{-1}B)^{-1} \end{gathered} \tag{57}$$

Proof. Let the inverted block matrix be given by submatrices $E, F, G,$ and H. By definition, it holds that

$$\begin{pmatrix} A & B \\ C & D \end{pmatrix}\begin{pmatrix} E & F \\ G & H \end{pmatrix} = \begin{pmatrix} I & O \\ O & I \end{pmatrix} \text{and} \begin{pmatrix} E & F \\ G & H \end{pmatrix}\begin{pmatrix} A & B \\ C & D \end{pmatrix} = \begin{pmatrix} I & O \\ O & I \end{pmatrix},$$

where I and O are the identity and zero matrix, respectively. A matrix multiplication of the first and second equality yields two blocks of equations:

$$\begin{aligned} AE + BG &= I \\ AF + BH &= O \\ CE + DG &= O \\ CF + DH &= I \end{aligned} \tag{58}$$

which we call *block A* and

$$\begin{aligned} EA + FC &= I \\ EB + FD &= O \\ GA + HC &= O \\ GB + HD &= I \end{aligned} \tag{59}$$

which we call *block B*. Resolving block A for $E, F, G,$ and H yields the first version of the inverted block matrix, whereas resolving block B yields the second version.

A.2. Product formula for Gaussian densities

For two Gaussian distributed random variables x and z, it holds that

$$N(x; y, P)N(z; Hx, R) = N(x; \overline{y}, \overline{P})N(z; \overline{z}, S) \tag{60}$$

where

$$\begin{aligned} \overline{y} &= \begin{cases} y + Wv \\ P^{-1}y + H^{\mathrm{T}}R^{-1}z \end{cases} \\ \overline{P} &= \begin{cases} P - WSW^{\mathrm{T}} \\ \left(P^{-1} + H^{\mathrm{T}}R^{-1}H\right)^{-1} \end{cases} \\ S &= HPH^{\mathrm{T}} + R \\ W &= PH^{\mathrm{T}}S^{-1} \\ v &= z - Hy \end{aligned} \tag{61}$$

A proof can be found, for instance, in [11].

Author details

Felix Govaers

Address all correspondence to: felix.govaers@fkie.fraunhofer.de

Fraunhofer FKIE, Wachtberg, Germany

References

[1] Chong CY. Hierarchical estimation. Proceedings of the 2nd MIT/ONR Workshop on Distributed Information and Decision Systems Motivated by Naval Command Control Communication Systems, Monterey, CA, USA. 1979

[2] Mahmoud MS, Khalid HM. Distributed Kalman Filtering: A Bibliographic Review. IET Control Theory & Applications; 2013

[3] Campbell ME, Ahmed NR. Distributed Data Fusion: Neighbors, Rumors, and the Art of Collective Knowledge. IEEE Control Systems; 2016

[4] Hall D, Chong C-Y, Llinas J, Martin Liggins II, editors. Distributed Data Fusion for Network-Centric Operations. CRC Press; 2013

[5] Govaers F, Koch W. An exact solution to track-to-track-fusion at arbitrary communication rates. In: IEEE Transactions on Aerospace and Electronic Systems. Vol. 48. 2012

[6] Chong CY, Mori S. Optimal Fusion For Non-Zero Process Noise. Proceedings of the 16th International Conference on Information Fusion, Istanbul. 2013

[7] Koch W, Govaers F. On accumulated state densities with applications to out-of-sequence measurement processing. In: IEEE Transactions on Aerospace and Electronic Systems. Vol. 47. 2011

[8] Reinhardt M. Linear Estimation in Interconnected Sensor Systems with Information Costraints. PhD thesis: KIT Scientific Publishing; 2014

[9] Govaers F, Chong CY, Mori S, Koch W. Comparison of Augmented State Track Fusion Methods for Non-Full-Rate Communication. 18th International Conference on Information Fusion, Washington, DC. 2015

[10] Pfaff F, Noack B, Hanebeck UD, Govaers F, Koch W. Information Form Distributed Kalman Filtering (IDKF) with Explicit Inputs. 20th International Conference on Information Fusion, Xi'an, China. 2017

[11] Koch W. Tracking and Sensor Data Fusion: Methodological Framework and Selected Applications. Springer; 2014:251

Chapter 14

Consensus-Based Distributed Filtering for GNSS

Amir Khodabandeh, Peter J.G. Teunissen and
Safoora Zaminpardaz

Additional information is available at the end of the chapter

http://dx.doi.org/10.5772/intechopen.71138

Abstract

Kalman filtering in its distributed information form is reviewed and applied to a network of receivers tracking Global Navigation Satellite Systems (GNSS). We show, by employing consensus-based data-fusion rules between GNSS receivers, how the consensus-based Kalman filter (CKF) of individual receivers can deliver GNSS parameter solutions that have a comparable precision performance as their network-derived, fusion center dependent counterparts. This is relevant as in the near future the proliferation of low-cost receivers will give rise to a significant increase in the number of GNSS users. With the CKF or other distributed filtering techniques, GNSS users can therefore achieve high-precision solutions without the need of relying on a centralized computing center.

Keywords: distributed filtering, consensus-based Kalman filter (CKF), global navigation satellite systems (GNSS), GNSS networks, GNSS ionospheric observables

1. Introduction

Kalman filtering in its decentralized and distributed forms has received increasing attention in the sensor network community and has been extensively studied in recent years, see e.g. [1–8]. While in the traditional centralized Kalman filter setup all sensor nodes have to send their measurements to a computing (fusion) center to obtain the state estimate, in the distributed filtering schemes the nodes only share limited information with their neighboring nodes (i.e. a subset of all other nodes) and yet obtain state estimates that are comparable to that of the centralized filter in a minimum-mean-squared-error sense. This particular feature of the distributed filters would potentially make the data communication between the nodes cost-effective and develop the nodes' capacity to perform *parallel* computations.

Next to sensor networks, distributed filtering can therefore benefit several other applications such as formation flying of aerial vehicles [9], cooperative robotics [10] and disciplines that concern the Global Navigation Satellite Systems (GNSS). The latter is the topic of this present contribution. The GNSS have been proven to be an efficient tool for determination of time varying parameters that are of importance for Earth science disciplines like positioning, deformation, timing and atmosphere [11, 12]. Parameter estimation in GNSS often relies on the data processing of a *network* of receivers that collect measurements from visible GNSS satellites. In the context of sensor networks, GNSS network receivers therefore serve as sensor nodes, providing their data to a computing center, thereby computing network-based parameter solutions in a (near) real-time manner. In this contribution we intend to demonstrate how consensus algorithms [13] and the corresponding consensus-based Kalman filter (CKF), as a popular means for distributed filtering, can take an important role in GNSS applications for which a network of receivers are to be processed. Although each single receiver can run its own local filter to deliver GNSS-derived solutions, the precision of such single-receiver solutions is generally much lower than its network-derived counterparts, see e.g. [14, 15]. It will be shown, through a CKF setup, that single-receiver parameter solutions can achieve precision performances similar to that of their network-based versions, provided that a sufficient number of iterative communications between the neighboring receivers are established. The importance of such consensus-based single-receiver solutions is well appreciated in the light of the recent development of new GNSS constellations as well as the proliferation of low-cost mass-market receivers [16–18]. With the increase in the number and types of GNSS receivers, many more GNSS users can establish their own measurement setup to determine parameters that suit their needs. By taking recourse to the CKF or other distributed filtering techniques, GNSS users can therefore potentially deliver high-precision parameter solutions without the need of having a computing center.

The structure of this contribution is as follows. We first briefly review the principles of the standard Kalman filter and its information form in Section 2. The additivity property of the information filter that makes this filter particularly useful for distributed processing is also highlighted. In Section 3 we discuss average consensus rules on which the sensor nodes agree to fuse each other information. Different consensus protocols are discussed and a 'probabilistic' measure for the evaluation of their convergence rates is proposed. Section 4 is devoted to the CKF algorithmic steps. Its two time-scale nature is remarked and a three-step recursion for evaluating the consensus-based error variance matrix is developed. In Section 5 we apply the CKF theory to a small-scale network of GNSS receivers collecting ionospheric observables over time. Conducting a precision analysis, we compare the precision of the network-based ionospheric solutions with those of their single-receiver and consensus-based counterparts. It is shown how the CKF of each receiver responses to an increase in the number of iterative communications between the neighboring nodes. Concluding remarks and future outlook are provided in Section 6.

2. Kalman filtering

Consider a time series of observable random vectors $y_1, \ldots, y_t$. The goal is to *predict* the unobservable random state-vectors $x_1, \ldots, x_t$. By the term 'prediction', we mean that the observables $y_1, \ldots, y_t$ are used to estimate *realizations* of the random vectors $x_1, \ldots, x_t$. Accordingly, the means of the state-vectors $x_1, \ldots, x_t$ can be known, while their unknown realizations still need to be guessed (predicted) through observed realizations of $y_1, \ldots, y_t$. In the following, to show on which set of observables prediction is based, we use the notation $\widehat{x}_{t|\tau}$ as the predictor of x_t when based on $y_{[\tau]} = \left[y_1^T, \ldots, y_\tau^T\right]^T$. The expectation, covariance and dispersion operators are denoted by E(.), C(.,.) and D(.), respectively. The capital Q is reserved for (co)variance matrices. Thus $\mathrm{C}(x_t, y_\tau) = Q_{x_t y_\tau}$.

2.1. The Kalman filter standard assumptions

To predict the state-vectors in an optimal sense, one often uses the minimum mean squared error (MMSE) principle as the optimality criterion, see e.g., [19–25]. In case no restrictions are placed on the class of predictors, the MMSE predictor $\widehat{x}_{t|\tau}$ is given by the conditional mean $E\left(x_t|y_{[\tau]}\right)$, known as the Best Predictor (BP). The BP is unbiased, but generally nonlinear, with exemptions, for instance in the Gaussian case. In case x_t and $y_{[\tau]}$ are jointly Gaussian, the BP becomes linear and identical to its linear counterpart, i.e. the Best Linear Predictor (BLP)

$$\widehat{x}_{t|\tau} = E(x_t) + Q_{x_t y_{[\tau]}} Q_{y_{[\tau]} y_{[\tau]}}^{-1} \left\{ y_{[\tau]} - \mathrm{E}\left(y_{[\tau]}\right) \right\} \tag{1}$$

Eq. (1) implies that (1) the BLP is unbiased, i.e. $\mathrm{E}\left(\widehat{x}_{t|\tau}\right) = \mathrm{E}(x_t)$, and that (2) the prediction error of a BLP is always uncorrelated with observables on which the BLP is based, i.e. $C\left(x_t - \widehat{x}_{t|\tau}, y_{[\tau]}\right) = 0$. These two basic properties can be alternatively used to uniquely specify a BLP [26].

The Kalman filter is a *recursive* BP (Gaussian case) or a recursive BLP. A recursive predictor, say $\widehat{x}_{t|t}$, can be obtained from the previous predictor $\widehat{x}_{t|t-1}$ and the newly collected observable vector y_t. Recursive prediction is thus very suitable for applications that require *real-time* determination of temporally varying parameters. We now state the standard assumptions that make the Kalman filter recursion feasible.

The dynamic model: The linear dynamic model, describing the time-evolution of the state-vectors x_t, is given as

$$x_t = \Phi_{t,t-1} x_{t-1} + d_t, \quad t = 1, 2, \ldots \tag{2}$$

with

$$\mathrm{E}(x_0) = x_{0|0}, \quad \mathrm{D}(x_0) = Q_{x_0 x_0} \tag{3}$$

and

$$\mathrm{E}(d_t) = 0, \quad \mathrm{C}(d_t, d_s) = S_t \delta_{t,s}, \quad \mathrm{C}(d_t, x_0) = 0 \tag{4}$$

for the time instances $t, s = 1, 2, \ldots$, with $\delta_{t,s}$ being the Kronecker delta. The nonsingular matrix $\Phi_{t,t-1}$ denotes the transition matrix and the random vector d_t is the system noise. The system noise d_t is thus assumed to have a zero mean, to be uncorrelated in time and to be uncorrelated with the initial state-vector x_0. The transition matrix from epoch s to t is denoted as $\Phi_{t,s}$. Thus $\Phi_{t,s}^{-1} = \Phi_{s,t}$ and $\Phi_{t,t} = I$ (the identity matrix).

The measurement model: The link between the observables y_t and the state-vectors x_t is assumed given as

$$y_t = A_t x_t + \varepsilon_t, \quad t = 1, 2, \ldots \tag{5}$$

with

$$\mathrm{E}(\varepsilon_t) = 0, \quad \mathrm{C}(\varepsilon_t, \varepsilon_s) = R_t \delta_{t,s}, \quad \mathrm{C}(\varepsilon_t, x_0) = 0, \quad \mathrm{C}(\varepsilon_t, d_s) = 0 \tag{6}$$

for $t, s = 1, 2, \ldots$, with A_t being the known design matrix. Thus the zero-mean measurement noise ε_t is assumed to be uncorrelated in time and to be uncorrelated with the initial state-vector x_0 and the system noise d_t.

2.2. The three-step recursion

Initialization: As the mean of x_0 is known, the best predictor of x_0 in the absence of data is the mean $\mathrm{E}(x_0) = x_{0|0}$. Hence, the initialization is simply given by

$$\widehat{x}_{0|0} = x_{0|0}, \quad P_{0|0} = Q_{x_0 x_0} \tag{7}$$

That the initial error variance matrix $P_{0|0} = \mathrm{D}(x_0 - \widehat{x}_{0|0})$ is identical to the variance matrix $Q_{x_0 x_0}$ follows from the equality $\mathrm{D}(x_0 - x_{0|0}) = \mathrm{D}(x_0)$.

Time update: Let us choose $\Phi_{t,t-1}\widehat{x}_{t-1|t-1}$ as a candidate for the BLP $\widehat{x}_{t|t-1}$. According to Eq. (1), the candidate would be the BLP if it fulfills two conditions: (1) it must be unbiased and (2) it must have a prediction error uncorrelated with the previous data $y_{[t-1]}$. The first condition, i.e. $\mathrm{E}(\Phi_{t,t-1}\widehat{x}_{t-1|t-1}) = \mathrm{E}(x_t)$, follows from Eq. (2) and the equalities $\mathrm{E}(\widehat{x}_{t-1|t-1}) = \mathrm{E}(x_{t-1})$ and $\mathrm{E}(d_t) = 0$. The second condition, i.e. $\mathrm{C}(x_t - \Phi_{t,t-1}\widehat{x}_{t-1|t-1}, y_{[t-1]}) = 0$, follows from the fact that the prediction error $x_t - \Phi_{t,t-1}\widehat{x}_{t-1|t-1}$ is a function of the previous BLP prediction error $x_{t-1} - \widehat{x}_{t-1|t-1}$ and the system noise d_t, i.e. (cf. 2)

$$x_t - \Phi_{t,t-1}\widehat{x}_{t-1|t-1} = \Phi_{t,t-1}\left(x_{t-1} - \widehat{x}_{t-1|t-1}\right) + d_t, \tag{8}$$

that are both uncorrelated with the previous data $y_{[t-1]}$. Hence, the time update is given by

$$\widehat{x}_{t|t-1} = \Phi_{t,t-1}\widehat{x}_{t-1|t-1}, \quad P_{t|t-1} = \Phi_{t,t-1} P_{t-1|t-1} \Phi_{t,t-1}^T + S_t \tag{9}$$

The error variance matrix $P_{t|t-1} = \mathrm{D}(x_t - \widehat{x}_{t|t-1})$ follows by applying the covariance propagation law to (8), together with $\mathrm{C}(x_{t-1} - \widehat{x}_{t-1|t-1}, d_t) = 0$.

Measurement update: In the presence of new data y_t, one may yet offer $\widehat{x}_{t|t-1}$ as a candidate for the BLP $\widehat{x}_{t|t}$. Such a candidate fulfills the unbiasedness condition $\mathrm{E}(\widehat{x}_{t|t-1}) = \mathrm{E}(x_t)$, but *not* necessarily the zero-correlation condition, that is, $\mathrm{C}(x_t - \widehat{x}_{t|t-1}, y_{[t]}) \neq 0$. Note, however, that $\mathrm{C}(x_t - \widehat{x}_{t|t-1}, y_{[t-1]}) = 0$. Thus the zero-correlation condition $\mathrm{C}(x_t - \widehat{x}_{t|t-1}, y_{[t]}) = 0$ would have been met if the most recent data y_t of $y_{[t]} = \left[y_{[t-1]}^T, y_t^T\right]^T$ would be a function of the previous data $y_{[t-1]}$, thereby *fully* predicted by $y_{[t-1]}$. Since an observable is its own best predictor, this implies that $y_t = A_t\widehat{x}_{t|t-1}$, where $A_t\widehat{x}_{t|t-1}$ is the BLP of y_t. But this would require the zero-mean quantity $v_t = y_t - A_t\widehat{x}_{t|t-1}$ to be identically zero which is generally not the case. It is therefore the presence of v_t that violates the zero-correlation condition. Note that v_t is a function of the prediction error $x_t - \widehat{x}_{t|t-1}$ and the measurement noise ε_t, i.e. (cf. 5)

$$v_t = A_t\left(x_t - \widehat{x}_{t|t-1}\right) + \varepsilon_t, \tag{10}$$

that are both uncorrelated with $y_{[t-1]}$. Therefore, v_t cannot be predicted by the previous data $y_{[t-1]}$, showing that v_t contains truly *new* information. That is why v_t is sometimes referred to as the *innovation* of y_t, see e.g. [27–29]. We now amend our earlier candidate $\widehat{x}_{t|t-1}$ by adding a linear function of v_t to it. It reads $\widehat{x}_{t|t} = \widehat{x}_{t|t-1} + K_t v_t$, with K_t being an unknown matrix to be chosen such that the zero-correlation condition is met. Such a matrix, known as the Kalman gain matrix, is uniquely specified by

$$K_t = P_{t|t-1}A_t^T Q_{v_t v_t}^{-1} \quad \Leftrightarrow \quad \mathrm{C}\left(x_t - \widehat{x}_{t|t-1} - K_t v_t, y_t\right) = 0 \tag{11}$$

since $\mathrm{C}\left(x_t - \widehat{x}_{t|t-1}, y_t\right) = P_{t|t-1}A_t^T$ and $\mathrm{C}\left(v_t, y_t\right) = Q_{v_t v_t}$. The measurement update reads then

$$\widehat{x}_{t|t} = \widehat{x}_{t|t-1} + K_t v_t, \quad \text{with} \quad P_{t|t} = P_{t|t-1} - K_t Q_{v_t v_t} K_t^T \tag{12}$$

The error variance matrix $P_{t|t} = \mathrm{D}\left(x_t - \widehat{x}_{t|t}\right)$ follows by an application of the covariance propagation law, together with $\mathrm{C}\left(x_t - \widehat{x}_{t|t-1}, v_t\right) = P_{t|t-1}A_t^T$. Application of the covariance propagation law to (10) gives the variance matrix of v_t as follows

$$Q_{v_t v_t} = A_t P_{t|t-1} A_t^T + R_t \tag{13}$$

since $\mathrm{C}\left(x_t - \widehat{x}_{t|t-1}, \varepsilon_t\right) = 0$.

2.3. A remark on the filter initialization

In the derivation of the Kalman filter one assumes the mean of the random initial state-vector x_0, in Eq. (3), to be known, see e.g. [30–37]. This is because of the BLP structure (1) that needs knowledge of the means $\mathrm{E}(x_t)$ and $\mathrm{E}(y_{[\tau]})$. Since in many, if not most, applications the means of the state-vectors $x_1, \ldots, x_t$ are unknown, such derivation is therefore not appropriate. As shown in Ref. [38], one can do away with this need to have both the initial mean $x_{0|0}$ and variance matrix $Q_{x_0x_0}$, given in Eq. (3), known. The corresponding three-step recursion would then follow the Best Linear Unbiased Prediction (BLUP) principle and not that of the BLP. The BLUP is also a MMSE predictor, but within a more restrictive class of predictors. It replaces the means $\mathrm{E}(x_t)$ and $\mathrm{E}(y_{[\tau]})$ by their corresponding Best Linear Unbiased Estimators (BLUEs). Within such BLUP recursion, the initialization Eq. (7) is revised and takes place at time instance $t = 1$ in the presence of the data y_1. Provided that matrix A_1 is of full column rank, the predictor $\widehat{x}_{1|1}$ follows from solving the normal equations

$$N_1 \widehat{x}_{1|1} = r_1, \quad \text{with} \quad N_1 = A_1^T R_1^{-1} A_1, \quad r_1 = A_1^T R_1^{-1} y_1 \tag{14}$$

Thus

$$\widehat{x}_{1|1} = N_1^{-1} r_1, \quad \text{and} \quad P_{1|1} = N_1^{-1} \tag{15}$$

The above error variance matrix $P_{1|1}$ is thus *not* dependent on the variance matrix of x_1, i.e. $Q_{x_1x_1} = \Phi_{1,0} Q_{x_0x_0} \Phi_{1,0}^T + S_1$. This is, however, not the case with the variance matrix of the predictor $\widehat{x}_{1|1}$ itself, i.e. $Q_{\widehat{x}_{1|1}\widehat{x}_{1|1}} = \mathrm{D}(\widehat{x}_{1|1})$. This variance matrix is given by [38]

$$Q_{\widehat{x}_{1|1}\widehat{x}_{1|1}} = Q_{x_1x_1} + P_{1|1} \tag{16}$$

showing that $P_{1|1} \neq Q_{\widehat{x}_{1|1}\widehat{x}_{1|1}}$. Matrices $P_{t|t}$ and $Q_{\widehat{x}_{t|t}\widehat{x}_{t|t}}$ $(t = 1, 2, \ldots)$ are used for two *different purposes*. The error variance matrix $P_{t|t} = \mathrm{D}(x_t - \widehat{x}_{t|t})$ is a measure of 'closeness' of $\widehat{x}_{t|t}$ to its target random vector x_t, thereby meant to describe the 'quality' of prediction, i.e. precision of the prediction error $(x_t - \widehat{x}_{t|t})$. The variance matrix $Q_{\widehat{x}_{t|t}\widehat{x}_{t|t}} = \mathrm{D}(\widehat{x}_{t|t})$ however, is a measure of closeness of $\widehat{x}_{t|t}$ to the nonrandom vector $\mathrm{E}(x_t)$, as $\mathrm{D}(\widehat{x}_{t|t}) = \mathrm{D}(\mathrm{E}(x_t) - \widehat{x}_{t|t})$. Thus $Q_{\widehat{x}_{t|t}\widehat{x}_{t|t}}$ does *not* describe the quality of prediction, but instead the precision of the predictor $\widehat{x}_{t|t}$.

The MMSE of the BLUP recursion is never smaller than that of the Kalman filter, as the Kalman filter makes use of additional information, namely, the known mean $x_{0|0}$ and variance matrix $Q_{x_0x_0}$. When the stated information is available, the BLUP recursion is shown to encompass the Kalman filter as a special case [39]. In the following we therefore assume that the means of the state-vectors $x_1, \ldots, x_t$ are unknown, a situation that often applies to GNSS applications.

2.4. Filtering in information form

The three-step recursion presented in Eqs. (7), (9) and (12) concerns the time-evolution of the predictor $\widehat{x}_{t|t}$ and the error variance matrix $P_{t|t}$. As shown in Eq. (15), both $P_{1|1}$ and $\widehat{x}_{1|1}$ can be determined by the normal matrix $N_1 = P_{1|1}^{-1}$ and the right-hand-side vector $r_1 = P_{1|1}^{-1}\widehat{x}_{1|1}$. One can therefore alternatively develop recursion concerning the time-evolution of $P_{t|t}^{-1}$ and $P_{t|t}^{-1}\widehat{x}_{t|t}$. From a computational point of view, such recursion is found to be very suitable when the inverse-variance or *information* matrices S_t^{-1} and R_t^{-1} serve as input rather than the variance matrices S_t and R_t. To that end, one may define [34]

$$\text{information vector} \quad i_{t|\tau} := P_{t|\tau}^{-1}\hat{x}_{t|\tau} \quad \text{and} \quad \text{information matrix } \mathcal{I}_{t|\tau} := P_{t|\tau}^{-1} \tag{17}$$

Given the definition above, the information filter recursion concerning the time-evolution of $i_{t|t}$ and $\mathcal{I}_{t|t}$ would then follow from the recursion Eqs. (15), (9) and (12), along with the following matrix-inversion equalities

$$\begin{array}{lll} \text{Time-update}: & (\Phi_{t,t-1}P_{t-1|t-1}\Phi_{t,t-1}^T + S_t)^{-1} & = M_t - M_t(M_t + S_t^{-1})^{-1}M_t \\ \text{Measurement-update}: & (P_{t|t-1} - P_{t|t-1}A_t^T Q_{v_t v_t} A_t P_{t|t-1})^{-1} & = P_{t|t-1}^{-1} + A_t^T R_t^{-1} A_t \end{array} \tag{18}$$

where $M_t = \Phi_{t-1,t}^T P_{t-1|t-1}^{-1}\Phi_{t-1,t}$.

The algorithmic steps of the information filter are presented in **Figure 1**. In the absence of data, the filter is initialized by the zero information $i_{1|0} = 0$ and $\mathcal{I}_{1|0} = 0$. In the presence of the data y_t, the corresponding normal matrix N_t and right-hand-side vector r_t are added to the time update information $i_{t|t-1}$ and $\mathcal{I}_{t|t-1}$ to obtain the measurement update information $i_{t|t}$ and $\mathcal{I}_{t|t}$.

Kalman filter in information form

- Step 1 - Initialization:

$$\boldsymbol{i}_{1|0} = 0, \qquad \mathcal{I}_{1|0} = 0$$

Set $t \mapsto 1$

- Step 2 - Measurement Update:

$$\overset{inputs}{\rightarrow} y_t,\ R_t^{-1},\ A_t$$

$$r_t = A_t^T R_t^{-1} y_t, \qquad N_t = A_t^T R_t^{-1} A_t$$

$$\boldsymbol{i}_{t|t} = \boldsymbol{i}_{t|t-1} + r_t, \qquad \mathcal{I}_{t|t} = \mathcal{I}_{t|t-1} + N_t$$

Set $t + 1 \mapsto t$

- Step 3 - Time Update:

$$\overset{inputs}{\rightarrow} \Phi_{t,t-1},\ S_t^{-1}$$

$$\begin{array}{lll} M_t & = & \Phi_{t,t-1}^{-T}\mathcal{I}_{t-1|t-1}\Phi_{t,t-1}^{-1} \\ \mathcal{I}_{t|t-1} & = & M_t - M_t(M_t + S_t^{-1})^{-1}M_t \\ \boldsymbol{i}_{t|t-1} & = & \mathcal{I}_{t|t-1}\Phi_{t,t-1}\mathcal{I}_{t-1|t-1}^{-1}\boldsymbol{i}_{t-1,t-1} \end{array}$$

Go to Step 2

Figure 1. Algorithmic steps of the information filter recursion concerning the time-evolution of the information vector $i_{t|t}$ and matrix $\mathcal{I}_{t|t}$.

The transition matrix $\Phi_{t,t-1}$ and inverse-variance matrix S_t^{-1} would then be used to time update the previous information $i_{t-1|t-1}$ and $\mathcal{I}_{t-1|t-1}$.

Singular matrix S_t: In the first expression of Eq. (18) one assumes the variance matrix S_t to be nonsingular and invertible. There are, however, situations where some of the elements of the state-vector x_t are nonrandom, i.e., the corresponding system noise is identically zero. As a consequence, the variance matrix S_t becomes singular and the inverse-matrix S_t^{-1} does not exist. An example of such concerns the presence of the GNSS carrier-phase ambiguities in the filter state-vector which are treated constant in time. In such cases the information time update in **Figure 1** must be generalized so as to accommodate singular variance matrices S_t. Let $\tilde{S}_t$ be an invertible sub-matrix of S_t that has the *same* rank as that of S_t. Then there exists a full-column rank matrix H_t such that

$$S_t = H_t \tilde{S}_t H_t^T \tag{19}$$

Matrix H_t can be, for instance, structured by the columns of the identity matrix I corresponding to the columns of S_t on which the sub-matrix $\tilde{S}_t$ is positioned. The special case

$$S_t = \begin{bmatrix} \tilde{S}_t & 0 \\ 0 & 0 \end{bmatrix} = \begin{bmatrix} I \\ 0 \end{bmatrix} \tilde{S}_t \begin{bmatrix} I \\ 0 \end{bmatrix}^T \quad \Rightarrow \quad H_t = \begin{bmatrix} I \\ 0 \end{bmatrix}, \tag{20}$$

shows an example of the representation (19). With Eq. (19), a generalization of the time update (**Figure 1**) can be shown to be given by

$$\mathcal{I}_{t|t-1} = M_t - M_t H_t \left[H_t^T M_t H_t + \tilde{S}_t^{-1} \right]^{-1} H_t^T M_t \tag{21}$$

Thus instead of S_t^{-1}, the inverse-matrix $\tilde{S}_t^{-1}$ and H_t are assumed available.

2.5. Additivity property of the information measurement update

As stated previously, the information filter delivers outcomes equivalent to those of the Kalman filter recursion. Thus any particular preference for the information filter must be attributed to the *computational* effort required for obtaining the outcomes. For instance, if handling matrix inversion requires low computational complexity when working with the input inverse-matrices S_t^{-1} and R_t^{-1}, the information filter appears to be more suitable. In this subsection we will highlight yet another property of the information filter that makes this recursion particularly useful for *distributed* processing.

As shown in **Figure 1**, the information measurement update is *additive* in the sense that the measurement information N_t and r_t is added to the information states $\mathcal{I}_{t|t-1}$ and $i_{t|t-1}$. We now make a start to show how such additivity property lends itself to distributed filtering. Let the measurement model Eq. (5) be partitioned as

$$y_t = A_t x_t + \varepsilon_t \quad \Rightarrow \quad \begin{bmatrix} y_{1,t} \\ \vdots \\ y_{i,t} \\ \vdots \\ y_{n,t} \end{bmatrix} = \begin{bmatrix} A_{1,t} \\ \vdots \\ A_{i,t} \\ \vdots \\ A_{n,t} \end{bmatrix} x_t + \begin{bmatrix} \varepsilon_{1,t} \\ \vdots \\ \varepsilon_{i,t} \\ \vdots \\ \varepsilon_{n,t} \end{bmatrix}, \quad t = 1, 2, \ldots \tag{22}$$

Accordingly, the observable vector y_t is partitioned into n sub-vectors $y_{i,t}$ $(i = 1, \ldots, n)$, each having its own design matrix $A_{i,t}$ and measurement noise vector $\varepsilon_{i,t}$. One can think of a network of n sensor nodes where each collects its own observable vector $y_{i,t}$, but aiming to determine a common state-vector x_t. Let us further assume that the nodes collect observables independently from one another. This yields

$$C(\varepsilon_{i,t}, \varepsilon_{j,t}) = R_{i,t}\delta_{i,j}, \quad \text{for} \quad i, j = 1, \ldots, n, \quad \text{and} \quad t = 1, 2, \ldots \tag{23}$$

Thus the measurement noise vectors $\varepsilon_{i,t}$ $(i = 1, \ldots, n)$ are assumed to be *mutually* uncorrelated. With the extra assumption Eq. (23), the normal matrix $N_t = A_t^T R_t^{-1} A_t$ and right-hand-side vector $r_t = A_t^T R_t^{-1} y_t$ can then be, respectively, expressed as

$$N_t = \sum_{i=1}^{n} N_{i,t}, \quad \text{and} \quad r_t = \sum_{i=1}^{n} r_{i,t} \tag{24}$$

where

$$N_{i,t} = A_{i,t}^T R_{i,t}^{-1} A_{i,t}, \quad \text{and} \quad r_{i,t} = A_{i,t}^T R_{i,t}^{-1} y_{i,t} \tag{25}$$

According to Eq. (24), the measurement information of each node, say $N_{i,t}$ and $r_{i,t}$, is individually added to the information states $\mathcal{I}_{t|t-1}$ and $i_{t|t-1}$, that is

$$\mathcal{I}_{t|t} = \mathcal{I}_{t|t-1} + \sum_{i=1}^{n} N_{i,t}, \qquad i_{t|t} = i_{t|t-1} + \sum_{i=1}^{n} r_{i,t} \tag{26}$$

Now consider the situation where each node runs its own *local* information filter, thus having its own information states $\mathcal{I}_{i,t|t}$ and $i_{i,t|t}$ $(i = 1, \ldots, n)$. The task is to recursively update the local states $\mathcal{I}_{i,t|t}$ and $i_{i,t|t}$ in a way that they remain equal to their *central* counterparts $\mathcal{I}_{t|t}$ and $i_{t|t}$ given in Eq. (26). Suppose that such equalities hold at the time update, i.e. $\mathcal{I}_{i,t|t-1} = \mathcal{I}_{t|t-1}$ and $i_{i,t|t-1} = i_{t|t-1}$. Given the number of contributing nodes n, each node just needs to be provided with the *average* quantities

$$\overline{N}_t = \frac{1}{n}\sum_{i=1}^{n} N_{i,t}, \quad \text{and} \quad \bar{r}_t = \frac{1}{n}\sum_{i=1}^{n} r_{i,t} \tag{27}$$

The local states $\mathcal{I}_{i,t|t-1}$ and $i_{i,t|t-1}$ would then be measurement updated as (cf. 26)

$$\mathcal{I}_{i,t|t} = \mathcal{I}_{i,t|t-1} + n\overline{N}_t, \qquad i_{i,t|t} = \quad i_{i,t|t-1} + n\overline{r}_t \tag{28}$$

that are equal to the central states $\mathcal{I}_{t|t}$ and $i_{t|t}$, respectively. In this way one has multiple distributed local filters $i = 1, \ldots, n$, where each recursively delivers results identical to those of a central filter.

To compute the average quantities $\overline{N}_t$ and $\overline{r}_t$, node i may need to receive all other information $N_{j,t}$ and $r_{j,t}$ $(j \neq i)$. In other words, node i would require direct connections to *all* other nodes $j \neq i$, a situation that makes data communication and processing power very expensive (particularly for a large number of nodes). In the following cheaper ways of evaluating the averages $\overline{N}_t$ and $\overline{r}_t$ are discussed.

3. Average consensus rules

In the previous section, we briefly discussed the potential applicability of the information filter as a tool for handling the measurement model Eq. (22) in a distributed manner. With the representation Eq. (28) however, one may be inclined to conclude that such applicability is limited to the case where the nodes $i = 1, \ldots, n$, have 'direct' communication connections to one another in order to receive/send their measurement information $N_{i,t}$ and $r_{i,t}$ $(i = 1, \ldots, n)$.

Instead of having direct connections, the idea is now to relax such a stringent requirement by assuming that the nodes are linked to each other *at least* through a 'path' so that information can flow from each node to all other nodes. It is therefore assumed that each node along the path plays the role of an *agent* transferring information to other nodes. To reach the averages $\overline{N}_t$ and $\overline{r}_t$, the nodes would then agree on specific 'fusion rules' or *consensus protocols*, see e.g. [6, 8, 40]. Note that each node exchanges information with neighboring nodes (i.e. those to which the node has direct connections) and *not* the entire nodes. Therefore, a *repeated* application of the consensus protocols is required to be carried out. The notion is made precise below.

3.1. Communication graphs

The way the nodes interact with each other to transfer information is referred to as the interaction topology between the nodes. The interaction topology is often described by a directed graph whose vertices and edges, respectively, represent the nodes and communication links [4]. The interaction topology may also undergo a *finite* number of changes over sessions $k = 1, \ldots, k_o$. In case of one-way links, the directions of the edges face toward the receiving nodes (vertices). Here we assume that the communication links between the nodes are two-way, thus having *undirected* (or bidirectional) graphs. Examples of such representing a network of 20 nodes with their interaction links are shown in **Figure 2**. Let an undirected graph at session k be denoted by $\mathcal{G}_k = (\mathcal{V}, \mathcal{E}_k)$ where $\mathcal{V} = \{1, \ldots, n\}$ is the vertex set and $\mathcal{E}_k \subset \{(i,j) |\ i, j \in \mathcal{V}\}$ is the edge set. We assume that the nodes remain unchanged over time, that is why the subscript k is omitted for $\mathcal{V}$. This is generally not the case with their interaction

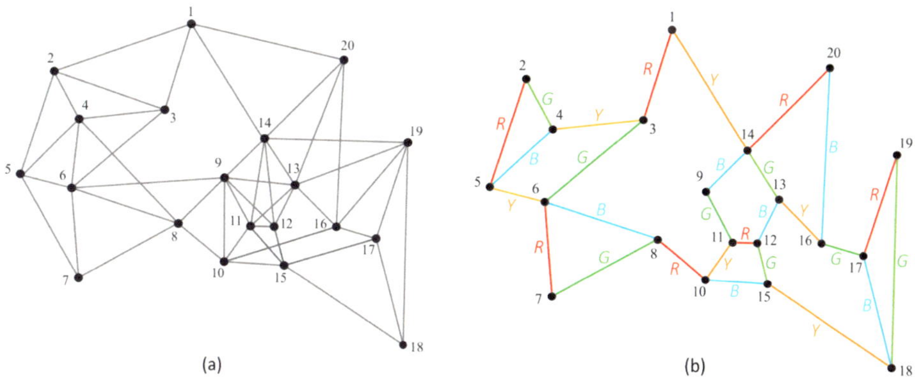

Figure 2. Communications graphs of 20 sensor nodes. The edges represent two-way communication links between the nodes. (a) Network with 49 links. (b) Network with different numbers of links over four sessions: 7 links in session 1 (R), 6 links in session 2 (Y), 8 links in session 3 (G) and 7 links in session 4 (B).

links though, i.e. the edge set $\mathcal{E}_k$ depends on k. As in **Figure 2 (b)**, the number of links between the nodes can be different for different sessions $k = 1, \ldots, k_o$. Each session represents a graph that may not be *connected*. In a 'connected' graph, every vertex is linked to all other vertices at least through one path. In order for information to flow from each node to all other nodes, the *union* of the graphs $\mathcal{G}_k$ ($k = 1, \ldots, k_o$), i.e.

$$\mathcal{G} = (\mathcal{V}, \mathcal{E}) \quad \text{with} \quad \mathcal{E} = \bigcup_{k=1}^{k_o} \mathcal{E}_k \quad (k_o : \text{a finite number}) \tag{29}$$

is therefore assumed to be connected. We define the *neighbors* of node i as those to which the node i has direct links. For every session k, they are collected in the set $\mathcal{N}_{i,k} = \{j \mid (j, i) \in \mathcal{E}_k\}$. For instance for network (a) of **Figure 2**, we have only one session, i.e. $k_o = 1$, in which $\mathcal{N}_{2,1} = \{1, 3, 4, 5\}$ represents the neighbors of node 2. In case of network (b) however, we have different links over four sessions, i.e. $k_o = 4$. In this case, the neighbors of node 2 are given by four sets: $\mathcal{N}_{2,1} = \{5\}$ in session 1 (red), $\mathcal{N}_{2,2} = \{\}$ in session 2 (yellow), $\mathcal{N}_{2,3} = \{4\}$ in session 3 (green) and $\mathcal{N}_{2,4} = \{\}$ in session 4 (blue).

3.2. Consensus protocols

Given the right-hand-vector $r_{i,t}$, suppose that node i aims to obtain the average $\bar{r}_t$ for which all other vectors $r_{j,t}$ ($\forall j \neq i$) are required to be available (cf. 27). But the node i only has access to those of its neighbors, i.e. the vectors $r_{j,t}$ ($j \in \mathcal{N}_{i,k}$). For the first session $k = 1$, it would then seem to be reasonable to compute a *weighted-average* of the available vectors, i.e.

$$r_{i,t}(1) = \sum_{j \in \{i, \mathcal{N}_{i,1}\}} w_{ij}(1) r_{j,t} \tag{30}$$

as an *approximation* of $\bar{r}_t$, where the scalars $w_{ij}(1)$ ($j \in \{i, \mathcal{N}_{i,1}\}$) denote the corresponding weights at session $k = 1$. Now assume that all other nodes $j \neq i$ agree to apply the fusion rule

Eq. (30) to those of their own neighbors. Thus the neighboring nodes $j \in \mathcal{N}_{i,k}$ also have their own weighted-averages $r_{j,t}(1)$. But they may have access to those to which the node i has *no* direct links. In other words, the weighted-averages $r_{j,t}(1)$ $(j \in \mathcal{N}_{i,1})$ contain information on the nodes to which the node i has no access. For the next session $k = 2$, it is therefore reasonable for the node i to *repeat* the fusion rule Eq. (30), but now over the new vectors $r_{j,t}(1)$ $(j \in \{i, \mathcal{N}_{i,2}\})$, aiming to improve on the earlier approximation $r_{i,t}(1)$. This yields the following iterative computations

$$r_{i,t}(k) = \sum_{j \in \{i, \mathcal{N}_{i,k}\}} w_{ij}(k) r_{j,t}(k-1), \quad k = 1, 2, \ldots \tag{31}$$

with $r_{j,t}(0) := r_{j,t}$. Choosing a set of weights $w_{ij}(k)$, the nodes $i = 1, \ldots, n$ agree on the consensus protocol (31) to iteratively fuse their information vectors $r_{i,t}(k)$. Here and in the following, we use the letter *'k'* for the *'session number'* $k = 1, \ldots, k_o$ (cf. 29) and for the *'number of iterative communications'* $k = 1, \ldots, k_n$ (cf. 34). The maximum iteration k_n is assumed to be not smaller than the maximum session number k_o, i.e. $k_n \geq k_o$.

The question that now comes to the fore is how to choose the weights $w_{ij}(k)$ such that the approximation $r_{i,t}(k)$ gets close to $\bar{r}_t$ through the iteration Eq. (31). More precisely, the stated iteration becomes favorable if $r_{i,t}(k) \to \bar{r}_t$ when $k \to \infty$ for all nodes $i = 1, \ldots, n$. To address this question, we use a multivariate formulation. Let p be the size of the vectors $r_{i,t}$ $(i = 1, \ldots, n)$. We define the higher-dimensioned vector $r = \left[r_{1,t}^T, \ldots, r_{n,t}^T\right]^T$. The multivariate version of Eq. (31) reads then

$$\begin{bmatrix} r_{1,t}(k) \\ \vdots \\ r_{i,t}(k) \\ \vdots \\ r_{n,t}(k) \end{bmatrix} = \begin{bmatrix} w_{11}(k)I_p & \ldots & w_{1i}(k)I_p & \ldots & w_{1n}(k)I_p \\ \vdots & \ddots & \vdots & \ddots & \vdots \\ w_{i1}(k)I_p & \ldots & w_{ii}(k)I_p & \ldots & w_{in}(k)I_p \\ \vdots & \ddots & \vdots & \ddots & \vdots \\ w_{n1}(k)I_p & \ldots & w_{ni}(k)I_p & \ldots & w_{nn}(k)I_p \end{bmatrix} \begin{bmatrix} r_{1,t}(k-1) \\ \vdots \\ r_{i,t}(k-1) \\ \vdots \\ r_{n,t}(k-1) \end{bmatrix}, \quad k = 1, 2, \ldots \tag{32}$$

or

$$r(k) = \left[W(k) \otimes I_p\right] r(k-1), \quad k = 1, 2, \ldots \tag{33}$$

The $n \times n$ weight matrix $W(k)$ is structured by $w_{ij}(k)$ $(j \in \{i, \mathcal{N}_{i,k}\})$ and $w_{ij}(k) = 0$.

$(j \notin \{i, \mathcal{N}_{i,k}\})$. The symbol $\otimes$ is the Kronecker matrix product [41]. According to Eq. (33), after k_n iterations the most recent iterated vector $r(k_n)$ is linked to the initial vector $r(0)$ by $\left[\prod_{k=1}^{k_n} W(k) \otimes I_p\right] r(0)$. Thus the vectors $r_{i,t}(k)$ $(i = 1, \ldots, n)$ converge to $\bar{r}_t$ when

$$L(k_n) := \prod_{k=1}^{k_n} W(k) \to \frac{1}{n} e_n e_n^T, \quad \text{as} \quad k_n \to \infty \tag{34}$$

where the n-vector e_n contains ones. If the condition Eq. (34) is met, the set of nodes $\{1, \ldots, n\}$ can *asymptotically* reach average consensus [4]. It can be shown that (34) holds if the weight matrices $W(k)$ $(k = 1, \ldots, k_o)$ have bounded nonnegative entries with positive diagonals, i.e. $w_{ij}(k) \geq 0$ and $w_{ii}(k) > 0$, having row- and column-sums equal to one, i.e. $\sum_{j=1}^{n} w_{ij}(k) = 1$ and $\sum_{i=1}^{n} w_{ij}(k) = 1$ $(i, j = 1, \ldots, n)$, see e.g. [3, 5, 40, 42, 43].

Examples of such consensus protocols are given in **Table 1**. As shown, the weights form a symmetric weight matrix $W(k)$, i.e. $w_{ji}(k) = w_{ij}(k)$. In all protocols presented, self-weights $w_{ii}(k)$ are chosen so that the condition $\sum_{j=1}^{n} w_{ij}(k) = 1$ is satisfied. The weights of Protocols 1 and 2 belong to the class of 'maximum-degree' weights, while those of Protocol 3 are referred to as 'Metropolis' weights [8]. The weights of Protocols 1 and 3 are driven by the degrees (number of neighbors) of nodes $i = 1, \ldots, n$, denoted by $dg_i(k) = \mathcal{N}_{i,k}$. For instance, in network (a) of **Figure 2** we have $dg_1(1) = 4$ as node 1 has 4 neighbors, while $dg_{14}(1) = 7$ as node 14 has 7 neighbors. Protocol 4 is only applicable to networks like (b) in **Figure 2**, i.e. when each node has *at most* one neighbor at a session [4]. In this case, each node exchanges its information to just one neighbor at a session. Thus for two neighboring nodes i and j we have $w_{ii}(k) = w_{jj}(k) = w_{ij}(k) = 0.5$, each averaging $r_{i,t}(k-1)$ and $r_{j,t}(k-1)$ to obtain $r_{i,t}(k) = r_{j,t}(k)$.

To provide insight into the applicability of the protocols given in **Table 1**, we apply them to the networks of **Figure 2**. Twenty values (scalars), say r_i $(i = 1, \ldots, 20)$, are generated whose average is equal to 5, i.e. $\bar{r} = 5$. Each value is assigned to its corresponding node. For network (a), Protocols 1, 2 and 3 are separately applied, whereas Protocol 4 is only applied to network (b). The corresponding results, up to 30 iterations, are presented in **Figure 3**. As shown, the iterated values $r_i(k)$ $(i = 1, \ldots, 20)$ get closer to their average (i.e. $\bar{r} = 5$), the more the number of iterative communications.

3.3. On convergence of consensus states

Figure 3 shows that the states $r_{i,t}(k)$ $(i = 1, \ldots, n)$ converge to their average $\bar{r}_t$, but with *different* rates. The convergence rate depends on the initial states $r_{i,t}(0) = r_{i,t}$ and on the consensus protocol employed. From the figure it seems that the convergence rates of Protocols 1 and 3 are about the same, higher than those of Protocols 2 and 4. Note that the stated results are

Protocols	$w_{ij}(k)$ $(\{i,j\} \in \mathcal{E}_k)$	$w_{ii}(k)$
Protocol 1	$\frac{1}{\max\limits_{u \in \{1,\ldots,n\}} \{dg_u(k)\}}$	$1 - \sum_{u \neq i} w_{iu}(k)$
Protocol 2	$\frac{1}{n}$	$1 - \sum_{u \neq i} w_{iu}(k)$
Protocol 3	$\frac{1}{1+\max\{dg_i(k), dg_j(k)\}}$	$1 - \sum_{u \neq i} w_{iu}(k)$
Protocol 4	$\frac{1}{2}$	$1 - \sum_{u \neq i} w_{iu}(k)$
otherwise $w_{ij}(k) = 0$		

The degree (number of neighbors) of node i is denoted by $dg_i(k) = \#\mathcal{N}_{i,k}$. Protocol 4 is only applicable when each node has *at most* one neighbor at a session.

Table 1. Examples of average-consensus protocols forming the weights $w_{ij}(k)$ in Eq. (31).

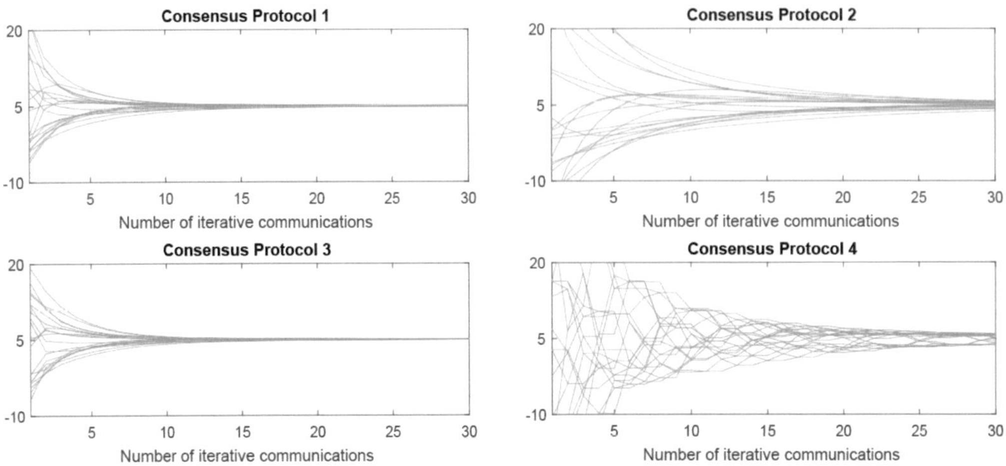

Figure 3. Performances of protocols 1, 2 and 3 (network (a) of **Figure 2**) and protocols 4 (network (b)) in delivering the average of 20 values (scalars). The iterated values get closer to their average (i.e. $\bar{r} = 5$), the more the number of iterative communications.

obtained on the basis of specific 'realizations' of $r_{i,t}$ $(i = 1, \ldots, n)$. Consider the states $r_{i,t}$ to be *random* vectors. In that case, can the results be still representative for judging the convergence performances of the protocols? To answer this question, let us define the difference vectors $dr_{i,t}(k) = r_{i,t}(k) - \bar{r}_t$ that are collected in the higher-dimensioned vector $dr(k) = \left[dr_{1,t}^T(k), \ldots, dr_{n,t}^T(k)\right]^T$. The more the number of iterations, the smaller the norm of $dr(k)$ becomes. According to Eq. (33), after k_n iterations the difference vector $d_r(k_n)$ is linked to $r = \left[r_{1,t}^T, \ldots, r_{n,t}^T\right]^T$ through

$$dr(k_n) = \left[\left(L(k_n) - \frac{1}{n}e_n e_n^T\right) \otimes I_p\right] r \tag{35}$$

Now let the initial states $r_{i,t}$ have the same mean and the same variance matrix $\mathrm{D}(r_{i,t}) = Q$ $(i = 1, \ldots, n)$, but mutually uncorrelated. An application of the covariance propagation law to (35), together with $L(k_n)e_n = e_n$, gives

$$\mathrm{D}(dr(k_n)) = \left(L^2(k_n) - \frac{1}{n}e_n e_n^T\right) \otimes Q \tag{36}$$

Thus the closer the squared matrix $L^2(k_n)$ to $(1/n)e_n e_n^T$, the smaller the variance matrix Eq. (36) becomes. In the limit when $k_n \to \infty$, the stated variance matrix tends to zero. This is what one would expect, since $dr(k_n) \to 0$. Under the conditions stated in Eq. (34), matrices $W(k)$ have $\lambda_n = 1$ as the largest absolute value of their eigenvalues [42]. A symmetric weight matrix W can then be expressed in its *spectral* form as

$$W = \sum_{i=1}^{n-1} \lambda_i u_i u_i^T + \frac{1}{n}e_n e_n^T \tag{37}$$

with the eigenvalues $\lambda_1 \leq \ldots \leq \lambda_{n-1} < \lambda_n = 1$, and the corresponding orthogonal unit eigenvectors $u_1, \ldots, u_{n-1}, u_n = (1/\sqrt{n})e_n$. By a repeated application of the protocol W, we get $L(k_n) = W^{k_n}$. Substitution into Eq. (36), together with Eq. (37), gives finally

$$\mathrm{D}(dr(k_n)) = \left(\sum_{i=1}^{n-1} \lambda_i^{2k_n} u_i u_i^T\right) \otimes Q \leq \lambda_{n-1}^{2k_n} [I_n \otimes Q] \tag{38}$$

The above equation shows that the entries of the variance matrix (36) are largely driven by the second largest eigenvalue of W, i.e. λ_{n-1}. The smaller the scalar $|\lambda_{n-1}|$, the faster the quantity $\lambda_{n-1}^{2k_n}$ tends to zero, as $k_n \to \infty$. The scalar $|\lambda_{n-1}|$ is thus often used as a measure to judge the convergence performances of the protocols [7]. For the networks of **Figure 2**, $|\lambda_{n-1}|$ of Protocols 1, 2 and 3 are about 0.92, 0.97, 0.91, respectively. As Protocol 3 has the smallest $|\lambda_{n-1}|$, it is therefore expected to have the best performance. Note, in Protocol 4, that the weight matrix $W(k)$ varies in every session, the performance of which cannot be judged by a single eigenvalue λ_{n-1}. One can therefore think of another means of measuring the convergence performance. Due to the randomness of the information vectors $r_{i,t}$ $(i = 1, \ldots, n)$, one may propose 'probabilistic' measures such as

$$\mathrm{Prob}\left(\max_i \|dr_{i,t}(k_n)\|_Q \leq q\right), \quad (q > 0) \tag{39}$$

to evaluate the convergence rates of the protocols, where $\|dr_{i,t}\|_Q^2 := dr_{i,t}^T Q^{-1} dr_{i,t}$. Eq. (39) refers to the probability that the maximum-norm of the difference vectors $dr_{i,t}(k_n) = r_{i,t}(k_n) - \bar{r}_t$ $(i = 1, \ldots, n)$ is not larger than a given positive scalar q for a fixed number of iterations k_n. The higher the probability Eq. (39), the better the performance of a protocol. For the scalar case $Q = \sigma^2$, Eq. (39) is reduced to

$$\mathrm{Prob}\left(\max_i |dr_{i,t}(k_n)| \leq q\sigma\right) \tag{40}$$

which is the probability that the absolute differences $|dr_{i,t}(k_n)|$ $(i = 1, \ldots, n)$ are not larger than q times the standard-deviation σ. For the networks of **Figure 2**, 100,000 normally-distributed vectors as samples of $r = [r_1, \ldots, r_{20}]^T$ are simulated to evaluate the probability (40). The results for Protocols 1, 2, 3 and 4 are presented in **Figure 4**. The stated probability is plotted as a function of q for three numbers of iterative communications $k_n = 10, 20$ and 30. As shown, Protocol 3 gives rise to highest probabilities, while Protocol 2 delivers lowest probabilities. After 10 iterations, the probability of having absolute differences smaller than one-fifth of the standard-deviation σ (i.e. $q = 0.2$) is about 80% for Protocol 1, whereas it is less than 5% for Protocol 2. After 30 iterations, the stated probability increases to 80% for Protocol 2, but close to 100% for Protocols 1 and 3.

Figure 4 demonstrates that the convergence performance of Protocol 4 is clearly better than that of Protocol 2, as it delivers higher probabilities (for the networks of **Figure 2**). Such a conclusion however, cannot be made on the basis of the results of **Figure 3**. This shows that

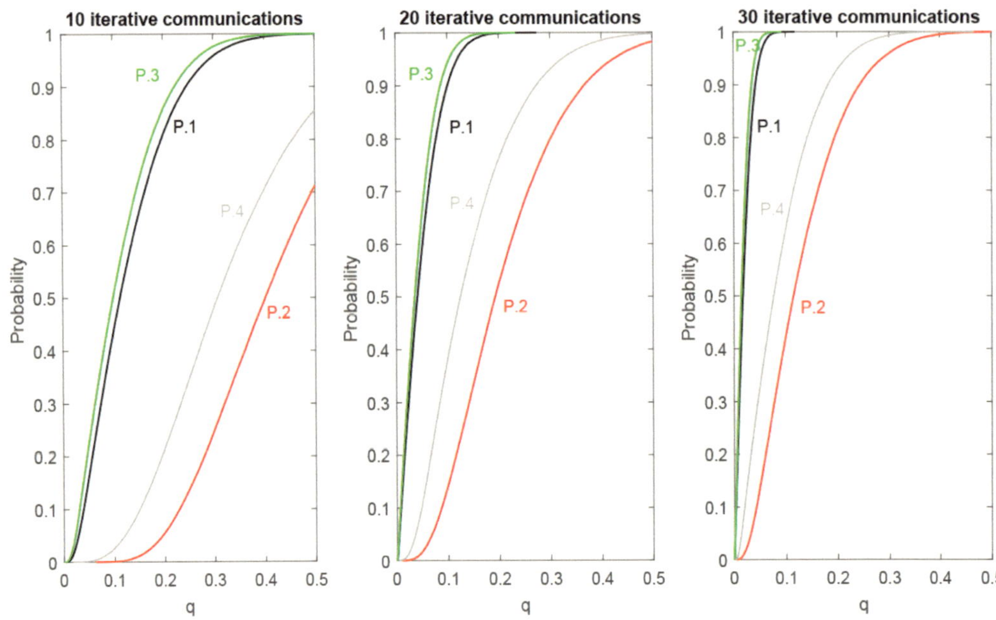

Figure 4. Probability Eq. (40) as a function of q for three numbers of iterative communications $k_n = 10, 20$ and 30 (protocols 1 (P.1), 2 (P.2), 3 (P.3) and 4 (P.4)). It refers to the probability that the absolute differences $|r_{i,t}(k_n) - \overline{r}_t|$ $(i = 1, \ldots, 20)$ are not larger than q times the standard-deviation σ (cf. **Figure 2**).

results obtained on the basis of specific 'realizations' of $r_{i,t}$ $(i = 1, \ldots, n)$ are not necessarily representative.

4. Consensus-based Kalman filters

4.1. Two time-scale approach

In Section 2.5 we discussed how the additivity property of the measurement update Eq. (26) offers possibilities for developing multiple distributed local filters $i = 1, \ldots, n$, each delivering local states $\mathcal{I}_{i,t|t}$ and $i_{i,t|t}$ equal to their central counterparts $\mathcal{I}_{t|t}$ and $i_{t|t}$. In doing so, each node has to evaluate the averages $\overline{N}_t$ and $\overline{r}_t$ at *every* time instance t. Since in practice the nodes do not necessarily have direct connections to each other, options such as the consensus-based fusion rules (cf. Section 3) can alternatively be employed to 'approximate' $\overline{N}_t$ and $\overline{r}_t$. As illustrated in **Figures 3** and **4**, such consensus-based approximation requires a number of iterative communications between the nodes in order to reach the averages $\overline{N}_t$ and $\overline{r}_t$. The stated iterative communications clearly require some time to be carried out and must take place during every time interval $[t, t+1]$ (see **Figure 5**). We distinguish between the *sampling* rate Δ and the *sending* rate δ. The sampling rate refers to the frequency with which the node i collects its observables $y_{i,t}$ $(t = 1, 2, \ldots)$, while the sending rate refers to the frequency with which the node i sends/receives information $N_{j,t}(k)$ and $r_{j,t}(k)$ $(k = 1, \ldots, k_n)$ to/from its neighboring nodes. As shown in **Figure 5**, the sending rate δ should therefore be reasonably smaller

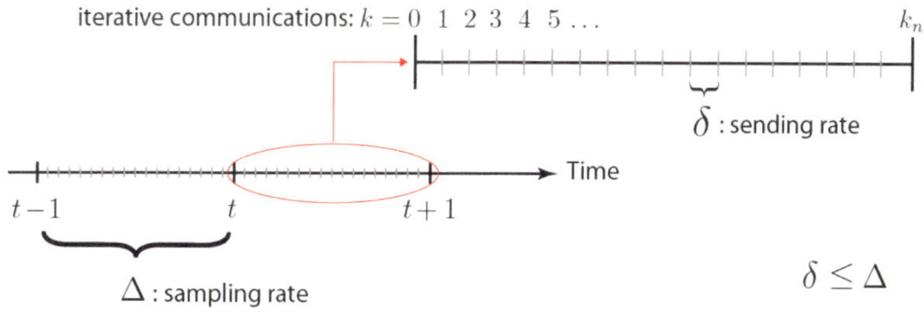

Figure 5. The two time-scale nature of a consensus-based Kalman filter (CKF): The data sampling time-scale $t = 1, 2, \ldots,$ versus the data sending time-scale $k = 1, \ldots, k_n$. The sending rate δ must be reasonably smaller than the sampling rate Δ so as to be able to incorporate consensus protocols into the CKF setup.

than the sampling rate Δ so as to be able to incorporate consensus protocols into the information filter setup. Such a consensus-based Kalman filter (CKF) would thus generally be of a two time-scale nature [2], the data sampling time-scale $t = 1, 2, \ldots,$ versus the data sending time-scale $k = 1, \ldots, k_n$. The CKF is a suitable tool for handling real-time data processing in a distributed manner for the applications in which the state-vectors x_t $(t = 1, 2, \ldots)$ change rather slowly over time (i.e. Δ can take large values) and/or for the cases where the sensor nodes transfer their data rather quickly (i.e. δ can take small values).

Under the assumption $\delta \leq \Delta$, the CKF recursion follows from the Kalman filter recursion by considering an extra step, namely, the 'consensus update'. The algorithmic steps of the CKF in information form are presented in **Figure 6**. Compare the recursion with that of the information filter given in **Figure 1**. Similar to the information filter, the CKF at node i is initialized by the zero information $\mathcal{I}_{i,1|0} = 0$ and $i_{i,1|0} = 0$. In the presence of the data $y_{i,t}$, node i computes its local normal matrix $N_{i,t}$ and right-hand-side vector $r_{i,t}$ to send them to its neighboring nodes $j \in \mathcal{N}_{i,k}$ $(k = 1, \ldots, k_n)$. In the consensus update, iterative communications between the neighboring nodes $\{i, \mathcal{N}_{i,k}\}$ are carried out to approximate the averages $\overline{N}_t$ and $\overline{r}_t$ by $N_{i,t}(k_n)$ and $r_{i,t}(k_n)$, respectively. After a finite number of communications k_n, the consensus states $N_{i,t}(k_n)$ and $r_{i,t}(k_n)$ are, respectively, added to the time update information $\mathcal{I}_{i,t|t-1}$ and $i_{i,t|t-1}$ to obtain their measurement update version $\mathcal{I}_{i,t|t}$ and $i_{i,t|t}$ at node i (cf. 28). The time update goes along the same lines as that of the information filter.

4.2. Time evolution of the CKF error covariances

With the consensus-based information filter, presented in **Figure 6**, it is therefore feasible to develop multiple distributed filters, all running in parallel over time. By taking recourse to an average-consensus protocol, not all the nodes are needed to be directly linked, thereby allowing non-neighboring nodes to also benefit from information states of each other. The price one has to pay for such an attractive feature of the CKF is that the local predictors

$$\widehat{x}_{i,t|t} = \mathcal{I}_{i,t|t}^{-1}\ i_{i,t|t}, \quad i = 1, \ldots, n, \tag{41}$$

will have a *poorer* precision performance than that of their central counterpart $\widehat{x}_{t|t}$. This is due to the fact that the consensus states $N_{i,t}(k_n)$ and $r_{i,t}(k_n)$ $(i = 1, \ldots, n)$ are just *approximations* of

Consensus-based information filter

- Step 1 - Initialization (for node i):

$$\boldsymbol{i}_{i,1|0} = 0, \qquad \mathcal{I}_{i,1|0} = 0$$

Set $t \mapsto 1$

- Step 2 - Consensus Update (for node i):

$$\overset{inputs}{\rightarrow} \; y_{i,t},\; R_{i,t}^{-1},\; A_{i,t}$$

$$r_{i,t} = A_{i,t}^T R_{i,t}^{-1} y_{i,t}, \qquad N_{i,t} = A_{i,t}^T R_{i,t}^{-1} A_{i,t}$$

$$r_{i,t}(0) \mapsto r_{i,t}, \qquad N_{i,t}(0) \mapsto N_{i,t}$$

(i) Set $k \mapsto 0$
(ii) Send $[N_{i,t}(k), r_{i,t}(k)]$ to neighboring nodes $j \in \mathcal{N}_{i,k}$
(iii) Receive $[N_{j,t}(k), r_{j,t}(k)]$ from neighboring nodes $j \in \mathcal{N}_{i,k}$

$$[N_{i,t}(k+1), r_{i,t}(k+1)] = \sum_{j \in \{i, \mathcal{N}_{i,k}\}} w_{i,j}(k)[N_{j,t}(k), r_{j,t}(k)]$$

(iv) Set $k \mapsto k+1$
(v) Go to (ii) while $k < k_n$

- Step 3 - Measurement Update (for node i):

$$\boldsymbol{i}_{i,t|t} = \boldsymbol{i}_{i,t|t-1} + n\, r_{i,t}(k_n), \qquad \mathcal{I}_{i,t|t} = \mathcal{I}_{i,t|t-1} + n\, N_{i,t}(k_n)$$

Set $t+1 \mapsto t$

- Step 4 - Time Update (for node i):

$$\overset{inputs}{\rightarrow} \; \Phi_{t,t-1},\; S_t^{-1}$$

$$\begin{aligned} M_{i,t} &= \Phi_{t,t-1}^{-T} \mathcal{I}_{i,t-1|t-1} \Phi_{t,t-1}^{-1} \\ \mathcal{I}_{i,t|t-1} &= M_{i,t} - M_{i,t}(M_{i,t} + S_t^{-1})^{-1} M_{i,t} \\ \boldsymbol{i}_{i,t|t-1} &= \mathcal{I}_{i,t|t-1} \Phi_{t,t-1} \mathcal{I}_{i,t-1|t-1}^{-1} \boldsymbol{i}_{i,t-1,t-1} \end{aligned}$$

Go to Step 2

Figure 6. Algorithmic steps of the CKF in information form concerning the time-evolution of the local information vector $i_{i,t|t}$ and matrix $\mathcal{I}_{i,t|t}$ of node i.

the averages $\overline{N}_t$ and $\overline{r}_t$. Although they reach the stated averages as $k_n \to \infty$, one of course always comes up with a *finite* number of communications k_n. As a consequence, while the inverse-matrix $\mathcal{I}_{t|t}^{-1}$ represents the error variance matrix $P_{t|t} = \mathrm{D}(x_t - \widehat{x}_{t|t})$ (cf. 17), the inverse-matrices $\mathcal{I}_{i,t|t}^{-1}$ $(i = 1, \ldots, n)$ do *not* represent the error variance matrices $P_{i,t|t} = \mathrm{D}(x_t - \widehat{x}_{i,t|t})$. To see this, consider the local prediction errors $(x_t - \widehat{x}_{i,t|t})$ which can be expressed as (**Figure 6**)

$$x_t - \widehat{x}_{i,t|t} = \mathcal{I}_{i,t|t}^{-1}\left\{\mathcal{I}_{i,t|t-1}\left(x_t - \widehat{x}_{i,t|t-1}\right) - n\left[r_{i,t}(k_n) - N_{i,t}(k_n)x_t\right]\right\} \tag{42}$$

Note that the terms $\left(x_t - \widehat{x}_{i,t|t-1}\right)$ and $\left[r_{i,t}(k_n) - N_{i,t}(k_n)x_t\right]$ are uncorrelated. With l_{ij} as the entries of the product matrix $L(k_n)$ in Eq. (34), one obtains4

$$\left[r_{i,t}(k_n) - N_{i,t}(k_n)x_t\right] = \sum_{j=1}^{n} l_{ij}\left(r_{j,t} - N_{j,t}x_t\right), \quad \Rightarrow \quad \mathrm{D}(\left[r_{i,t}(k_n) - N_{i,t}(k_n)x_t\right]) = \sum_{j=1}^{n} l_{ij}^2 N_{j,t} \tag{43}$$

since $\mathrm{D}\left(r_{j,t} - N_{j,t}x_t\right) = N_{j,t}$. With this in mind, an application of the covariance propagation law to (42) results in the error variance matrix

$$P_{i,t|t} = \mathcal{I}_{i,t|t}^{-1}\left\{\mathcal{I}_{i,t|t-1}P_{i,t|t-1}\mathcal{I}_{i,t|t-1} + n^2\sum_{j=1}^{n} l_{ij}^2 N_{j,t}\right\}\mathcal{I}_{i,t|t}^{-1} \tag{44}$$

that is not necessarily equal to $\mathcal{I}_{i,t|t}^{-1}$ (see the following discussion on Eqs. (47) and (48)).

In **Figure 7** we present the three-step recursion of the error variance matrix $P_{i,t|t}$ (for node i). As shown, the node i would need an *extra* input, i.e., the term $\sum_{j=1}^{n} l_{ij}^2 N_{j,t}$ in order to be able to compute $P_{i,t|t}$. In practice however, such additional information is absent in the CKF setup. This means that the node i does not have enough information to evaluate the error variance matrix $P_{i,t|t}$. Despite such restriction, it will be shown in Section 5 how the recursion of $P_{i,t|t}$ conveys useful information about the performance of the local filters $i = 1, \ldots, n$, thereby allowing one to *a-priori* design and analyze sensor networks with different numbers of iterative communications.

To better appreciate the recursion given in **Figure 7**, let us consider a special case where a *stationary* state-vector x_t is to be predicted over time. Thus $\Phi_{t,t-1} = I$ and $S_t = 0$ $(t = 1, 2, \ldots)$. Moreover, we assume that all nodes deliver the same normal matrices $N_{i,t} = N$ $(i = 1, \ldots, n)$.

The central error variance matrix $P_{t|t}$ would then simply follow by inverting the sum of all normal matrices over n nodes and t time instances. Collecting observables up to and including time instance t, the stated variance matrix reads $P_{t|t} = (1/tn)N^{-1}$. We now compare $P_{t|t}$ with its consensus-based local counterpart at node i, i.e. $P_{i,t|t}$. The aforementioned assumptions, together with $\sum_{j=1}^{n} l_{ij} = 1$, give

$$N_{i,t}(k_n) = \sum_{j=1}^{n} l_{ij}N_{j,t} = N, \quad \text{and} \quad n^2\sum_{j=1}^{n} l_{ij}^2 N_{j,t} = \alpha n N \tag{45}$$

in which the scalar α is given by

$$\alpha := n\sum_{j=1}^{n} l_{ij}^2 \tag{46}$$

Substitution into the stated recursion provides us with the time-evolution of the error variance matrix $P_{i,t|t}$ as follows (**Figure 7**)

(Co)Variance propagation of CKF prediction errors

- Step 1 - Initialization (for node i):

$$\mathcal{I}_{i,1|0} = 0, \qquad F_{i,1} = 0$$

Set $t \mapsto 1$

- Step 2 - Measurement Update (for node i):

$$\overset{inputs}{\rightarrow} N_{i,t}(k_n),\ \sum_{j=1}^{n} l_{ij}^2 N_{j,t}$$

$$\begin{aligned}
\mathcal{I}_{i,t|t} &= \mathcal{I}_{i,t|t-1} + nN_{i,t}(k_n) \\
P_{i,t|t} &= \mathcal{I}_{i,t|t}^{-1} [F_{i,t} + n^2 \sum_{j=1}^{n} l_{ij}^2 N_{j,t}] \mathcal{I}_{i,t|t}^{-1}
\end{aligned}$$

Set $t+1 \mapsto t$

- Step 3 - Time Update (for node i):

$$\overset{inputs}{\rightarrow} \Phi_{t,t-1},\ S_t^{-1}$$

$$\begin{aligned}
P_{i,t|t-1} &= \Phi_{t,t-1} P_{i,t-1|t-1} \Phi_{t,t-1}^T + S_t \\
M_{i,t} &= \Phi_{t,t-1}^{-T} \mathcal{I}_{i,t-1|t-1} \Phi_{t,t-1}^{-1} \\
\mathcal{I}_{i,t|t-1} &= M_{i,t} - M_{i,t}(M_{i,t} + S_t^{-1})^{-1} M_{i,t} \\
F_{i,t} &= \mathcal{I}_{i,t|t-1} P_{i,t|t-1} \mathcal{I}_{i,t|t-1}
\end{aligned}$$

Go to Step 2

Figure 7. The three-step recursion of the error variance matrix $P_{i,t|t} = \mathrm{D}(x_t - \widehat{x}_{i,t|t})$ for node i. The *extra* term $\sum_{j=1}^{n} l_{ij}^2 N_{j,t}$ would be required to compute $P_{i,t|t}$. The entries of the product matrix $L(k_n)$ in Eq. (34) are denoted by l_{ij}.

$$\begin{array}{ccc}
\mathcal{I}_{i,1|0} = 0, \quad F_{i,1} = 0 & & \\
\downarrow & & \\
\mathcal{I}_{i,1|1} = nN, & \Rightarrow & P_{i,1|1} = \alpha \dfrac{1}{n} N^{-1} \\
\downarrow & & \\
\mathcal{I}_{i,2|1} = \mathcal{I}_{i,1|1}, \quad F_{i,2} = \alpha nN & \Rightarrow & P_{i,2|1} = P_{i,1|1} \\
\vdots & \vdots & \vdots \\
\mathcal{I}_{i,t|t-1} = \mathcal{I}_{i,t-1|t-1}, \quad F_{i,t} = \alpha(t-1)nN & \Rightarrow & P_{i,t|t-1} = P_{i,t-1|t-1} \\
\downarrow & & \\
\mathcal{I}_{i,t|t} = tnN, & \Rightarrow & P_{i,t|t} = \alpha \dfrac{1}{tn} N^{-1}
\end{array} \tag{47}$$

This shows that the consensus-based error variance matrix $P_{i,t|t}$ is α times its central counterpart $P_{t|t} = (1/tn)N^{-1}$. With the vector $l{:=}[l_{i1}, \ldots, l_{in}]^T$, application of the Cauchy-Schwarz inequality gives the lower-bound

$$\alpha = \left(e_n^T e_n\right)\left(l^T l\right) \geq \left(l^T e_n\right)^2 = 1 \tag{48}$$

as $l^T e_n = 1$. Thus scalar α is never smaller than 1, i.e. $P_{i,t|t} \geq P_{t|t}$, showing that the performance of the consensus-based predictor $\widehat{x}_{i,t|t}$ is never better than that of its central version $\widehat{x}_{t|t}$. The lower-bound Eq. (48) is reached when $l = (1/n)e_n$, i.e. when $l_{ij} = 1/n$ $(j = 1, \ldots, n)$. According to Eq. (34), this can be realized if $L(k_n) \rightarrow (1/n)e_n e_n^T$, for which the number of iterations k_n might be required to be reasonably large. The conclusion reads therefore that the local filters at nodes $i = 1, \ldots, n$, generate information matrices $\mathcal{I}_{i,t|t}$, the inverse of which are different from the actual error variance matrices of the predictors $\widehat{x}_{i,t|t}$, i.e. $\mathcal{I}_{i,t|t}^{-1} \neq P_{i,t|t}$.

5. Applications to GNSS

The purpose of this section is to demonstrate how the CKF theory, discussed in Section 4, can play a pivotal role in applications for which the GNSS measurements of a *network* of receivers are to be processed in a real-time manner. In a GNSS network setup, each receiver serves as a sensor node for receiving observables from visible GNSS satellites to determine a range of different parameters such as positions and velocities in an Earth-centered Earth-fixed coordinate system, atmospheric delays, timing and instrumental biases, see e.g. [11, 12]. As the observation equations of the receivers have satellite specific parameters in common, the receivers' observables are often integrated through a computing (fusion) center to provide network-derived parameter solutions that are more precise than their single-receiver versions. Now the idea is to deliver GNSS parameter solutions without the need of having a computing center, such that their precision performance is still comparable to that of network-derived solutions.

As previously discussed, consensus-based algorithms and in particular the CKF can be employed to process network data in a distributed filtering scheme, i.e. no computing center is required. In order to illustrate such applicability, we simulate a network of 13 GNSS receivers located in Perth, Western Australia (**Figure 8**). As shown in the figure, each node (white circle) represents a receiver having data links (red lines) to its neighbors with inter-station distances up to 4 km. We therefore assume that the receivers receive each other data within the ranges not longer than 4 km. For instance, receiver 1 is directly connected to receivers 2 and 6, but not to receiver 3 (the inter-station distance between receivers 1 and 3 is about 8 km).

5.1. GNSS ionospheric observables: Dynamic and measurement models

Although the GNSS observables contain information on various positioning and non-positioning parameters, here we restrict ourselves to *ionospheric observables* of the GPS *pseudo-range* measurements only [44]. One should however bear in mind that such restriction is made just for the sake of presentation and illustration of the theory discussed in Sections 3 and 4.

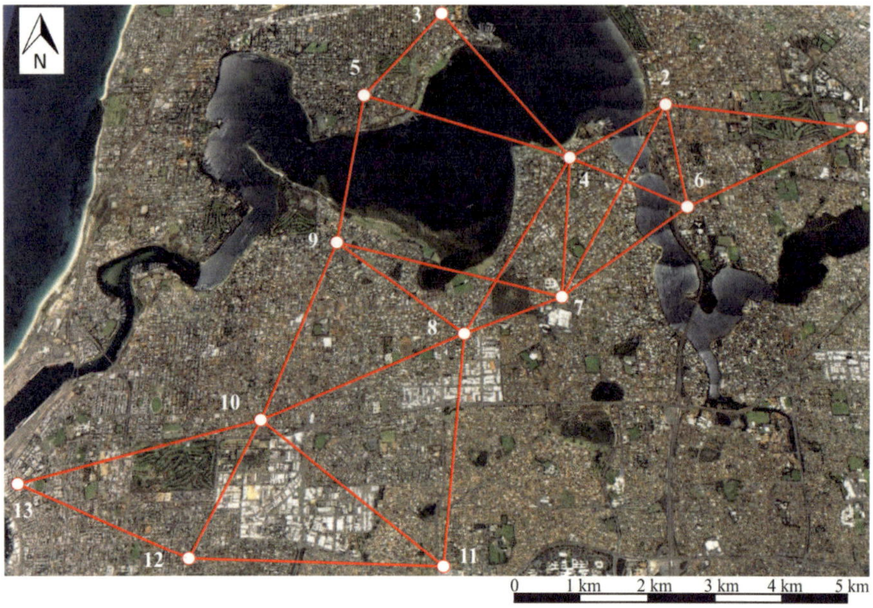

Figure 8. A network of 13 GNSS receivers simulated over Perth, Western Australia. Each node (white circle) represents a receiver tracking GNSS satellites. The receivers have data links to their neighbors with inter-station distances up to 4 km. The data links are shown by red lines.

Would one, for instance, make use of the very precise carrier-phase measurements and/or formulate a multi-GNSS measurement setup, solutions of higher precision levels are therefore expected.

Let the scalar $y^s_{i,t}$ denote the pseudo-range ionospheric observable that the receiver i collects from satellite s at time instance t. The corresponding measurement model, formed by the between-satellite differences $y^{ps}_{i,t}{:=}y^s_{i,t} - y^p_{i,t}$ $(s \neq p)$, reads (cf. 5)

$$y^{ps}_{i,t} = \left\{a^{ps}_{i,t;o}\,\nu_{o,t} + a^{ps}_{i,t;\phi}\,\nu_{\phi,t} + a^{ps}_{i,t;\psi}\,\nu_{\psi,t}\right\} - b^{ps}_t + \varepsilon^{ps}_{i,t} \tag{49}$$

where the term within $\{.\}$ refers to the first-order slant ionospheric delays, and b^{ps}_t denotes the between-satellite differential code biases (DCBs). We use a regional single-layer model [45, 46] to represent the slant ionospheric delays in terms of 1) $\nu_{o,t}$ as the vertical total electron content (TEC), 2) $\nu_{\phi,t}$ and 3) $\nu_{\psi,t}$ as the south-to-north and west-to-east spatial gradient of $\nu_{o,t}$, respectively. The corresponding known coefficients follow from Ref. [47]

$$a^s_{i,t;o} = \frac{1}{\cos\left(z^s_{i,t}\right)}, \quad a^s_{i,t;\phi} = \frac{1}{\cos\left(z^s_{i,t}\right)}\left(\phi^s_{i,t} - \phi_{o,t}\right), \quad a^s_{i,t;\psi} = \frac{1}{\cos\left(z^s_{i,t}\right)}\cos\left(\phi^s_{i,t}\right)\left(\psi^s_{i,t} - \psi_{o,t}\right) \tag{50}$$

with $(.)^{ps}{:=}(.)^s - (.)^p$. The angles $\psi^s_{i,t}$ and $\phi^s_{i,t}$, respectively, denote the longitude and latitude of the ionospheric piercing points (IPPs) corresponding to the receiver-to-satellite line-of-sight

$i-s$ (see **Figure 9**). They are computed with respect to those of the reference IPP at time instance t, i.e. $\psi_{o,t}$ and $\phi_{o,t}$. The angle $z^s_{i,t}$ denotes the zenith angle of the IPPs. These angles are computed based on the mean Earth's radius 6378.137 km and height of layer 450 km. The measurement noises $\varepsilon^s_{i,t}$ are assumed to be mutually uncorrelated with the dispersion (cf. 6)

$$D\left(\varepsilon^s_{i,t}\right) = \frac{1.02^2}{\left(0.02 + \sin\left(\theta^s_{i,t}\right)\right)^2}\sigma^2 \tag{51}$$

forming the variance matrices R_t in Eq. (6), where $\theta^s_{i,t}$ is the satellite elevation angle. The scalar σ is set to $\sigma \approx 65.6$ cm as the zenith-referenced standard-deviation of the GPS 'geometry-free' pseudo-range measurements [48].

Suppose that m number of satellites $s = 1, \ldots, m$, are tracked by the network receivers $i = 1, \ldots, n = 13$, during the observational campaign. The state-vector sought is structured as

$$x_t = \left[\nu_{o,t}, \nu_{\phi,t}, \nu_{\psi,t}, b^{p1}_t, b^{p2}_t, \ldots, b^{pm}_t\right]^T \tag{52}$$

Thus the state-vector x_t contains three TEC parameters $\nu_{o,t}$, $\nu_{\phi,t}$, $\nu_{\psi,t}$ and $(m-1)$ between-satellite DCBs b^{ps}_t $(s \neq p)$. The dynamic model is assumed to be given by (cf. 2, 4 and 21)

$$\begin{bmatrix} \nu_{o,t} \\ \nu_{\phi,t} \\ \nu_{\psi,t} \end{bmatrix} = \begin{bmatrix} \nu_{o,t-1} \\ \nu_{\phi,t-1} \\ \nu_{\psi,t-1} \end{bmatrix} + \begin{bmatrix} d_o \\ d_\phi \\ d_\psi \end{bmatrix}, \quad \text{and} \quad b^{ps}_t = b^{ps}_{t-1} \quad s \neq p \tag{53}$$

Thus the DCBs b^{ps}_t are assumed constant in time, while the temporal behavior of the TEC parameters $\nu_{o,t}$, $\nu_{\phi,t}$, $\nu_{\psi,t}$ is captured by a random-walk process. The corresponding zero-mean

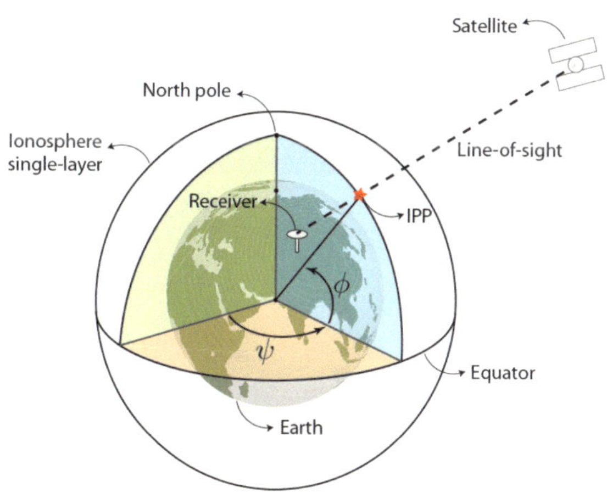

Figure 9. Longitude (ψ) and latitude (ϕ) of an ionospheric piercing point (IPP) corresponding to a receiver-to-satellite line-of-sight. The distance scales are exaggerated.

process noises are assumed to be mutually uncorrelated, having the standard-deviations $\sigma_{d_o} = 1$ mm/$\sqrt{sec}$ and $\sigma_{d_\phi} = \sigma_{d_{psi}} = 5$ mm/rad/$\sqrt{sec}$ [49].

5.2. Observational campaign

The network receivers $i = 1, \ldots, n$ ($n = 13$), shown in **Figure 8**, are assumed to track GPS satellites over 16 hours from 8:00 to 24:00 Perth local time, on 02-06-2016. The observation sampling rate is set to $\Delta = 1$ minute. Thus the number of observational epochs (time instances) is 960. As to the data sending rate δ (cf. 5), we assume three different sending rates $\delta = 5, 10$ and 15 seconds. Thus the number of iterative communications between the neighboring receivers takes the values $k_n = 4, 6$ and 12. The consensus protocol 3 (**Table 1**) is applied to the CKF of each receiver.

As the satellites revolve around the Earth, not all of which are simultaneously visible to the 'small-scale' network of **Figure 8**. Their visibility over time is shown in **Figure 10** (left panel) in which the satellites with elevation angles smaller than 10 degrees are excluded. There are 31 GPS satellites (i.e. $m = 31$), with PRN 4 absent (PRN refers to the satellite identifier). PRN 22 has the maximum duration of visibility, while PRN 21 has the minimum duration of visibility. Note also that PRNs 2, 6, 16, 17, 19, 26 and 32 disappear (set) and reappear (re-rise). That is why their visibility is shown via two separate time intervals. **Figure 10** (right panel) shows the trajectories of the ionospheric pierce points on the ionospheric single layer that are made by receiver-to-satellite line-of-sight paths. It is the spatial distribution of these points that drives the coefficients $a^{ps}_{i,t;o}$, $a^{ps}_{i,t;\phi}$, $a^{ps}_{i,t;\psi}$ in Eq. (49).

In the following we present precision analyses on the *measurement update* solutions of x_t in Eq. (52), given the network and satellite configurations shown in **Figures 8** and **10**, respectively. Throughout the text, PRN 10 is chosen as the pivot satellite p (cf. (49)). By the term 'standard-deviation', we mean the square-root of prediction errors' variance.

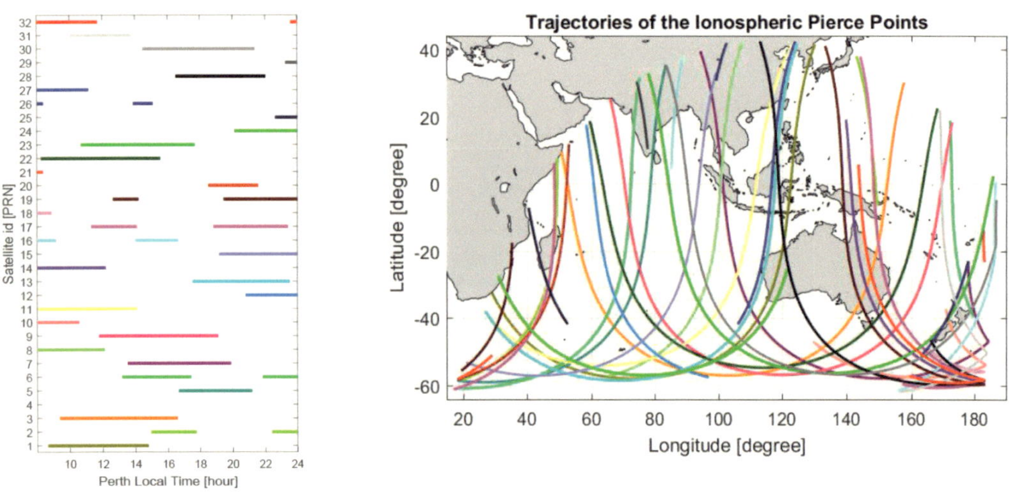

Figure 10. *Left*: GPS satellites visibility over time, viewed from Perth, Western Australia. *Right*: Trajectories of the corresponding IPPs made by receiver-to-satellite line-of-sight paths. The satellites are indicated by different colors.

http://dx.doi.org/10.5772/intechopen.71138

5.3. Central (network-based) versus local (single-receiver) solutions

Before discussing the precision performance of the CKF solutions, we first compare the network-based (central) TEC solutions with the solutions that are obtained by the data of one single-receiver only (referred to as the local solutions). At the filter initialization, the standard-deviations of the local TEC solutions are $\sqrt{13} \approx 3.6$ times larger than those of the central TEC solutions (i.e. square-root of the number of nodes). This is because of the fact that each of the 13 network receivers independently provides equally precise solutions. In that case, the central solution follows then by averaging all the 13 local solutions. Due to the common dynamic model Eq. (53) however, the local solutions become correlated over time. After the filter initialization, the central solution would therefore *not* follow the average of its local versions. The standard-deviation results, after one hour of the filter initialization, are presented in **Figure 11**. Only the results of the receiver 1 are shown as local solutions (in red). As shown, the standard-deviations get stable over time as the filters reach their steady-state. On the right panel of the figure, the local-to-central standard-deviation ratios are also presented. In case of the vertical TECs $\nu_{o,t}$, the ratios vary from 1.5 to 3. For the horizontal gradients $\nu_{\phi,t}$ and $\nu_{\psi,t}$, the ratios are about 2 and 2.5, respectively.

5.4. Role of CKF in improving local solutions

With the results of **Figure 11**, we observed that the central TEC solutions considerably outperform their local counterparts in the sense of delivering more precise outcomes, i.e. the local-to-central standard-deviation ratios are considerably larger than 1. We now employ the

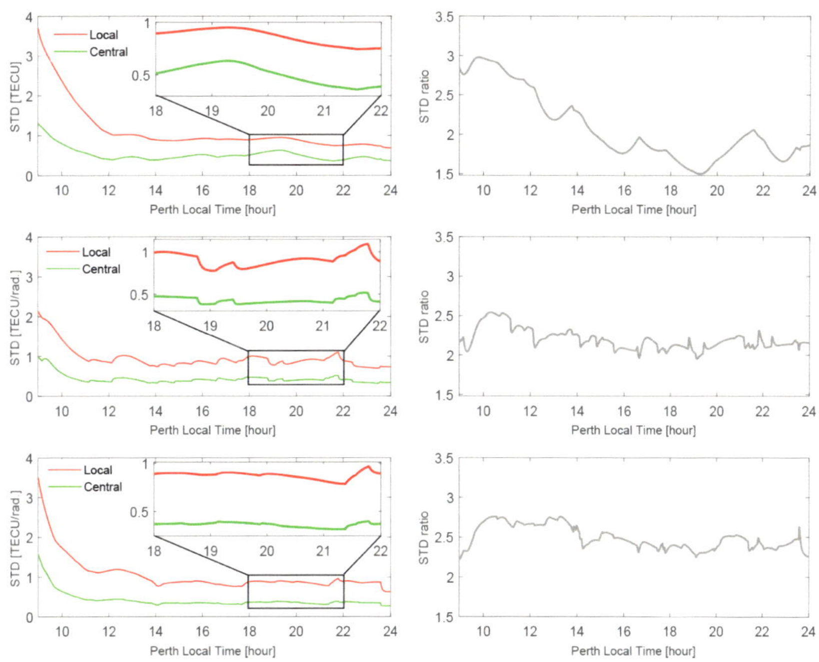

Figure 11. *Left*: Standard-deviation of the central (green) and local (red) solutions of the TEC parameters $\nu_{o,t}$ (top), $\nu_{\phi,t}$ (middle) and $\nu_{\psi,t}$ (bottom) as functions of time. *Right*: The corresponding local-to-central standard-deviation ratios.

CKF for each node (receiver) $i = 1, \ldots, 13$, to *improve* the local solutions' precision performance via consensus-based iterative communications between the receivers. In doing so, we make use of the three-step recursion given in **Figure 7** to evaluate the error variance matrices $P_{i,t|t}$ ($i = 1, \ldots, 13$), thereby computing the CKF-to-central standard-deviation ratios. The stated ratios are presented in **Figure 12** for two different data sending rates $\delta = 15$ seconds (left panel) and $\delta = 5$ seconds (right panel). In both cases, the CKF-to-central standard-deviation ratios are smaller than their local-to-central versions shown in **Figure 11** (right panel), illustrating that employing the CKF does indeed improve the local solutions' precision. Since more iterative communications take place for $\delta = 5$, the corresponding ratios are very close to 1. In that case, the CKF of each receiver is expected to have a similar precision performance to that of the central (network-based) filter. For the case $\delta = 15$ however, the CKF performance of each receiver does very much depend on the number of the receiver's neighbors. This is because of the fact that only 4 iterative communications between the receivers take place (i.e. $k_n = 4$). The receivers with the minimum number of neighbors, i.e. receivers 1, 3 and 13 (**Figure 8**), have the worst precision performance as the corresponding ratios take largest values. On the other hand, the receivers with the maximum number of neighbors, i.e. receivers 4, 7, 9 and 8, have the best performance as the corresponding ratios are close to 1.

Next to the solutions of the TEC parameters $\nu_{o,t}$, $\nu_{\phi,t}$ and $\nu_{\psi,t}$, we also analyze CKF solutions of the between-satellite DCBs b_t^{ps} ($s \neq p$) in Eq. (52). Because of the difference in the satellites visibility over time (cf. **Figure 10**), the DCBs' standard-deviations are quite distinct and very much depend on the duration of the satellites visibility. The more a pair of satellites $p - s$ are

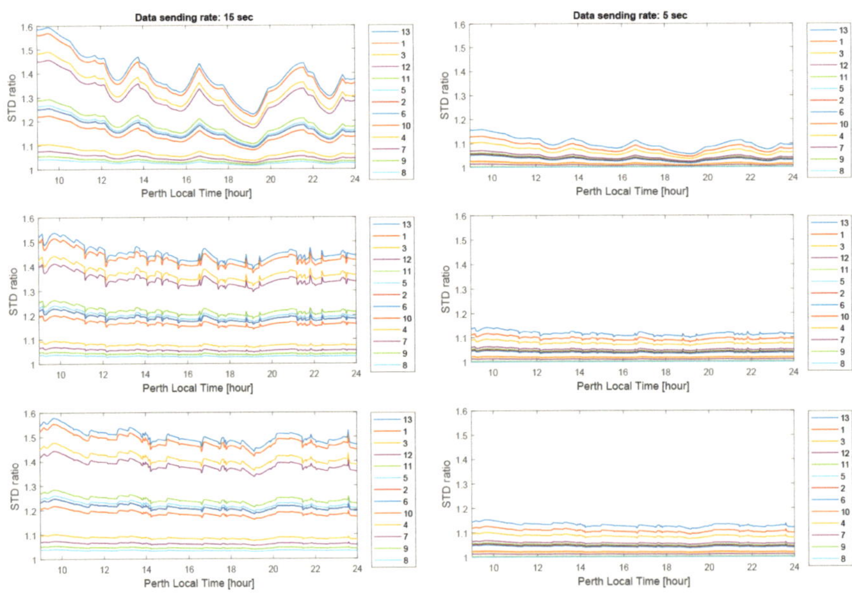

Figure 12. Time-series of the CKF-to-central standard-deviation ratios corresponding to the TEC parameters $\nu_{o,t}$ (top), $\nu_{\phi,t}$ (middle) and $\nu_{\psi,t}$ (bottom). *Left*: The data sending rate is set to $\delta = 15$ seconds (i.E. 4 iterative communications). *Right*: The data sending rate is set to $\delta = 5$ seconds (i.E. 12 iterative communications). The results of the nodes (receivers) are indicated by different colors.

visible, the smaller the standard-deviation is expected. We now consider the required time to have between-satellite DCBs solutions with standard-deviation smaller than 0.5 nanoseconds. Because of the stated difference in the standard-deviations, each between-satellite DCB corresponds to a different required time. For the central filter, the minimum value of such required time is 7 minutes, with the 25th percentile as 12, median as 38, 75th percentile as 63 and the maximum as 84 minutes. Thus after 84 minutes of the filter initialization, all central DCB solutions have standard-deviations smaller than 0.5 nanoseconds. Such percentiles can be represented by a 'boxplot'. We compute the stated percentiles for all the CKF solutions and compare their boxplots with the central one in **Figure 13**. The results are presented for three

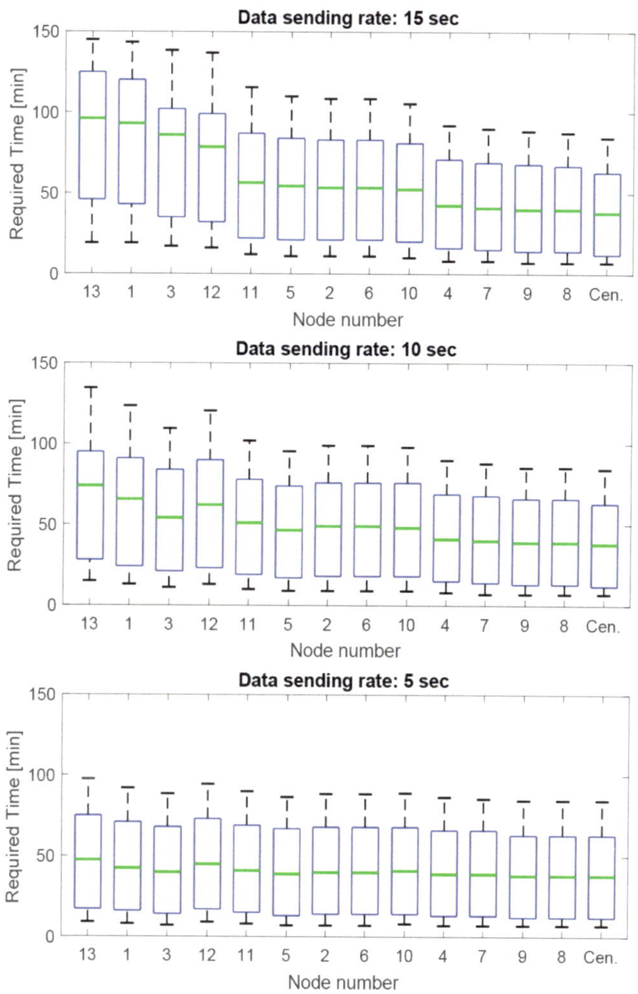

Figure 13. Boxplots of the required time (minutes) to have between-satellite DCBs solutions with standard-deviation smaller than 0.5 nanoseconds. The performance of the CKF of each node (receiver) is compared to that of the central filter (Cen.). In each boxplot, the horizontal lines from bottom to top show the minimum (black), 25th percentile (blue), median (green), 75th percentile (blue) and maximum (black) of the stated required time. The data sending rate is set to *Top*: $\delta = 15$ seconds (i.E. 4 iterative communications), *Middle*: $\delta = 10$ seconds (i.E. 6 iterative communications), and *Bottom*: $\delta = 5$ seconds (i.E. 12 iterative communications).

different data sending rates $\delta = 15$ seconds (top), $\delta = 10$ seconds (middle) and $\delta = 5$ seconds (bottom). As shown, the more the number of iterative communications, the more similar the boxplots becomes, i.e. the nodes (receivers) are reaching consensus. Similar to the TEC solutions, the DCB precision performance of the CKF corresponding to the receivers 4, 7, 9 and 8 is almost similar to that of the central one, irrespective of the number of iterative communications. This follows from the fact that the stated receivers have the maximum number of neighbors (**Figure 8**), thus efficiently approximating the averages $\overline{N}_t$ and $\overline{r}_t$ in Eq. (28) after a few iterations. On the other hand, the receivers with the minimum number of neighbors require more number of iterative communications in order for their CKF precision performance to get similar to that of the central filter.

6. Concluding remarks and future outlook

In this contribution we reviewed Kalman filtering in its information form and showed how the additive measurement update (28) can be realized by employing average-consensus rules, even when *not* all nodes are directly connected, thus allowing the sensor nodes to develop their own distributed filters. The nodes are assumed linked to each other *at least* through a 'path' so that information can flow from each node to all other nodes. Under this assumption, average-consensus protocols can deliver consensus states $[N_{i,t}(k_n), r_{i,t}(k_n)]$ as an approximation of the averages $[\overline{N}_t, \overline{r}_t]$ in Eq. (28) at every time instance $t = 1, 2, \ldots$, thus allowing one to establish a CKF recursion at every node $i = 1, \ldots, n$. To improve the stated approximation, the neighboring nodes have to establish a number of iterative data communications to transfer and receive their consensus states. This makes the CKF implementation applicable only for the applications in which the state-vectors change rather slowly over time (i.e. the sampling rate Δ can take large values) and/or for the cases where the sensor nodes transfer their data rather quickly (i.e. the sending rate δ can take small values).

We developed a three-step recursion of the CKF error variance matrix (**Figure 7**). This recursion conveys useful information about the precision performance of the local filters $i = 1, \ldots, n$, thereby enabling one to a-priori design and analyze sensor networks with different numbers of iterative communications. As an illustrative example, we applied the stated recursion to a small-scale network of GNSS receivers and showed the role taken by the CKF in improving the precision of the solutions at each single receiver. In near future the proliferation of low-cost receivers will give rise to an increase in the number of GNSS users. Employing the CKF or other distributed filtering techniques, GNSS users can therefore potentially deliver high-precision parameter solutions without the need of having a computing center.

Acknowledgements

The second author is the recipient of an Australian Research Council Federation Fellowship (project number FF0883188). This support is gratefully acknowledged.

Author details

Amir Khodabandeh[1]*, Peter J.G. Teunissen[1,2] and Safoora Zaminpardaz[1]

*Address all correspondence to: amir.khodabandeh@curtin.edu.au

1 Curtin University of Technology, Australia

2 Delft University of Technology, The Netherlands

References

[1] Cattivelli FS, Sayed AH. Diffusion strategies for distributed Kalman filtering and smoothing. IEEE Transactions on Automatic Control. 2010;**55**(9):2069-2084

[2] Das S, Moura JMF. Consensus+innovations distributed Kalman filter with optimized gains. IEEE Transactions on Signal Processing. 2017;**65**(2):467-481

[3] Jadbabaie A, Lin J, Stephen Morse A. Coordination of groups of mobile autonomous agents using nearest neighbor rules. IEEE Transactions on Automatic Control. 2003;**48**(6):988-1001

[4] Kingston DB, Beard RW. Discrete-time average-consensus under switching network topologies. In: American Control Conference, 2006. IEEE; 2006. pp. 3551-3556

[5] Moreau L. Stability of multiagent systems with time-dependent communication links. IEEE Transactions on Automatic Control. 2005;**50**(2):169-182

[6] Olfati-Saber R, Murray RM. Consensus problems in networks of agents with switching topology and time-delays. IEEE Transactions on Automatic Control. 2004;**49**(9):1520-1533

[7] Scherber DS, Papadopoulos HC. Locally constructed algorithms for distributed computations in ad-hoc networks. In: Proceedings of the 3rd International Symposium on Information Processing in Sensor Networks. ACM; 2004. pp. 11-19

[8] Xiao L, Boyd S, Lall S. A scheme for robust distributed sensor fusion based on average consensus. In: Proceedings of the 4th International Symposium on Information Processing in Sensor Networks. IEEE Press; 2005. pp. 63-70

[9] Rigatos GG. Distributed filtering over sensor networks for autonomous navigation of UAVs. Intelligent Service Robotics. 2012;**5**(3):179-198

[10] Sugar TG, Kumar V. Control of cooperating mobile manipulators. IEEE Transactions on Robotics and Automation. 2002;**18**(1):94-103

[11] Hofmann-Wellenhof B, Lichtenegger H, Wasle E. GNSS: Global Navigation Satellite Systems: GPS, Glonass, Galileo, and More. New York: Springer; 2008

[12] Teunissen PJG, Montenbruck O, editors. Springer Handbook of Global Navigation Satellite Systems. Switzerland: Springer; 2017

[13] Casbeer DW, Beard R. Distributed information filtering using consensus filters. In: American Control Conference, 2009. ACC'09. IEEE; 2009. pp. 1882-1887

[14] Khodabandeh A, Teunissen PJG. An analytical study of PPP-RTK corrections: Precision, correlation and user-impact. Journal of Geodesy. 2015;**89**(11):1109-1132

[15] Li W, Nadarajah N, Teunissen PJG, Khodabandeh A, Chai Y. Array-aided single-frequency state-space RTK with combined GPS, Galileo, IRNSS, and QZSS L5/E5a observations. Journal of Surveying Engineering. 2017;**143**(4):04017006

[16] Li X, Ge M, Dai X, Ren X, Fritsche M, Wickert J, Schuh H. Accuracy and reliability of multi-GNSS real-time precise positioning: GPS, GLONASS, BeiDou, and Galileo. Journal of Geodesy. 2015;**89**(6):607-635

[17] Odolinski R, Teunissen PJG. Single-frequency, dual-GNSS versus dual-frequency, single-GNSS: A low-cost and high-grade receivers GPS-BDS RTK analysis. Journal of Geodesy. 2016;**90**(11):1255-1278

[18] Zaminpardaz S, Teunissen PJG, Nadarajah N. GLONASS CDMA L3 Ambiguity Resolution and Positioning. Berlin, Heidelberg: GPS Solutions; 2016. pp. 1-15

[19] Brammer K, Siffling G. Kalman-Bucy Filters. Berlin: Artech House; 1989

[20] Candy JV. Signal Processing: Model Based Approach. New Jersey: McGraw-Hill, Inc.; 1986

[21] Gelb A. Applied Optimal Estimation. London: MIT press; 1974

[22] Gibbs BP. Advanced Kalman Filtering, Least-Squares and Modeling: A Practical Handbook. New Jersey: Wiley; 2011

[23] Jazwinski AH. Stochastic Processes and Filtering Theory. Maryland: Dover Publications; 1991

[24] Kailath T. Lectures on Wiener and Kalman Filtering. Number 140. Vienna: Springer; 1981

[25] Kalman RE. A new approach to linear filtering and prediction problems. Journal of Basic Engineering. 1960;**82**(1):35-45

[26] Teunissen PJG. Best prediction in linear models with mixed integer/real unknowns: Theory and application. Journal of Geodesy. 2007;**81**(12):759-780

[27] Bode HW, Shannon CE. A simplified derivation of linear least square smoothing and prediction theory. Proceedings of the IRE. 1950;**38**(4):417-425

[28] Kailath T. An innovations approach to least-squares estimation–part I: Linear filtering in additive white noise. Automatic Control, IEEE Transactions on. 1968;**13**(6):646-655

[29] Zadeh LA, Ragazzini JR. An extension of Wiener's theory of prediction. Journal of Applied Physics. 1950;**21**(7):645-655

[30] Anderson BDO, Moore JB. Optimal Filtering. Vol. 11. New Jersey: Prentice-hall Englewood Cliffs; 1979

[31] Bar-Shalom Y, Li XR. Estimation and Tracking- Principles, Techniques, and Software. Vol. 1993. Norwood, MA: Artech House, Inc; 1993

[32] Grewal MS, Andrews AP. Kalman Filtering; Theory and Practice Using MATLAB. 3rd ed. New Jersey: John Wiley and Sons; 2008

[33] Kailath T, Sayed AH, Hassibi B. Linear Estimation. New Jersey: Prentice-Hall; 2000

[34] Maybeck PS. Stochastic Models, Estimation, and Control. Vol. 1. Academic Press; 1979. Republished 1994

[35] Simon D. Optimal State Estimation: Kalman, H [Infinity] and Nonlinear Approaches. New Jersey: John Wiley and Sons; 2006

[36] Sorenson HW. Kalman filtering techniques. In: Leondes CT editor. Advances in Control Systems Theory and Applications, Vol. 3; 1966. pp. 219-292

[37] Stark H, Woods JW. Probability, Random Processes, and Estimation Theory for Engineers. Englewood Cliffs, New Jersey: Prentice-Hall; 1986

[38] Teunissen PJG, Khodabandeh A. BLUE, BLUP and the Kalman filter: Some new results. Journal of Geodesy. 2013;**87**(5):461-473

[39] Khodabandeh A, Teunissen PJG. A recursive linear MMSE filter for dynamic systems with unknown state vector means. GEM - International Journal on Geomathematics. 2014; **5**(1):17-31

[40] Ren W, Beard RW. Consensus of information under dynamically changing interaction topologies. In: American Control Conference, 2004. Proceedings of the 2004, Vol. 6. IEEE; 2004. pp. 4939-4944

[41] Henderson HV, Pukelsheim F, Searle SR. On the history of the Kronecker product. Linear and Multilinear Algebra. 1983;**14**(2):113-120

[42] Horn RA, Johnson CR. Matrix Analysis. New York: Cambridge UP; 1985

[43] Wolfowitz J. Products of indecomposable, aperiodic, stochastic matrices. Proceedings of the American Mathematical Society. 1963;**14**(5):733-737

[44] Blewitt G. An automatic editing algorithm for GPS data. Geophysical Research Letters. 1990;**17**(3):199-202

[45] Mannucci AJ, Wilson BD, Yuan DN, Ho CH, Lindqwister UJ, Runge TF. A global mapping technique for GPS-derived ionospheric total electron content measurements. Radio Science. 1998;**33**(3):565-582

[46] Schaer S. Mapping and predicting the Earth's ionosphere using the global positioning system. PhD thesis. Bern, Switzerland: University of Bern; 1999

[47] Brunini C, Azpilicueta FJ. Accuracy assessment of the GPS-based slant total electron content. Journal of Geodesy. 2009;**83**(8):773-785

[48] Khodabandeh A, Teunissen PJG. Array-aided multifrequency GNSS Ionospheric sensing: Estimability and precision analysis. IEEE Transactions on Geoscience and Remote Sensing. 2016;**54**(10):5895-5913

[49] Julien O, Macabiau C, Issler JL. Ionospheric delay estimation strategies using Galileo E5 signals only. In: GNSS 2009, 22nd International Technical Meeting of The Satellite Division of the Institute of Navigation, Savannah; 2009. pp. 3128-3141